D1261851

THE BALANCE BETWEEN INDUSTRY AND AGRICULTURE IN ECONOMIC DEVELOPMENT
Volume 2: SECTOR PROPORTIONS

This volume is IEA conference volume no. 87

THE BALANCE BETWEEN INDUSTRY AND AGRICULTURE IN ECONOMIC DEVELOPMENT

Volume 1 BASIC ISSUES
Kenneth J. Arrow (*editor*)

Volume 2 SECTOR PROPORTIONS
Jeffrey G. Williamson and Vadiraj R. Panchamukhi (*editors*)

Volume 3 MANPOWER AND TRANSFERS
Sukhamoy Chakravarty (*editor*)

Volume 4 SOCIAL EFFECTS
Irma Adelman and Sylvia Lane (*editors*)

Volume 5 FACTORS INFLUENCING CHANGE
Nurul Islam (*editor*)

These volumes are, respectively, nos 86–90 in the IEA/Macmillan series

IEA Conference volume series

The Balance between Industry and Agriculture in Economic Development

Proceedings of the Eighth World Congress of the International Economic Association, Delhi, India

Volume 2
SECTOR PROPORTIONS

Edited by
Jeffrey G. Williamson
and Vadiraj R. Panchamukhi

MACMILLAN
PRESS

in association with the
INTERNATIONAL ECONOMIC
ASSOCIATION

First published 1989

Published by
THE MACMILLAN PRESS LTD
Houndmills, Basingstoke, Hampshire RG21 2XS
and London
Companies and representatives
throughout the world

Printed in Hong Kong

British Library Cataloguing in Publication Data
International Economic Association. *World Congress (8th: Delhi, India)*
The balance between industry and agriculture in economic development: proceedings of the Eighth World Congress of the International Economic Association, Delhi, India.——(Macmillan/International Economic Association Series).
Vol. 2: Sector proportions
I. Title II: Williamson. Jeffrey G. (Jeffrey Gale), *1935* – III. Panchamukhi, Vadiraj R.
330.9
ISBN 0–333–46716–7

Contents

Preface

The Eighth World Congress of the International Economic Association was held in Delhi from 1 to 5 December 1986, presided over by Professor Kenneth J. Arrow, President of the IEA from 1983 to 1986. The subject of the Congress was 'The Balance between Industry and Agriculture in Economic Development'.

Participation in the Congress was broadly based in terms both of geography and of the types of economy from which participants came; market orientated and centrally planned; developed and developing; mainly agricultural and predominantly industrial.

The Congress included a number of plenary sessions, but much of the work of the Congress was undertaken in eighteen specialised meetings. The volume of papers was too large for them all to be published, but the five volumes in this group, together with a volume on the Indian economy being published separately in India, represent the major viewpoints expressed. The volumes generally contain reports on the discussions which took place during the specialised sessions.

The volumes are:

1. *Basic Issues*, edited by Kenneth J. Arrow
2. *Sector Proportions*, edited by Jeffrey G. Williamson and Vadiraj R. Panchamukhi
3. *Manpower and Transfers*, edited by Sukhamoy Chakravarty
4. *Social Effects*, edited by Irma Adelman and Sylvia Lane
5. *Factors Influencing Change*, edited by Nurul Islam

The Indian volume is edited by Dr P. R. Brahmananda and Dr S. Chakravarty under the title *The Indian Economy: Balance between Industry and Agriculture* and will be published by Macmillan (India).

This volume contains selected papers from four sessions of the Congress, as follows:

Session 1–Sector Proportions and Economic Development: Theory and Cross-section Comparisons, organised by Professor Hollis B. Chenery

Session 2–Sector Proportions and Economic Development: Country Experience, organised by Dr Kemal Dervis

Session 4–Overall Growth Rates and Sector Growth Rates, organised by Professor P. R. Brahmananda
Session 5–Effects of Sector Growth on Real Magnitudes, organised by Professor Bruce F. Johnston

The International Economic Association

A non-profit organisation with purely scientific aims, the International Economic Association (IEA) was founded in 1950. It is in fact a federation of national economic associations and presently includes fifty-eight such professional organisations from all parts of the world. Its basic purpose is the development of economics as an intellectual discipline. Its approach recognises a diversity of problems, systems and values in the world and also takes note of methodological diversities.

The IEA has, since its creation, tried to fulfil that purpose by promoting mutual understanding of economists from the West and the East as well as from the North and the South through the organisation of scientific meetings and common research programmes and by means of publications on problems of current importance. During its thirty-seven years of existence, it has organised seventy-nine round-table conferences for specialists on topics ranging from fundamental theories to methods and tools of analysis and major problems of the present-day world. Eight triennial World Congresses have also been held, and these have regularly attracted the participation of a great many economists from all over the world. The proceedings of all these meetings are published by Macmillan.

The Association is governed by a Council, composed of representatives of all member associations, and by a fifteen-member Executive Committee which is elected by the Council. The present Executive Committee (1986–89) is composed as follows:

President:	Professor Amartya Sen, India
Vice-President:	Professor Béla Csikós-Nagy, Hungary
Treasurer:	Professor Luis Angel Rojo, Spain
Past President:	Professor Kenneth J. Arrow, USA
Other members:	Professor Edmar Lisboa Bacha, Brazil
	Professor Ragnar Bentzel, Sweden
	Professor Oleg T. Bogomolov, USSR
	Professor Silvio Borner, Switzerland
	Professor P. R. Brahmananda, India

Professor Phyllis Deane, United Kingdom
Professor Luo Yuanzheng, China
Professor Edmond Malinvaud, France
Professor Luigi Pasinetti, Italy
Professor Don Patinkin, Israel
Professor Takashi Shiraishi, Japan

Adviser: Professor Tigran S. Khachaturov, USSR
Secretary-General: Professor Jean-Paul Fitoussi, France
General Editor: Mr Michael Kaser, United Kingdom
Adviser to General Editor: Professor Sir Austin Robinson, United Kingdom
Conference Editor: Dr Patricia M. Hillebrandt, United Kingdom

The Association has also been fortunate in having secured the following outstanding economists to serve as President:

Gottfried Haberler (1950–3), Howard S. Ellis (1953–6), Erik Lindahl (1956–9), E. A. G. Robinson (1959–62), G. Ugo Papi (1962–5), Paul A. Samuelson (1965–8), Erik Lundberg (1968–71), Fritz Machlup (1971–4), Edmond Malinvaud (1974–7), Shigeto Tsuru (1977–80), Victor L. Urquidi (1980–3), Kenneth J. Arrow (1983–86).

The activities of the Association are mainly funded from the subscriptions of members and grants from a number of organisations, including continuing support from UNESCO.

Acknowledgements

The host for the Eighth World Congress of the International Economic Association was the Indian Economic Association and all Congress participants are in its debt for the organisation of the Congress itself and for the welcome given to economists from all over the world. The preparation for such a gathering, culminating in a week of lectures, discussions and social activities, is an enormous undertaking. The International Economic Association wishes to express its appreciation on behalf of all participants.

Both the Indian and the International Economic Associations are grateful to the large number of institutions and organisations, including many states, banks, business firms, research and trade organisations which provided funds for the Congress. They particularly wish to thank the following Indian Government Departments and other Official agencies:

Ministry of Finance
Ministry of External Affairs
The Reserve Bank of India
The State Bank of India
The Industrial Development Bank of India
The Indian Council of Social Science Research
The Industrial Credit and Investment Corporation of India
The Industrial Finance Corporation of India
The National Bank for Agriculture and Rural Development
The Industrial Reconstruction Bank of India
The Punjab National Bank
The Canara Bank
Tata Group of Industries
The Government of Uttar Pradesh
The Government of Karnataka
The Government of Kerala
The Government of Madhya Pradesh

Valuable support was given by the Ford Foundation and the International Development Research Centre. The Research and Information System for Non-aligned and Other Developing Countries and the Institute of Applied Manpower Research provided valuable assistance in staffing and in the infrastructure of the Congress.

The social events of the Congress provided a useful opportunity for informal discussion as well as being a source of great enjoyment. The hospitality of the Indian Economic Association, the Export–Import Bank of India, the Federation of Indian Chambers of Commerce and Industry, the Punjab Haryana and Delhi (PHD) Chambers of Commerce and Industry and the DCM Ltd created memorable occasions. Thanks go to the Indian Council for Cultural Relations, which organised a cultural evening. In addition there were many small social gatherings which stressed the international flavour of the Congress.

Lastly, and vitally important, was the contribution of the members of the IEA Organising Committee and the Indian Steering Committee listed overleaf; in particular, Dr Manmohan Singh, Chairman of the Steering Committee, Professor S. Chakravarty, President of the Indian Economic Association, and then Vice-President of the International Economic Association, and Dr V. R. Panchamukhi, Convenor of the Steering Committee; the authors; the discussants; the rapporteurs and the ever-helpful students. The International Economic Association wishes to thank them all for the success of the Congress in Delhi in 1986.

Thanks are expressed to the International Social Science Council under whose auspices the publications programme is carried out, and to UNESCO for its financial support.*

* Under the auspices of the International Social Science Council and with a grant from UNESCO (1986–87/DG/7.6.2./SUB. 16 (SHS)).

The IEA Programme Committee

Irma Adelman
Kenneth J. Arrow
P. R. Brahmananda
Sukhamoy Chakravarty
Béla Csikós-Nagy
Shigeru Ishikawa
Nurul Islam
Bruce Johnston
Paolo Sylos-Labini

Indian Steering Committee

Dr Manmohan Singh (*Chairman*)
Professor S. Chakravarty (*President, Indian Economic Association, then Vice-President of the International Economic Association*)
Dr V. R. Panchamukhi (*Convenor*)

Dr Malcolm S. Adiseshaiah
Dr M. S. Ahluwalia
Dr D. S. Awasthi
Dr Mahesh Bhatt
Professor P. R. Brahmananda
Shri M. Dubey
Professor Alak Ghosh
Professor P. D. Hajela
Dr Bimal Jalan
Professor A. M. Khusro
Professor D. T. Lakdawala
Dr M. Madaiah
Professor Gautam Mathur
Professor M. V. Mathur
Professor Iqbal Narain
Professor D. L. Narayana
Dr D. D. Narula
Professor Kamta Prasad
Dr C. Rangarajan
Dr N. K. Sengupta
Professor Shanmugasundaram
Dr R. K. Sinha

List of the Contributors

Professor Irma Adelman, Department of Agricultural and Resource Economics, University of California at Berkeley, USA

Dr Isher Judge Ahluwalia, Centre for Policy Research, New Delhi, India

Dr Jean-Marc Bourniaux, OECD, Paris, France

Professor P. R. Brahmananda, Economics Department, University of Bombay, India

Professor Hollis B. Chenery, Economics Department, Harvard University, Cambridge, Mass., USA

Professor Dr Csaba Csáki, Department of Agricultural Economics, Karl Marx University of Economic Sciences, Budapest, Hungary

Dr Kemal Dervis, Industrial and Policy Division, The World Bank, Washington, DC, USA

Professor Dr Wolfgang Heinrichs, Economic Association at the Academy of Sciences of the GDR, Berlin, GDR

Professor Erh-Cheng Hwa, Country Analysis and Projections Division The World Bank, Washington, DC, USA

Dr C. J. Jepma, Faculty of Economics, State University, Gröningen, Netherlands

Professor Bruce F. Johnston, Food Research Institute, Stanford University, USA

Professor Hiromitsu Kaneda, Economics Department, University of California at Davis, USA

Dr Peter Kilby, Economics Department, Wesleyan University, Middletown, Connecticut, USA

Dr Danny M. Leipziger, Korea and South East Asia Division, The World Bank, Washington, DC, USA

Professor Carl Liedholm, Economics Department, Michigan State University, E. Lansing, Mich.,USA

Dr Rajesh Mehta, Research and Information System for Non-Aligned and other Developing Countries, New Delhi, India

Dr John W. Mellor, International Food Policy Research Institute, Washington, DC, USA

Dr R. G. Nambiar, Research and Information System for Non-Aligned and other Developing Countries, New Delhi, India

Professor L. V. Nikiforov, Institute of Economics of the Academy of Sciences of the USSR, Moscow, USSR

Dr Mieko Nishimizu, Industrial Strategy and Policy Division, The World Bank, Washington, DC, USA

Professor T. Ademola Oyejide, University of Ibadan, Nigeria

Dr John M. Page, Jr., Industrial Strategy and Policy Division, The World Bank, Washington, DC, USA

Dr Vadiraj R. Panchamukhi, Research and Information System for the Non-Aligned and Other Developing Countries, New Delhi, India

Professor Peter A. Petri, Brandeis University, Watham, Mass., USA

Dr C. Rangarajan, Reserve Bank of India, Bombay, India

Professor Klaus Steinitz, Economic Association at the Academy of Sciences of the GDR, Berlin, GDR

Professor Moshe Syrquin, Economics Department, Bar-Ilan University, Israel and Harvard

Professor Lance Taylor, Department of Economics, Massachusetts Institute of Technology, Cambridge, Mass., USA

Professor Shujiro Urata, School of Social Sciences, Waseda University, Tokyo, Japan and World Bank

Professor Jean Waelbroeck, Centre d'Economie Mathématique et d'Econometrie, Brussels, Belgium

Professor Jeffrey G. Williamson, Economics Department, Harvard University, Cambridge, Mass., USA

Abbreviations and Acronyms

ADLI	Agricultural development-led industrialisation
AIC	Agro-industrial complex
CCCN	Customs Cooperation Council Nomenclature
CES	Constant elasticity of substitution
CGE	Computable general equilibrium
CMEA	Council for Mutual Economic Assistance
CMS	Constant market share analysis
CPSU	Communist Party of the Soviet Union
CRS	Constant returns to scale
CSO	Central Statistical Office
ESCAP	Economic and Social Commission for Asia and the Pacific
FAO	Food and Agriculture Organisation (of the United Nations)
FAP	Food and Agriculture Program Group (of IIASA)
GATT	General Agreement on Tariffs and Trade
GDP	Gross domestic product
GMF	Grain Management Fund (Korea)
GNP	Gross national product
HCI	Heavy chemical industry
IBRD	International Bank for Reconstruction and Development
ICOR	Incremental capital output ratio
IIASA	International Institute for Applied Systems Analysis
ILO	International Labour Organisation
IRRI	International Rice Research Insitute
ISIC	International Standard Industrial Classification
KDI	Korean Development Institute
LDC	Less developed country
LPG	Landwirtschaftliche Produktiongenossenschaft
MSU	Michigan State University
NDP	Net domestic product
NIC	Newly-industrialising country
NIF	National Investment Fund (Korea)

OECD	Organisation for Economic Cooperation and Development
OLS	Ordinary least squares
OPEC	Organisation of Petroleum Exporting Countries
PL	Public Law (USA)
R & D	Research and Development
RUNS	Rural–urban, North–South (model)
SBC	Social benefit cost (ratios)
SITC	Standard International Trade Classification
TFP	Total Factor Productivity

Introduction

Jeffrey G. Williamson
and Vadiraj R. Panchamukhi

OVERVIEW

The eighteen paper included in this volume certainly fit the theme of the Eight International Economic Association Congress, 'The Balance between Industry and Agriculture in Economic Development'. The papers were collected from four of the sessions on 'Sector Proportions and Economic Development: Theory and Cross-Section Comparisons', organised by Hollis B. Chenery; 'Sector Proportions and Economic Development: Country Experience', organised by Kemal Dervis; 'Overall Growth Rates and Sector Growth Rates', organised by P. R. Brahmananda; and 'Effects of Sector Growth on Real Magnitudes', organised by Bruce F. Johnston. The four parts of this volume roughly, but not exactly, conform to the structure of those sessions.

PART I

Thirty years ago, the late Simon Kuznets initiated a programme of historical and cross-sectional studies of the main features of the structural transformation. Kuznets characterised the transformation as Modern Economic Growth, and searched for the stylised 'patterns of development' which would offer a yardstick against which to gauge the performance of any developing country. Except for the theoretical survey by Lance Taylor, the papers in Part I are all in the Kuznets tradition. Four of these were presented at Delhi in the session organised by Hollis B. Chenery. The fifth paper, by V. R. Panchamukhi, R. G. Nambiar and R. Mehta, was presented in another session but clearly fits best here.

In its simplest form, the Kuznets research programme measures the association between increases in per capita income and structural change. Erh-Cheng Hwa, Panchamukhi *et al.*, Moshe Syrquin, and

Jeffrey G. Williamson augment the list of explanatory variables to include population, size, resource endowment, trade policy and other factors thought to be central to the transformation. Thus, the econometric models used here are far richer than those used by Kuznets in his pioneering work. Furthermore, these papers reflect a much larger data base than that which was available to Kuznets. Syrquin and Panchamukhi *et al.*, for example, are able to exploit a data base of about 100 nations across three decades of post-Second World War experience, and Williamson is able to report the application of the Chenery-Syrquin model to fifteen developing countries in nineteenth century Europe. The vastly augmented data base makes it possible, of course, to explore much more elaborate econometric specifications.

Despite these advances, controversy surrounds the 'patterns of development' approach. The debate centres on the interpretation of the patterns. In effect, the regressions are reduced form equations that could be derived from any number of competing models of development. This is especially apparent of any 'patterns' specification in which per capita income is an exogenous variable. After all, the latter can be argued to serve as a proxy for both supply side factors, such as accumulation, as well as for demand side factors, such as Engel's Law. Furthermore, the 'patterns' approach is not well equipped to assign roles to policy, institutional failure, or commodity price shocks in accounting for any country's departure from the stylised patterns.

Despite these limitations, the patterns approach appears to have many effective uses in comparative historical studies. The most obvious is to relate the study of a given country to the typical development experience. For this purpose, the patterns regressions can be treated as a way of summarising historical experience. When an individual country departs from the typical patterns, the analysis serves to motivate an assessment of those conditions which served to make that country's experience different. Williamson, for example, illustrates this approach in his analysis of the First Industrial Revolution in Britain, where the application of a computable general equilibrium model makes it possible to explore the impact of some institutional factors which may have been unique to British experience.

Although the structural transformation has been widely modelled for individual countries, there is no general theory of transformation. Indeed, as Parts III and IV of this volume suggest, we are not even sure whether agricultural-led development is the quickest and most efficient road to transformation. Some of the reasons for this state of affairs are illustrated in Taylor's survey of theories of sectoral balance. Taylor compares several theories, starting with Schumpeter's concept of the circular flow. The competing models shed light on medium-term phenomena, such as an economy's response to commodity booms ('Dutch disease'), which are the main source of exceptions to the universal tendency of developing countries to industrialise.

It is notable that conventional neoclassical theory has little to say about the directions of structural change and its relation to growth, although dualistic models certainly do. So long as conventional neoclassical assumptions of competitive markets and perfect foresight hold, equilibrium is maintained in commodity and factor markets, and thus the real reallocation of labour and capital among sectors has little effect on growth and accumulation. Williamson shows, however, that departures from the neoclassical assumptions during Britain's industrial revolution may have had a substantial impact on the allocation of resources between industry and agriculture, on the 'patterns of development' and on the overall rate of accumulation and growth.

What do these studies tell us about policy? In earlier interpretations there was a tendency to leap from the observation that all advanced countries were industrialised, to the policy conclusion that all countries should industrialise in order to develop. More recent work has stressed the differences in the feasible patterns of development and the alternative interpretations that can be given to a country's experience. Indeed, Part IV devotes considerable attention to the treatment of agriculture, assessing the advantages of concurrent agricultural development (or 'balanced growth') or even agricultural-led development.

From this starting point, Taylor classifies multisectoral models according to whether they are primarily 'supply-driven', as in neoclassical systems, or 'demand-driven', as in neo-Keynesian systems. The assumptions which 'close' the model reflect the analyst's views of the functioning of different markets and of the policy choices available. Since it is difficult to test with historical data the relative superiority of one closure rule over another, it is essential that we explore the sensitivity of policy conclusions to different assumptions about closure.

These five papers stress the difficulties of making causal inferences from statistical correlations, and illustrate the effects of alternative assumptions on how development takes place, historical or contemporary. Regarding the question of central concern to the Delhi Congress, it cannot be concluded from these papers that neglecting agriculture is an effective policy. On the contrary, models that fit the experience of successful countries illustrate the need for raising agricultural productivity if the great transformation of Modern Economic Growth is to continue unimpeded.

PART II

The papers in Part II of this volume were all presented at the session organised by Kemal Dervis. While they raise the same issues which appear in Part I, they enrichen the analysis by offering country studies of India, Korea and Nigeria, by looking more closely at export performance of all developing countries, and by applying decomposition analysis to five very large economies and nine NICs.

Korea, of course, offers the supreme example of successful development and, according to the excellent paper by Danny M. Leipziger and Peter A. Petri, of relatively painless structural transition and modest tension between agriculture and industry. The structural transition was dramatic. Korean industry experienced an unusually rapid expansion after the mid-1950s, even by the standards of Chenery-Syrquin 'patterns' predictions. Leipziger and Petri show us that agriculture's share in GNP was about 45 per cent at the start of Korea's period of miraculous industrialisation – far larger than the 30 per cent predicted by Chenery-Syrquin regressions. Today, the much smaller agricultural share is only 2 per cent higher than predicted. Clearly, the Korean success story entailed a transformation of an economy which was unusually heavily committed to agriculture to one which roughly conforms to standard 'patterns'. The remarkable outward-looking industrial revolution in Korea reflects in part a redressing of immediate post-Korean War distortions. This massive shift in the economic structure of the Korean economy implied migration and urbanisation of comparable magnitude.

Industrialisation at this speed created economic tensions between industry and agriculture. Lagging outmigration and government neglect of agriculture had important distributional implications: it implied a widening gap between farm and city, much like it has appeared in East Europe (Part III), a result which we have come to expect since Simon Kuznets brought it to our attention. After the mid-1970s, however, the gap declined, partly due to continued rapid outmigration and partly due to explicit government intervention. By now, the gap has pretty much disappeared.

This is a remarkable story and Leipziger and Petri tell it well. They also augment the quantitative 'patterns of development' analysis with an effective overview of the policies which helped implement these trends, including the recent favourable treatment of agriculture. Indeed, it is striking to note that while industrial policy has moved towards liberalisation, the opposite has been the case for agriculture, so that the latter now closely resembles Japan and Western Europe, and for much the same reasons – distributional goals. It also presents the same problems – high food costs, chronic surpluses, and massive budget outlays.

Nigeria's post-colonial experience offers a striking contrast to Korea's. While her endowment differed – much larger, less literate, but greater land abundance – Nigeria also adopted a very different development strategy – import-substitution, and was beset by oil price shocks. An excellent paper by T. Ademola Oyejide attempts, with considerable success, to sort out each of those influences as they served to push Nigeria off the path of the standard 'patterns of development'. Oyejide uses the Chenery-Syrquin model to explore the impact of import-substitution strategies up to 1970, and then does the same for the 1970–84 period using the 'Dutch disease' model. This quantitative effort is then augmented with a detailed account of the policies and price shocks which pushed Nigeria off the standard patterns path. The result is a marvellous complement to the Korea country study.

India has developed since 1960 in yet another distinctly different environment. The largest of the three countries, the least involved in international trade, and undistributed by massive commodity price shocks, India raises somewhat different problems. Isher J. Ahluwalia and C. Rangarajan therefore focus on somewhat different issues, but ones that make sense for an economy relatively closed to trade and in which agriculture is so large. The two central issues in the paper are (1), the impact of the domestic terms of trade between agriculture and industry on the demand for industrial goods and (2), the impact of the terms of trade on investment allocation between sectors. These issues take on even greater relevance when we appreciate that input-output relationships between these two sectors strengthened dramatically between the late 1960s and the late 1970s.

The authors develop a sophisticated macroeconometric model to explore these issues, and estimate it on the two decades following 1960. Subsequent simulations with the model serve to increase our understanding of the interaction between these two sectors by paying special attention to the terms of trade. It is an analysis well worth reading.

The remaining two papers deal with large samples, but with issues that are closely related to many which have already been raised in this introduction. Shujiro Urata compares the sources of growth in five large countries – China, India, the USSR, Indonesia and Brazil, with nine smaller NICs. Domestic demand dominated the performance of the large nations – who also tended to pursue import-substitution strategies, while export expansion dominated the performance of the smaller countries – who also tended to pursue export-promotion strategies. The paper by C. J. Jepma follows quite naturally since it focuses on the export performance of fifty-nine developing countries who had significant trade with the OECD area. What determined export performance? Was it luck – that is, was the successful country's exports fortuitously in those commodities which enjoyed strong demand? Or was it performance – that is, competitiveness? It appears to have been the latter, at least between 1966 and 1980.

PART III

The papers in Part III of this volume were taken from the session organised by P. R. Brahmananda. One of these appears in Part I (Panchamukhi), one appears in Part IV (Adelman *et al.*), and four which deal with India are published elsewhere (Brahmananda and Chakravarty, forthcoming).[1] Three papers remain which are essential to give balance to this volume. All three deal with East European economies.

The USSR, the German Democratic Republic and Hungary offer three very different development environments compared with the country observations which underlie the 'patterns' analysis in Part I or the country studies in Part II. All have undergone post-war development under state planning. All have pursued pro-industrialisation policies. And all have adopted semi-autarkic and inward-looking development strategies. The distinctive policy environment offers an opportunity to make comparative assessments directly relevant to the theme of the Delhi Congress. While none of the three papers respond fully to this challenge, they do lay the groundwork for what should be an exciting research agenda for the future.

L. V. Nikiforov distinguishes four stages of development in the USSR, where input-output relationships between agriculture and industry are stressed. The first is the pre-revolutionary stage, where agriculture was largely subsistence and the links with industry weak. The second, during the late 1920s and early 1930s, is characterised by the rise of agricultural mechanisation and the increased demand for output from the domestic capital goods sector. The third stage covers the post-war period up to the mid-1960s, years of additional mechanisation, application of chemical fertilisers, and the rise of agricultural processing in industry. The fourth stage, since the mid-1960s, has seen the rise of an impressive agro-industrial complex. Throughout these stages, we know that agriculture lagged behind and that policy favoured industry. What we would truly like to know, however, is the quantitative impact of that policy compared with other large nations at similar levels of income. Would we find the USSR to be an outlier in a 'patterns' analysis? If so, which of these stages would exhibit the greatest pro-industrial bias? And if the evidence of a pro-industrial bias were confirmed by the patterns analysis, does the historical evidence since the First Five Year Plan suggest that these policies were most effective in achieving Modern Economic Growth? If the answer is yes, how would we reconcile it with the opposite findings for the unplanned economies discussed in Part IV? These are all very old questions that were raised long ago in the famous Soviet industrialisation debates of the 1920s. One can only hope that the methodologies contained elsewhere in this volume can be used to help get the answers in future research.

The paper on Hungary by Csaba Csáki and the paper on the German Democratic Republic by Wolfgang Heinrichs and Klaus Steinlitz raise similar issues. Polish economist J. Winiecki attributes economic problems 'in the small European socialist countries to the disproportionate size of industry and to inappropriate participation in the international division of labour' (Csáki, section 1.3). If these two economies in the post-war period can be described as semi-autarkic and inward-looking, we should see it reflected in any comparative assessment of the 'patterns of development'. Furthermore, if Central Europe's agricultural exports have been driven out of their historic markets by EEC discrimination, then evidence confirming that hypothesis should emerge when the methodologies of this volume are applied. How much of Hungary's export problems can be attributed to policy-induced supply side forces, and how much to demand side problems caused by discrimination in the West? Is the GDR's intended policy of foodgrain self-sufficiency in the 1980s and the 1990s an efficient strategy?

PART IV

Part IV of this volume contains papers which were presented to the session organized by Bruce F. Johnston, and one (by Irma Adelman, Jean-Marc Bourniaux and Jean Waelbroeck) which was presented to the session organised by P. R. Brahmananda, but which seemed to fit best here in Part Four. The common themes uniting these five important papers include: the link between agriculture and industry, the importance of rural nonfarm activities, and the effectiveness of a balanced growth or even an agricultural-led development strategy. As we pointed out above in the discussion of Part One of this volume, the latter issue was of central concern to the Delhi Congress, and these papers here in Part Four offer some useful answers.

Hiromitsu Kaneda's paper on rural resource mobility and intersectoral balance in Japan's early modern growth provides historical background for the remaining papers whose focus is on the contemporary less developed countries. Kaneda stresses that concurrent growth of agriculture and manufacturing (and other non-farm activities) was central in facilitating Japanese economic development from the late nineteenth century onwards. Relative to European countries on the eve of industrialisation, the share of the labour force engaged in agriculture in Meiji Japan was extremely large – about 80 per cent, and therefore similar to many LDCs in the early post-Second World War period. Since agriculture was already quite intensive, Japan had to find non-farm activities to absorb additions to the labour force. While it is true that labour force growth was modest in Meiji Japan (but about equal to European nineteenth century experience), and thus much lower than the growth rates which have characterised the Third World in the past three decades or so, Japan's non-farm employment growth was favoured by rapid output growth in those sectors, by an output mix favouring labour-intensive products, and by the adoption of relatively labour-intensive techniques. In addition, Kaneda points out that agriculture helped absorb labour since technological change on the farm was labour-using. Furthermore, expanded use of fertilisers and other yield-increasing inputs were crucial in achieving impressive output growth in spite of limited arable land; and expansion of sericulture and tea cultivation were also important in achieving fuller utilisation of a farm work force that was large relative to the farmland available for cultivation.

The balance of Kaneda's paper is devoted to two related themes. First, a two-sector model, suggested some time ago by Herbert Simon, is elaborated and tested against Japan's experience. Secondly, the sources and rates of growth in rice production per farm household are examined. Kaneda finds that a major source of output growth was the expansion of rice cultivation into the less developed regions, coupled with technological progress that steadily narrowed regional yield differentials. Since the capital and labour requirements for that increase in agricultural output were modest, Japan could achieve a concurrent expansion of agriculture and industry, an important lesson of history.

In an impressive empirical paper, Carl Liedholm and Peter Kilby draw on extensive research by the authors and collaborators in a number of developing countries to provide new evidence on the size and potential contribution of the rural nonfarm sector. The sector is very large. For example, manufacturing employment in rural areas as a share of total manufacturing employment ranged from 30 per cent in Korea to 86 per cent in Sierra Leone, with India (1967) 57 per cent and the Philippines (1976) 61 per cent about average for their sample of thirteen countries. Liedholm and Kilby also document the fact that the contribution of nonfarm income is especially large in the poorest rural households, thereby making a significant contribution to the alleviation of rural poverty. These findings by Liedholm and Kilby on the contemporary Third World are quite consistent, of course, with the economic historians' stress on cottage industries during what they call proto-industrialisation. In any case, many readers may be even more interested in the Liedholm and Kilby estimates of social benefit-cost ratios comparing large urban firms and small rural firms in a dozen African and Latin American industries. In eight of the twelve cases examined, social benefit-cost ratios are higher in the small rural firms. Moreover, for the only country (Sierra Leone) in which it was possible to compare social benefit-cost ratios at world prices, the relative efficiency of small rural firms is even larger. These are important findings, with obvious policy implications.

In their international comparison of productivity change in industry and agriculture, Mieko Nishimizu and John Page examine a large body of evidence to shed light on another question central to the theme of the Delhi Congress: Are there any identifiable relationships between agricultural and industrial output and total factor productivity growth? Along the way, they also supply some very useful stylised facts.

One interesting general finding is that the variance in total factor productivity growth (TFPG) rates is negatively correlated with income levels. That is, as income per capita rises, sectoral TFPG rates converge. At low levels of development, unbalanced TFPG tends to exacerbate the imbalance between agriculture and industry, while these forces tend to dissipate as modern economic growth unfolds. Within industry, Nishimizu and Page show that TFPG by country falls into three categories. The 'developed country pattern' is characterised by low input and output growth, but where TFPG accounts for sizeable shares of output growth. The 'typical LDC pattern' is characterised by more rapid output growth, but where TFPG accounts for a smaller share of output growth. In 'typical' LDCs, therefore, conventional capital deepening plays a more important role. However, the 'high productivity LDC pattern' is characterised by rapid growth of output, inputs, and productivity, and TFPG accounts for a high share of output growth.

The relationship between TFPG in agriculture and industry differs between developed and less developed countries. In high income countries TFPG in agriculture exceeds that of industry, and agricultural TFPG contributes more to total agricultural output growth. Furthermore, differences between developed and less developed countries in TFPG are significant in agriculture but not in industry. Nevertheless, for industry there was more variation across countries within individual industries than across industries within countries. This is an important finding which confirms the importance of country-specific factors, including policy, in accounting for productivity change.

The Nishimizu-Page paper also points to some conclusions relevant for policy. For example, export growth is positively associated with TFPG, while regimes which rely on quantitative import restrictions have lower rates of TFPG. Furthermore, agricultural output growth is positively associated with industrial TFPG in regimes with few quantitative restrictions on trade. In quota constrained economies, the opposite relationship holds. That is, the ability to export and the maintenance of close links between international and domestic prices appears to have significant positive effects on TFPG.

These papers relate well to John Mellor's effort to generalise agriculture's role in economic growth, emphasising those factors that determine whether agriculture can serve as a 'principal engine of growth'. A major proposition is that acceleration of agricultural growth through technological change facilitates employment growth by relaxing the wage goods constraint on the supply of labour. Accelerated agricultural growth also increases the demand for labour. However, Mellor also notes that agriculturally-led growth tends to be less effective in Latin America than Asia due to the highly skewed distribution of agricultural land and income. Hence, growth in effective demand associated with increases in agricultural income is biased towards products that are capital-intensive and import-intensive in Latin America. In either case, investment in rural infrastructure and education are of critical importance in fostering the growth of rural-based industries characterised by low capital intensity. Indeed, as Mellor stresses, heavy taxation of agriculture to finance capital-intensive (mostly urban) public expenditures is often a major obstacle to an agriculture-led, employment-oriented development strategy.

Since Mellor is thinking in general equilibrium, it is extremely useful to have Part IV conclude with a paper by Irma Adelman, Jean-Marc Bourniaux, and Jean Waelbroeck. The authors assess precisely the issues raised by Mellor, namely the impact of 'Agricultural Development-Led Industrialisation' (ADLI) strategy. However, they make this assessment at the global level, by use of a large multi-country model which stresses sector interdependence. This very important paper shifts attention from the individual country to the global impact of ADLI strategies when they are introduced simultaneously in all less developed countries. In their words, 'would the inelasticity of world demand lead to a sufficiently large decline in world prices for agricultural commodities to more than counterbalance the favourable domestic effects of the agricultural strategy?'

The ADLI strategy is implemented by allocating a greater share of the investment pool to agriculture, and to food production in particular, while maintaining an open-economy regime. Simulations of the ADLI strategy are based on a model underlying the IBRD's *World Development Reports*. The key finding is that the ADLI policy does generalise to all developing countries as a group: world real incomes are raised with the benefit split about equally between developed and developing countries. However, ADLI is most beneficial to the middle-income NICs, somewhat less to low-income Asia, and in

Africa, the policy cannot be implemented without external aid. This brief summary only skims the surface of a stimulating paper.

These papers make an enormous contribution to the theme of the Delhi Congress. They serve to remind us just how far development economics has come since 'The Balance between Industry and Agriculture in Economic Development' was raised as one of the critical issues of debate in the 1950s.

Note

1. Brahmananda, P. R., and Chakravarty, S. (eds) *The Indian Economy: Balance between Industry and Agriculture* (forthcoming, Macmillan, India).

Part I

Sector Proportions and Economic Development: Theory and Evidence

1 Theories of Sectoral Balance

Lance Taylor
MASSACHUSETTS INSTITUTE OF TECHNOLOGY

Separating sectors in the economy means they have structure – something differs between them. For centuries economists have set up models incorporating sectors with diverse natures – subsistence *v.* modern, agriculture *v.* non- agriculture, traded *v.* non-traded – to ask how they reach balance in the short run and grow over time. Much work on economic development is concerned with such issues. An attempt is made in this review to give the flavour (and some of the substance) of this literature. Its volume is enormous; any summary must be selective at best. The emphasis here is on models with practical relevance to policy issues in the Third World.

Sectoral disaggregations are based on prior notions about why certain parts of the economy are being separated and how they affect each other. Their linkages and differences can be described in terms of national income or inter-industry accounts. The more common demarcations are set out in section 1.

Once they are defined, the next question is how sectors interact macroeconomically. Mechanisms for their mutual relationships can be classified according to the economy's macro 'closure' or (to borrow the Schumpeterian metaphor) its pattern of circular flow. Section 2 outlines six kinds of circular flow, differing as to whether basic causality comes from the side of demand or supply, and the means via which macro balance (or savings-investment equilibrium) is attained. All six variants appear in the development literature.

Once their accounting and causal structures are specified, multi-sectoral models should fit the (stylised) facts. There are well-known regularities to the growth process, described in section 3. With structure, macro closure and relevance in hand, we can next address how theories work out in practice. Section 4 takes up models where circular flow is driven by conditions of farm labour and/or food supply – the emphasis is on agriculture/industry linkages which affect resource growth and allocation. Aggregate demand enters the discussion in section 5 – food or non-traded goods limit activity in other sectors and the terms-of-trade or the real exchange rate becomes the key adjustment variable. Other commodities – capital goods or foreign resources – may constrain the system, as in section 6. Demand composition figures in section 7 where feedbacks between income distribution and sectoral growth are considered. Section 8 is about extensions of sectoral schemes to deal with new institutions and technologies for production and use of goods. Conclusions appear in section 9.

1. SECTORAL ACCOUNTING SCHEMES

Formal models put most detail on the parts of the economy that their designers want to emphasise. Sectors may be split off according to numerous criteria:

1. Consumption patterns for their products may differ, in terms of budget shares from different income flows or elasticities. Distribution models emphasises shares; elasticities underlie generalisations like Engel's Law.
2. Investment demands may differ. Capital and non-capital goods producing sectors are often distinguished.
3. The insertion of a sector into the international system can take several forms. Its product may be non-traded, with an internal price determined according to various possible rules (some discussed below). Even if the sector's output is traded, there may be negligible or weak links between its internal and external prices, with the producers or the state pocketing or paying the difference. Contrariwise, the product may be imported or exported 'competitively' in the sense that its external price determines (via the Law of One Price), or at least strongly influences, its internal price. In another specification, the direction of causation is reversed – the external prices of some exports (typically of manufactures) may follow from the internal ones after the effects of taxes and subsidies are taken into account. Finally, with fixed export supply, the foreign price will be determined by demand. The rest of the world's products obey similar distinctions.
4. Production of traded goods may be more or less intimately linked with the rest of the economy. Much trade may be 'complemen-

tary' in the sense that export products are not extensively used within the country (copper in Chile, coffee in Tanzania) while the bulk of imports cannot be produced at home. External prices of imports may determine their internal ones competitively, or the internal price may follow other rules if import levels are fixed, say by external factors or a quota that binds.

5. Sectors may differ in their technologies as measured by input-output and/or factor use coefficients; their supply parameters such as elasticities of substitution need not be the same. Rates of change of the coefficients (or 'technical progress') can vary across sector boundaries.
6. Price formation rules differ. The most popular recent variants are 'flex-prices' determined in a flow equilibrium with inelastic supply, prices based on mark-up rules or cost functions with all input prices predetermined, and prices fixed by public authorities.
7. Factor payments may be fixed in nominal terms or indexed to some commodity bundle. For example, in surplus labour models the real wage in terms of a consumption basket dominated by food is held constant.
8. In different specifications, factor inputs such as labour or machines may move between sectors flexibly, sluggishly according to behavioural rules, or not at all.
9. Claims on sectoral levels of value-added by different social groups (workers, capitalists, small proprietors, peasants, landlords) may be specified with more or less detail.
10. Rules for transforming incomes generated in different parts of the economy into demands for all sectors' products may be more or less complex.
11. Sectoral expansion may itself feed back into parameters – technical change may be influenced by past growth, for example. More directly, production cost in major sectors may fall as output expands, illustrating the importance of economies of scale.

2. PROCESSES OF CIRCULAR FLOW

Accounting balances go a long way toward describing a model's structure. However, degrees of freedom beyond double-entry bookkeeping's adding-up identities remain. Behavioural restrictions must be imposed to close the model algebraically, or define its 'closure' in economic terms (Sen, 1963; Taylor, 1983). Broadly speaking, setting closure amounts to saying whether the macro system is driven largely from the side of demand or supply, followed by a description of the means via which equilibrium is attained. Different schools of economists have strong views about how the macro system functions, i.e., they prefer distinct patterns of circular flow. At least six can be distinguished in the development literature:

1. Neoclassical models treat growth as supply-determined. Resources are fully utilised. Their intrinsic productivity determines the level of income; thrift (or saving) then sets the overall rate of accumulation and growth. Solow (1956) presents the canonical formulation. One production input (conveniently called 'labour') expands at a given rate. Other means of production – fixed and circulating capital – are reproducible within the system. Their growth rates in steady state settle down to that of labour supply. In a multisectoral version, sectoral shares of output adjust to meet the pattern of consumer demand.
2. Instead of a fixed labour supply, the real wage can be predetermined in terms of a utility level or consumption basket. A share of non-labour income is saved and (as in the full development growth model) transformed without hitches into investment demand. Sectoral proportions adjust to demand, permitting the economy to grow at a steady rate.
3. The output of one sector may have a growth rate exogenous to the rest. Favourite examples are capital goods, food, non-traded goods or capital inflows and/or exports which provide foreign exchange. Overall growth converges to the rate for the key commodity. In some models its relative price, e.g., the agricultural terms of trade or the real exchange rate, is the fulcrum for adjustment of the macro system. Other models implicitly allow underutilisation of some input, e.g., productive capacity of labour.
4. So far we have not considered independent 'injections' of final demand, e.g., fiscal spending or investment on the part of the state or entrepreneurs. With resources fully utilised, a jump in demand (possibly financed by credit creation by the banks) must drive prices up. Income flows associated with prices that do not rise as fast as most (for example, non-indexed money wage) will decline in real terms. The resulting cuts in real spending – 'enforced lacking' in Dennis Robertson's phrase, or more commonly 'forced saving' – release resources to meet the demand injection. As

Kaldor (1955) recognised, such distributional shifts under full resource use violate marginal productivity rules for production input demands. Independent investment demand functions and neoclassical distributional theory are mutually incompatible under full employment.

5. The fifth variant differs from the first four in allowing the overall level of activity to respond to demand – investment determines saving by changing output levels and composition. Sectoral rates of capacity utilisation become endogenous, and converge to a steady state configuration generating enough saving to satisfy investment (which itself may respond to sector outputs or profit rates in a further feedback loop).
6. Finally, generalising the concept of steady state, there is the idea that development consists of continual introduction of new commodities as demand for old ones stagnates. Observed demand composition does not maintain constant shares, even though the weighted average of sectoral expansion rates may hold steady. If commodities and technologies appear in regular fashion, the waves of new activity superimpose themselves into a growth process.

How does one select the relevant pattern of circular flow? In practice, hypotheses regarding model closure are difficult to test – with sufficient ingenuity any one of the six structures just discussed can be made to fit any country's historical time series. The choice therefore boils down to a combination of institutional assessment and the case a particular author wants to make. The parable of the elephant and the blind men applies; more generally, all forms of circular flow can be parables for the same economic system. Economists' views about which stories fit best are pronounced, but a survey should treat them on the same footing. Value judgements are postponed to section 9.

3. STYLISED FACTS

Over the years, generalisation about the changing roles of different sectors in the growth process have built themselves up. They usually refer to national income or inter-industry accounting aggregates, which is a weakness. As discussed above and in section 8, the commodities and technologies that comprise a sector like 'agriculture' are themselves in constant flux; the sectoral shell conceals diversity within. Bearing the problem in mind, we can still say something about relationships between broad sectors; see Syrquin (1986) for a summary complementary to the one here.

1. If we distinguish three composites – agriculture, industry, and services – the share of total value added of the first falls, and the second rises, as per capita income goes up. The service share may hold fairly steady, probably increasing at very high income levels. The agriculture/industry shift is well known, and amply documented by Kuznets (1966) and Chenery (1979).
2. Clark's (1957) work shows that the agricultural labour share also drops. Only part of the released labour moves to industry, with the balance going to services. The decline in the employment share exceeds that of value-added. Hence, agricultural labour productivity increases relative to that of the rest of the economy. Kuznets computed indexes of productivity to show that, at least in the early stages of development, differentials across sectors narrow.
3. This convergence is consistent with other considerations. The importance of food products in the aggregate consumption basket declines with real income, i.e., Engel's Law. Technical change in agriculture usually involves increased use of inputs from outside the sector – fuels and fertilisers as intermediates and machines as capital goods. Correspondingly, the sector's share of value-added in gross output and its labour/output ratio go down. All these changes underlie the tendencies in employment and GDP shares just noted. They are complemented by a reduction in the relative importance of agricultural and other primary inputs in industry's intermediate demand basket.
4. Gerschenkron (1951) observed trends in commodity prices opposite to those of production levels. Industrial products may be relatively expensive in poor countries and cheap in rich ones, due to cumulative effects of scale economies, learning, and incorporation of better and best practice techniques. As development proceeds, relative prices rise for labour-intensive commodities – agriculture in the early stages and especially services.
5. The rise in the service share of GDP in rich countries is consistent with these price shifts and also consumer demand patterns. Though productivity advance is notoriously difficult to measure in services, the consensus is that it is slow. Hence, as Rothschild

(1986) shows, cross-sectoral labour productivity differentials tend to widen in rich countries, reversing Kuznets's narrowing at lower income levels (point 2 above).

6. The share of value-added imputed to wage labour rises is most sectors, in part because independent family-owned business wanes. In consequence, the contribution of new capital stock to output increase as measured in conventional total factor productivity (TFP) calculations is higher in poor countries than rich. Industrial TFP growth exceeds that in agriculture in absolute terms, but relative to growth of other inputs is less.
7. International trade becomes less complementary to the production structure as development proceeds (Chenery, 1975). Primary exports fall off in importance (McCarthy, Taylor, and Talati, 1987) while imports shift from technically obligate purchases of intermediate and capital goods toward products more similar to those produced at home. At the sectoral level, a period of import substitution may precede export expansion (Urata, 1986). Depending on policy, trade may become more competitive in the sense that the law of one price links imported and nationally produced goods. This observation is consistent with the rise in service prices in rich countries noted in point 4 – services are typically non-traded so that their prices cannot be held down by foreign competition (Balassa, 1964).
8. Kuznets (1985) conjectured that income inequality first widens and then narrows during the course of economic development – such a shift is consistent with labour productivity rising in an initially large agricultural sector while its GDP share falls. The hypothesis has been amply debated, but holds up weakly in cross-country comparisons. Whether it will continue to be true in rich countries as low productivity service sectors expand remains to be seen.
9. Investment and national saving shares of GDP increase as per capita income goes up. How much of the saving increase can be attributed to rising average rates of household savings, movements in the personal income distribution, and shifts among household, business, and government saving has never been adequately explored. But the overall increase serves as a base for a number of growth models.
10. Finally, subsectoral structural change can be discerned. Hoffman (1958), Chenery (1979) and others point out trends in the GDP shares of manufacturing sectors. Food products and textiles

dominate at low income levels; then intermediates, more technically complex consumer products, and capital goods come in. The process is accelerated in large economies in which GDP shares of foreign trade are relatively low and those of industrial sectors high. These phenomena are usually attributed to the interaction of economies of scale in production with relatively large internal markets, and form a major component of the industrial strategy lore.

4. SUPPLY-LIMITED INDUSTRIAL GROWTH

Labour 'supply' conditions in growth models can be stated in either quantity or price terms. The classical economists preferred the latter, setting a real wage from forces outside the economic system – Malthus's population dynamics and Marx's reserve army are well-known examples. Both represent variants on the second type of circular flow discussed in section 2.

Marx provies a useful introduction. In stripped-down form, his model is built around pervasive unemployment (the reserve army) and technical progress or choice of technique (the organic composition of capital). The scheme can be extended to two or more industrial sectors, as we will see in section 6, but for the moment we stick with one-sector dynamics only. In an upswing, employment rises and the wage is bid up from the level consistent with social norms. A turning point is passed, beyond which capitalists undertake technical innovation or substitution of machines for labour. The organic composition rises, setting the stage for a downturn. Resources are tied up in the production process, so that demand does not grow rapidly enough to absorb a growing supply of commodities; a realisation crisis ensues. Long-term expansion of the capitalist system involves a sequence of such cycles, perhaps of increasing severity as time goes on.

Similar themes entered the development literature in the two-sector model of Lewis (1954, 1958). In his story the 'reserve army' becomes the 'subsistence sector', but retains the function of providing labour at a socially-determined wage until unemployment dries up. Meanwhile, modern sector capitalists accumulate and employ additional labour from subsistence. As the mass of capital grows, so do total profits and saving therefrom – the national saving rate will rise in line with stylised facts listed above. Beyond a turning point, the labour supply curve slopes upward. Capital-labour substitution in industry becomes important with real wage determined by labour demand and supply. Lewis's vision of growth has less drama than Marx's, but the common elements stemming from the classical heritage of both are clear.

Despite his mellow transition from subsistence to a modern economic structure, Lewis was soon subject to fundamentalist attack. First, as Lluch (1977) emphasises, his subsistence sector made up of farmers, casual workers, petty traders, retainers, and women, was transformed by the profession into 'agriculture'. This reinterpretation set the stage for a purely academic debate among proponents of different agriculture/industry models. Dixit (1973) and Kanbur and McIntosh (1986) provide surveys.

In the best-known contribution, Fei and Ranis (1964) set up a model with two turning points – when food supply begins to decline as labour is withdrawn from agriculture, and when the marginal product of rural labour rises to the institutionally fixed urban wage. The parallels with Marx went unstated, as did the similarity of Fei and Ranis's landlords to Marx's capitalists. In Dixit's words, the landlord ' . . . should be eager to save. He should sell his surplus to industry and should transfer his savings to industrial entrepreneurs. He should be eager to innovate, and thereby improve the technology in agriculture'. The social origins of this master of manipulation do not flow naturally from the model.

Jorgenson (1961, 1967) was more technocratic still, switching focus from labour to food supply. He worked out a formula emphasising that productivity increase in agriculture is required to generate sufficient food to satisfy urban/industrial demand (in line with sylised facts above). Subsequent work continued to stress technical change in agriculture, to be elicited by adequate farm-gate prices. Schultz (1964, 1978) is the central figure; his work underlies the current mainstream approach to agricultural development.

The intellectual time trend in these models went from class distinctions and saving growth mediated by distributional change to technical advance triggered by price policy – marginal conditions enveloped Lewis's mild radicalism. Both early and later models addressed issues of sectoral balance, but from divergent perspectives. This line of thought exhausted itself by the late 1960s, though the neoclassicised synthesis has had major policy significance since.

Other work with supply-determined models has emphasised three themes – labour allocation across sectors, urban unemployment, and economic dualism. Sectoral migration models date at least to Simon (1947). He showed that if labour is the only production input with full employment and equal rates of productivity growth in two sectors, then workers will move toward the sector with income-elastic demand. Baumol (1967) argued that a non-progressive sector with low productivity growth will lose all its labour in the long run if its demand is sufficiently price-elastic (or the recipient sector's demand price-inelastic). Similarly, if agriculture faces price-inelastic demand, farmers' incomes will be hurt by rapid technical advance (Houthakker, 1976). Kaneda (1986) generalises these results – which all boil down to applying Slutsky demand decompositions to full employment income flows in a one-input model – to emphasise the importance of agricultural technical advance in liberating resources for industrial growth in Meiji Japan (the 'Ohkawa thesis'). The moral resembles Jorgenson's discussed above, but focuses more on the market for labour than food.

The second issue is urban unemployment. Its characterisation as the outcome of a search process by Todaro (1969) and Harris and Todaro (1970) spawned an enormous literature. Their closure is the first type mentioned in Section 2 above (labour supply limited) with the twist that some people in urban areas are unemployed as they seek jobs with a less than unitary probability of success.

Two state variables describe dynamics in a general equilibrium version of this model – allocation of capital stock between the sectors and a migration function whereby people move according to the differential between the rural wage and the expected urban wage (actual wage times probability of finding a job). Following Bartlett (1983), Kanbur and McIntosh (1986) point out that equilibrium is only saddlepoint stable. If the initial rural population lies above a certain level:

> its rate of growth will dominate the change in the unemployment rate which provides the signal for the sectoral reallocation of labour . . . The share of population in agriculture will then *rise*, contrary to the stylised facts. However, if the share of agriculture and the unemployment rate are lower than certain critical values, then the population share agriculture will go to zero, contrary to the observation that this share typically stabilises around a low value.

As a theory of transition, the second path may make sense, though it is hard to believe that the economic growth process is driven solely by a search for urban jobs. The migration pattern is consistent with Kuznets's conjecture about a U-shaped relationship between income equality and development (stylised fact number 8 above). The Kuznets story can also be rationalised in a full employment context by appropriate assumptions about technology in the two sectors (specifically, significantly lower capital-labour elasticities of substitution in industry than agriculture) as in Taylor (1979, ch. 11).

Finally, one can ask how asymmetry or 'dualism' between agricultural and industrial sectors affects their mutual balance. Technological differences creating a 'factor proportions problem' were stressed by Eckaus (1955). Chichilnisky (1981) elaborates the story in a model with a fixed-coefficients technology, assuming that labour supply is fully employed but elastic to the real wage, both sectors utilise labour and a given stock of shiftable capital, and agriculture is labour-intensive. A movement in the terms of trade toward agriculture will bid up the real wage, by the Stolper-Samuelson (1941) theorem. Labour supply will increase, and the production mix will shift toward food by the Rybczynski (1955) theorem. All appears harmonious, except that consumption of food products by labour may rise enough to make the marketable or exportable surplus increase only slightly or even decline.

This 'perverse' price response of excess food supply is the consequence of simultaneous dualism in production structure, labour supply, and patterns of demand. It can be reproduced with fixed labour supply and substitution in production (Taylor, 1979), and helps rationalise econometric evidence that the macro response of marketed food surpluses to price is often small or negative. Lele and Mellor (1981) work through a similar model emphasising technical change. If new technology raises the labour share, then the resulting increase in food demand may make marketed surplus problems worse. On the other hand, biased technical advance that increases food output can ease an urban wage constraint. Lele and Mellor don't ask how extra food sales are to realised in an economy in which full employment is not presumed – state intervention to subsidise both consumption and stock accumulation may be required (McCarthy and Taylor, 1980). The problem is likely to be practically important, as argued in the following section.

Distributional problems in dual economies can also be acute, as both Eckaus and Chichilnisky point out. Indeed, Roemer (1982) constructs a theory of exploitation based precisely on dualism in production technology, property ownership, and demand. The mathematics becomes ponderous, but as a sidelight does generate Lenin's and Mao's famous five-way classification of peasants – pure landlords who don't work at the top, pure landless who hire out all their labour at the bottom, and three intermediate classes (those who do some work on their land and hire in, those who hire neither in nor out, and those who hire out part of their labour) in between.

Most of the recent work on dualism has not been put into growth model form, but the effort would be worthwhile. Its emphasis could be either on adding social content to the neoclassical machines, or sharpening Lewis's mild class distinctions without sacrificing them to technique.

5. VARYING TERMS OF TRADE

The models just discussed are all closed from the side of supply. They do not cast much light on the intersectoral terms of trade, which are intimately linked with aggregate demand. Demand pressure against a sector with inelastic supply will swing relative prices in its direction, in the third kind of circular flow from section 2. Models stressing this linkage have recently been worked out for agriculture and non-agriculture, for non-traded and traded goods, and (at the world economy level) for the 'South' as a primary product exporter to an industrialised 'North'.

Terms of trade models date at least to the controversy between Ricardo and Malthus about whether the English Corn Laws limiting grain imports should or should not be repealed. A sketch of an agriculture-limited economy emphasising the distributional role of food prices was Preobrazhenski's (1965) contribution to the Soviet debate about industrialisation in the 1920s. Kalecki (1976) gave a version in lectures in Mexico City in 1953 which was adopted or reinvented by many others, e.g., Sylos-Labini (1984), Kaldor (1976), Taylor (1983). In the recent versions, the 'industrial' sector has prices formed by a mark-up over prime costs, while the price of 'agricultural' goods varies in the short run to clear the market with supply fixed or not very responsive to price changes.

Output in the industrial sector is demand-driven. It produces a commodity used both for consumption and capital formation. Agriculture produces food, and partly finances overall investment through savings flows. Urban forced saving (see Section 2 item 4) is central to the workings of the model, since the urban wage is not fully indexed to the food price. On the other hand, agricultural savings need not play a major role. Empirically, Quisumbing and Taylor (1986) argue that the share is only a few per cent of GDP except in countries where large agro-export flows can be easily tapped for investment finance.

Suppose that the terms of trade shift toward agriculture, say from a reduction in food imports. Manufactured goods output can either rise or fall. It will be pushed up by increased demand from higher rural income, but also held down by reduced real urban spending power (real wages decline). The implication is that if the farm sector is relatively small, a cheap food policy based on higher imports will increase industrial output. The urban sector is doubly benefited – food prices fall and production expands. Though he used a different model to argue his point, this outcome is consistent with Ricardo's advocacy of the Corn Laws.

The alternative view was espoused by Malthus who reasoned (more or less cogently) on aggregate demand lines. Agriculture is responsible for a large proportion of consumption demand. Farmers (or landlords for Malthus) hit by adverse terms of trade will cut spending enough to reduce overall economic activity. Which distributional configuration applies in developing countries today is highly relevant for policy. The answer seems to go either way in practice (McCarthy and Taylor, 1980; Londono, 1985).

Besides import policy, other state interventions can affect the terms of trade. Fiscal expansion or increased investment, for example, will bid up the flexible price by injecting aggregate demand. As noted above, forced saving on the part of the urban groups whose incomes are not fully indexed to the food prices will occur. Ellman (1975) argues that food price-induced forced saving supported the Soviet investment push of the early 1930s, in contrast to Preobrazhenski's suggestion that the terms of trade be shifted *against* agriculture to extract an investable surplus. Along political economy lines, Mitra (1977) stresses distributional consequences of shifts in relative prices between two staples in India – rice and wheat – and industrial goods.

Models with a flex-price sector apply to issues besides the agrarian question. They are frequently stated with one sector producing traded goods (with prices determined by a mark-up or the Law of One Price) and the other sector making non-traded goods with a variable internal price. Formalisation dates at least to Salter (1959) and Swan (1960); another observer of Australia, Cairns (1921), used the same set-up to discuss the political economy of gold booms over a century ago. The latest incarnation is the 'Dutch Disease', e.g., Corden (1984).

The key price is between traded and non-traded goods – the 'real exchange rate'. When overall demand rises, say from attempts to spend the proceeds of better external prices for a primary export or from capital inflows, part will be directed toward non-tradeds. If supply is inelastic, their price will rise or the real exchange rate (the ratio of foreign to home goods prices) will appreciate. Industrial exports sensitive to the exchange rate will decline; demand for home products similar to importables may stagnate as well. In Boutros-Ghali's (1981) useful terminology, both internal and external diversification (of the production and export baskets respectively) are reduced.

A third application of the terms of trade model is in global macroeconomics built around the South as a primary product exporter and an industrialised North. The South is described as a Lewis-type economy while the North may be demand-driven or obey supply-limited circular flow. The South is triply dependent in that terms of trade for its exports depend on activity in the North, it imports a substantial fraction of its capital goods, and (barring unlimited capital flows) its investment is ultimately limited by its own saving plus whatever transfers come from the other party. Comparative static exercises can illustrate the implications of dependency. As in Houthakker's (1976) model discussed above, a productivity increase in the South will reduce its income and/or slow its rate of growth because its export commodity is sold in a flex-price market subject to low income elasticities of demand in the North. Also, patterns of convergence to steady state growth for the two regions can be complex, but the South is likely to suffer doubly (slower export sales and lower terms of trade) from adverse shocks to the system. Ocampo (1986) and Kanbur and McIntosh (1986) survey the growing literature.

A final point is that the terms of trade model, already a potted version of the agrarian question, is susceptible to simplistic optimisation games. An example is treating Preobrazhenski's idea that the terms of trade should be shifted against agriculture as a problem in optimal taxation – the latest of two decades's worth of papers on the topic is by Sah and Stiglitz (1984). Manipulations of Lagrangian functions to find optimal tax rates for the peasantry are mildly illuminating so long as the results are not taken as serious policy advice. But if they are, then one has to ask if they make any sense in a given institutional context, e.g., Bukharin's (1971) observation that if Preobrazhenski's scheme were rigorously imposed in the 1920s Soviet economic system, the kulaks would simply abscond. (Neither author foresaw Stalin's drastic modifications of the system a few years later; both suffered greatly thereby.) Lines of policy also depend on the macro closures their proposers presuppose. Rattso (1986) observes that Sah and Stiglitz, like all optimal taxationists, prefer a neoclassical closure – savings determines investment and full employment is maintained. With demand-driven circular flow, their distributional results can be shifted to those reported by Ellman (1975) – the Soviet state as the agent of accumulation benefits from forced saving induced by inflation acting on a working class whose numbers are increased and political leverage dramatically reduced by the decapitalisation campaign. The political implications cannot begin to be addressed by economic models but are crucial to those who seek to apply them. All 'optimal' policy recommendations suffer such problems, as we will see in the capital- and foreign exchange-constrained models discussed in the following section.

6. SCARCE MACHINES OR FOREIGN EXCHANGE

Sismondi following the Physiocrats, and Marx, were among the first economists to formulate coherent models of multisectoral growth. Sismondi's demand side analysis is taken up in the following section. The reproduction schemes in volume II of *Capital* lead naturally here to models of development which stress limits to growth stemming from scarce products of key sectors.

To a low order approximation, Marx's 'departments' can be called sectors. Their rules for balance in simple or expanded reproduction (zero or positive growth) can be set out on intuitive grounds. Department I must reproduce enough constant capital (or 'machines') for itself. The rest of its costs – variable capital or goods-in-process plus labour payments, and surplus value or profits – must be shipped to Department II. There, the imports will be used as constant capital. From these relationships, balanced departmental proportions can be calculated in numerical examples as by Marx or via Laplace transforms (dated labour inputs discounted at the rate of steady growth) as in an elegant formulation by Foley (1986).

Not surprisingly, a sectoral structure based on the Departments underlay planning analysis during the Soviet industrialisation debate.

G. A. Feldman set up a two-sector capital and consumer goods model to address the balance question; a few decades later it was restated for North Atlantic economists by Domar (1957). More or less independently, the Indian statistician turned planner, Mahalanobis (1953), reproduced the same formulation.

Analytically, the story is simple; the social implications run substantially more deep. In contemporary accounting emphasising resource flows rather than Marxian value magnitudes, Sector I makes capital goods and Sector II consumer products. Both use capital goods as inputs and, once installed, machines cannot be moved. Growth overall is limited by the fraction of capital goods output directed to new investment in that sector. It is easy to show (Taylor, 1979) that with a constant reinvestment share, Sector II growth will converge to Sector I's steady rate.

The model is not fully specified macroeconomically. In his reproduction schemes, Marx was concerned with sectoral consistency and (implicitly) saving-investment balance. Mahalanobis did not go so far; nor did he allow a possible role for foreign trade. Successors set up optimisation exercises to find dynamic solutions to an already incomplete description of the economy. The usual result is that, to maximise a discounted integral of utility from consumption, it is optimal initially to push the reinvestment rate to the highest feasible level, and later switch to balanced growth. During the initial phase, the level of consumption goods output declines if machines installed in that sector depreciate rapidly enough.

The recommendation scarcely inspires confidence – putting all her eggs in some model's 'optimal' big-push basket is something that a prudent planner in a state with non-Stalinist levels of political control would never do. Other difficulties are that the models (and their disaggregated input-output cousins, for that matter) do not give enough detail for disaggregated resource allocation, and ignore the issue of reaching saving-investment balance.

The two-gap model of Chenery and Bruno (1962) takes up the saving-investment issue when the economy's potential growth is constrained by foreign exchange. A 'complementary' trade structure along the lines of point 4 of section 2 is presupposed. Imports are required as an unavoidable proportion of capital formation (most equipment must be imported) and as intermediate inputs into domestic production. Most goods made within the country are essentially non-traded. Export volumes (perhaps from an enclave) are subject to strict upper bounds, representing practical difficulties in selling more primary products or breaking into markets for non-primary ones. The main analytical question is what happens when net capital inflows (also pre-determined) change. Bacha (1984) sets the potential responses out.

If saving determines investment (circular flow, see Section 2 item 1), capital formation rises one-for-one with the capital inflow. The part of the increased current account deficit not devoted to capital goods can go to imports of intermediates or consumer products. The gap between the saving and trade gaps is covered by trade, as Findlay (1971) observes in dismissing the two-gap dilemma. However, suppose that dollar inflows decline. Then imports must fall or exports rise to maintain foreign balance. Beyond a point, exports are limited from above and import reduction will require output contraction to absorb fewer intermediate goods. Investment and/or domestic economic activity are cut to meet the external constraint.

In contrast, suppose that any extra dollars are devoted to capital formation – investment can rise more than one-for-one since it also draws on domestic production. A quantum of saving beyond that realised through extra machinery imports is required from domestic sources. The obvious possibilities are distributional change underlying forced saving, or capacity adjustment. Either may not be easy to arrange. The potential gap between the gaps persists, whether investment responds to available capital goods imports or to saving.

Longer-term prospects can be analysed either by extrapolation of the 'more binding' of the internal and external balance restrictions over time, or optimisation of some dynamic welfare function subject to the balances as constraints. Results of optimisation parallel those in the capital/consumption goods model. The utility-maximising solution is to specialise initially in activities like export promotion or import substitution which generate foreign exchange (to import capital goods) and later relax into balanced growth. Sensible advice, possibly, but easier said than done. In binding gap projections, scarce saving gives rise to a modified Harrod-Domar formula. If the trade gap binds, computation centres on import propensities. Either sort of extrapolation is obviously incomplete in ignoring which macro variable(s) adjust to make the gaps equal *ex post*. In practice, the two-gap dilemma is often 'resolved' by restrictions on state policy action. The model serves a useful purpose in directing attention at that fact.

7. DEMAND-DRIVEN GROWTH

Modern development economics began with the writings of Nurkse and Rosenstein-Rodan during and just after the Second World War. They emphasised causes of slow output expansion originating on the side of demand. Rosenstein-Rodan's (1943) shoe factory is the epitome of demand-constrained circular flow. The owner does not think it worthwhile to step up his operations because he knows that not all the income so generated will flow back into increased purchases of shoes – there must be economy-wide expansion to ratify his own investment plans. Nurkse's (1953) 'vicious circles' describe a similar problem, taking Schumpeter's metaphor for granted. Basu (1984) provides a formalisation involving a fixed money wage and firms in two sectors which select production levels depending on wage costs and anticipated marginal revenues from expanding sales down kinked demand curves. They can easily arrive at an equilibrium in which not all available workers find jobs. Fuller employment could be attained by increasing production in both sectors in line with consumers' tastes. Such expansion is at the core of 'balanced growth'.

This diagnosis was debated vigorously in the 1950s. A blow-by-blow recapitulation cannot be attempted here, but two controversies can be noted:

1. Adherents of balanced growth stressed the virtues of investment planning to keep sectoral expansion rates in line. Critics like Hirschman (1958) and Streeten (1959) wanted 'unbalanced growth' to shock the system from low-level circular flow. In their view, the development process is characterised by uneven advance of different sectors, disproportions and disequilibria, with inflationary and balance-of-payments tensions arising at different points. Under such conditions, policy should be oriented to investment strategies having the best chance of being self-propelling, i.e., of being able to induce further investments to correct imbalances created at previous stages. Despite their penchant for evocative labels for ill-defined economic effects (linkages, cumulative processes, growth poles, etc.), economists in this school were more practical than advocates of balanced growth. The latter never really specified how their plans for expansion were to be formulated, let alone put into practice. Input-output models and shadow-pricing of investments were their chosen instruments to

lead to balance, but neither could fulfil the detailed resource-allocation task.

2. A question of resource mobilisation to meet ambitious growth targets arises. Increasing output requires investment, with the saving counterpart to come from somewhere in the system. Rosenstein-Rodan (1961) laid great emphasis on the potential output gains implicit in indivisibilities and economies of scale. The saving for his investment 'big push' might come from profits generated by decreasing costs. And as Rodan's London School mentor Young (1928) pointed out, expansion of output makes room not only for more, but for more sophisticated capital equipment. The theme reoccurs in contemporary discussions of export-led growth.

Apart from the balanced/unbalanced growth controversy, demand-driven models have appeared in other contexts over the years. One example, already noted, is Sismondi's argument that growth in post-Napoleonic France could be stimulated by income redistribution toward workers. They would buy more textiles, leading to investment and growth in that 'leading sector'. Similar themes appeared in the structuralist analysis of problems of industrialisation in Latin America, reviewed by Lustig (1980). A contemporary debate is unfolding in India about slow industrialisation there.

To understand the issues, it is useful to begin with growth stagnation in a model with one sector, and then bring in multisectoral complications. Dutt (1984), formalising a literature beginning with Kalecki (1971) and Steindl (1952), points out that if investment responds to increased capacity utilisation as in an accelerator, then a shift in the income distribution against high-consuming wage-earners can lead to slower growth. The fall in wage income reduces consumption demand, and capital formation slows down.

Now suppose that in addition to changing the economy-wide marginal propensity to save, income shifts between classes also modify consumption demands by sector. In a two-sector model, Taylor (1985) argues that incomes policies under contemporary circumstances may foster neither redistribution nor growth. The wealthy may prefer to consume services and sophisticated manufactures, for example, and poor people simple industrial products and food. The implication is that consumption from profit income flows may be labour-intensive. By the Rybczynski (1955) theorem, a tax-

cum-transfer aimed at shifting the distribution toward labour will then reduce the wage share in a two-sector system. Effects on growth depend on whether investment demand in an accelerator formulation responds more strongly to demands for commodities preferred by rentiers or workers. If the former, overall growth and capacity utilisation can decline. The transition between steady states will be stable under the Harrodian condition that the inducement to invest created by new capacity utilisation (appropriately averaged across sectors) exceeds the marginal propensity to save.

Latin American structuralists like Tavares (1972) and Furtado (1972) more or less viewed the world along such lines, but with causality reversed. If industrialisation beyond production of simple goods like food and textiles is to occur, they said, then income concentration to sustain demand for more sophisticated commodities (especially those with production technologies subject to economies of scale) is unavoidable under present social conditions. Taylor and Bacha (1976) provide a formalisation. Investment responds to increased consumer demand for 'luxuries', leading to forced saving in a full capacity use specification. The distributional shift stimulates luxury sales, and further investment growth. The economy goes through a transition between steady states without and with luxury goods production.

The Indian debate follows Sismondi more closely, though with emphasis on the role of public capital formation and the state enterprises. Demand composition may indeed underlie slow industrial growth, e.g., but Dutt (1984) and Nayyar (1978). Chakravarty (1984) further cautions, like the Latins, that unless steps are taken to preclude regressive distributional shifts, more rapid growth may benefit only the top segment of the income spectrum. A counterargument is that stagnation may result from limited capacity in infrastructure (Ahluwalia, 1986), along the lines of the last section. The situation is said to be exacerbated by inefficient public enterprise and bureaucratic meddling which dampens private sector animal spirits.

Attempts at empirical analysis of this complex of ideas have been incomplete at best. A spate of studies in the 1960s, partly surveyed by Clark (1975), found no great variation in *marginal* propensities to consume by sector – the relevant parameters – across income classes. Input-output computations thus showed that income redistribution would not change sectoral output composition very much. Directions of effects also varied across economies. In some cases redistribution toward the rich would stimulate employment via the service sectors;

in others not. The major problem with this empirical work is that it did not consider feedbacks of changes in demand to investment and productivity gains – this is a promising area of research. It can be extended to examine the effects of trade interventions. For example, Buffie (1986) and Taylor (1986) show that currently modish 'equal incentives' trade policies need not maximise overall output growth in investment-driven models.

8. PATTERNS OF GROWTH

Stylised facts about how sectors behave over the course of development amount to a catalogue of phenomena without an integrating theory. Economists lack an analytical framework able to contain the demand, supply, and distributional shifts that underlie observed patterns of growth. The Walrasian system, with its dogged insistence that prices will adjust to bring overall full employment, is inappropriate. Can one do better? Several possibilities are open for theoretical and empirical work:

1. Pasinetti (1981) uses a dynamic input-output model to ask when stable growth with differing sectoral demand patterns and rates of technical change may prove possible. One conclusion – first stated by Hawkins (1948) – is that conditions on sectoral rates of demand and productivity growth required to assure full employment of capacity and labour are very strong – so strong that they are virtually impossible to satisfy without institutional changes in labour force participation rates or the length of the working day.
2. If Engels phenomena are restated in terms of initial increases, saturation and final decline of consumption of specific goods, then it is apparent that maintaining full employment requires the continual introduction of new commodities. In advanced economies, think of products in common use in the 1980s (Big Mac hamburgers, Walkperson tape players, video cassettes) that were unheard of a decade before. A role is naturally opened for entrepreneurs, and absorption of new production technologies could give rise to distributional problems.
3. Another sort of demand shift that can be captured in an input-output framework stems from changes in social relationships and the role of the state. An old point raised by Kuznets (1966) and elaborated by Rothschild (1987) is that much state activity –

regulation and environmental protection, for example – is functional to the production process under prevailing political norms. The same applies to many consumer demands, e.g., for transport services. Rothschild suggests the use of 'extramediate' matrices to explore the strength and stability of these linkages. Institutional change during the development process would surely shift the relative magnitudes of coefficients and flows.

4. Changes in demand patterns and input-output coefficients as functions of per capita GDP can be fed into the usual Leontief balance equation to generate shifts in sectoral value-added shares along the lines of points 1 and 3 of section 3 (Chenery, Shishido, and Watanabe, 1962). Such decompositions reveal regularities of Engels phenomena and coefficient changes as development proceeds; they can be extended toward country typologies (Chenery and Syrquin, 1986). Though it illuminates patterns of change in the past that may persist in the future, the open Leontief model used in these studies does not tie income generation back to demand in full circular flow. Analyses based on general equilibrium models under different closures are an obvious potential extension of this work.
5. The input-output investigations could naturally be extended to deal with changes in trading patterns, from complementary to non-complementary external activity, for example. They could also incorporate the relative price shifts pointed out by Gerschenkron (point 4 in section 3), but again a closure rule or explicit theory of price determination would have to be specified.
6. Finally, relative backwardness may determine both the sector-specific pattern of industrialisation and the degree of state intervention required to support it – Gerschenkron's (1962) other famous hypothesis. Beyond historical studies for Europe and Japan, the notion has not been elaborated empirically. How well does it apply to the NIC's?

To these input-output questions, one should add the problem of timing. Basing her theory on Marx's circuits of capital, Luxemburg (1968) pointed out that, with lags in the system, realisation of all potential labour power in the economy could prove impossible – an anticipation of Hawkins's more technology-oriented point. Such difficulties underlie Marx's cyclical crises discussed in section 4, and are formalised with specific lag structures for production, realisation of demand, and recommitment of sales proceeds to production by

Foley (1986). His accounting relies on integral equations, but for statistical application it would have to be restated in terms of lagged Leontief current and capital input matrices. The similarities with both the Pasinetti scheme and sectoral planning models like that of Eckaus and Parikh (1968) would then be more apparent.

A last area of investigation involves distribution and demand composition. On theoretical and empirical grounds respectively, Pasinetti (1981) and Burns (1934) argue that profit rates in 'new' rapidly-growing sectors are likely to be high – the firms need the cash flow for reinvestment. But if the fast-growing sectors don't generate all their own demand, how will sales for their products stay strong, e.g., how did Italy after the Second World War, and Brazil in the 1960s, sustain demand for rapid expansion of production of cars? The questions raised by the Latin American structuralists and Rosenstein-Rodan have yet to be adequately addressed.

9. CONCLUSIONS

Maintained hypotheses about directions of macroeconomic causality are key to the results of most models discussed herein. Whether the models adequately address empirical generalisations about the development process is another question. Sensible economics should incorporate both elements.

On the whole the models that do best at reproducing the data are the input-output simulations discussed in the previous section. But since they do not fully close the income-expenditure nexus, they provide little insight into dynamics or distributional change. Thinking about these issues requires strong hypotheses about directions of macro causality. On the terms of trade in the agrarian question, for example, Preobrazhenski and Malthus used models based substantially on the same equilibrium relationships, but came to opposite conclusions on the basis of their politics and judgements about directions of causality. Malthus thought cheap food would reduce economic activity by curtailing landlords' spending; Preobrazhenski thought the state could direct their income loss toward increased capital formation. At a given time and place, both could not be right. Both could be wrong if the institutional rules were to change, as under Stalin.

Similar conclusions apply to other models. Agriculture/industry balance when savings drives investment may pivot on distributive shares (Lewis), technological progress (Ohkawa), or intersectoral migration (Harris-Todaro). But if investment drives saving, the terms of trade and activity levels become central as in structuralist theory. Incomes policies are likely to matter little in full employment neoclassical circular flow – redistribution at most affects growth by altering the savings supply. But depending on factor intensities, marginal propensities to consume by sector, and investment responses, redistribution could trigger either rapid or slow cumulative growth processes in investment-driven models. In the first case, foreign trade policies generating equal incentives across sectors will be near-optimal; they could brake expansion in the second.

There is no philosophers' stone to cull these theories – two centuries of economists' debate attest to that. In a particular circumstance one causal structure may be more appropriate than another. However, the decision is mostly external to the models. Any description of circular flow can have enough epicycles added to fit the numbers. Criteria such as Occam's Razor, historical relevance, and political implications have to be used to select which one best fits the situation at hand. What *is* surprising is that some economists – principally neoclassicals at this juncture – really think that they can come up with a broadly applicable uniform model. One clear lesson from the history of economic change is that such a reductionist approach is bound to fail. What may serve better is a renewed, classically-based, attempt to formulate development theories on the basis of institutions, macroeconomic power in the sense of being able to impose one's demands upon the system, and sectoral and class relationships that do (or are likely to) exist.

References

Ahluwalia, I. J. (1986) 'Industrial Growth in India: Performance and Prospects', *Journal of Development Economics*, vol. 23, pp.1–8.

Bacha, E. L. (1984) 'Growth with Limited Supplies of Foreign Exchange: A Reappraisal of the Two-Gap Model', in Syrquin, M., Taylor, L. and Westphal, L. (eds) *Economic Structure and Performance: Essays in Honor of Hollis B. Chenery* (New York: Academic Press).

Balassa, B. (1964) 'The Purchasing-Power Parity Doctrine: A Reappraisal', *Journal of Political Economy*, vol. 72, pp. 584–96.

Bartlett, W. (1983) 'On the Dynamic Instability of Induced-Migration Unemployment in a Dual Economy', *Journal of Development Economics*, vol. 13, pp. 85–95.

Basu, K. (1984) *The Less Developed Economy* (Oxford: Basil Blackwell).

Baumol, W. J. (1967) 'Macroeconomics of Unbalanced Growth: The Anatomy of the Urban Crisis', *American Economic Review*, vol. 57, pp. 185–96.

Boutros-Ghali, Y. (1981) *Essays on Structuralism and Development*, Department of Economics, Massachusetts Institute of Technology, unpublished PhD dissertation.

Buffie, E. (1986) 'Commercial Policy, Growth, and the Distribution of Income in a Dynamic Trade Model', Department of Economics, University of Pennsylvania.

Bukharin, N. I. (1971) *Economics of the Transformation Period* (New York: Bergman Publishers).

Burns, A. F. (1934) *Production Trends in the United States since 1870* (New York: National Bureau of Economic Research).

Cairns, J. E. (1921) 'The Australian Episode', reprinted from an 1859 paper in F. W. Taussig (ed.) *Selected Readings in International Trade and Tariff Problems* (New York: Ginne Co).

Chakravarty, S. (1984) 'India's Development Strategy for the 1980's', *Economic and Political Weekly*, vol. 21, pp. 845–52.

Chenery, H. B. (1975) 'The Structuralist Approach to Development Policy', *American Economic Review* (Papers and Proceedings), vol. 65, pp. 310–6.

Chenery, H. B. (1979) *Structural Change and Development Policy* (New York: Oxford University Press).

Chenery, H. B. and Bruno, M. (1962) 'Development Alternatives in an Open Economy: The Case of Israel', *Economic Journal*, vol. 72, pp. 79–103.

Chenery, H. B., Shishido, S. and Watanabe, T. (1962) 'The Pattern of Japanese Growth, 1914–1954', *Econometrica* vol. 30, pp. 98–139.

Chenery, H. B., and Syrquin, M. (1986) 'Typical Patterns of Transformation', in Chenery, H. B., Robinson, S. and Syrquin, M. *Industrialization and Growth* (New York: Oxford University Press).

Chichilnisky, G. (1981) 'Terms of Trade and Domestic Distribution: Export-Led Growth with Abundant Labor', *Journal of Development Economics*, vol. 8, pp. 163–92.

Clark, C. (1957) *Conditions of Economic Progress* (3rd edn) (London: Macmillan).

Clark, P. B. (1975) 'Intersectoral Consistency and Macroeconomic Planning', in Blitzer, C., Clark, P. and Taylor, L. (eds) *Economy-Wide Models and Development Planning* (New York: Oxford University Press).

Corden, W. M. (1984) 'Booming Sector and Dutch Disease Economics: Survey and Consolidation', *Oxford Economic Papers*, vol. 36, pp. 359–80.

Dixit, A. (1973) 'Models of Dual Economies', in Mirrlees, J. A. and Stern, N. H. (eds) *Models of Economic Growth* (New York: Wiley).

Domar, E. (1957) 'A Soviet Model of Growth', in *Essays in the Theory of Economic Growth* (New York: Oxford University Press).

Dutt, A. (1984) 'Stagnation, Income Distribution, and Monopoly Power', *Cambridge Journal of Economics*, vol. 8, pp. 25–40.

Eckaus, R. (1955) 'The Factor-Proportions Problem in Underdeveloped Areas', *American Economic Review*, vol. 45, pp. 539–65.

Eckaus, R. and Parikh, K. (1968) *Planning for Growth: Multisectoral, Intertemporal Models Applied to India* (Cambridge, Mass.: MIT Press).

Ellman, M. (1975) 'Did the Agricultural Surplus Provide the Resources for the Increase in Investment in the USSR during the First Five Year Plan?' *Economic Journal*, vol. 85, pp. 844–64.

Fei, J. C. H. and Ranis, G. (1964) *Development of the Labor Surplus Economy*, (Homewood, Ill.: Irwin).

Findlay, R. (1971) 'The Foreign Exchange Gap', in Bhagwati, J. N. *et al.* (eds) *Trade, Balance of Payments and Growth: Papers in International Economics in Honor of Charles P. Kindleberger* (Amsterdam: North-Holland).

Foley, D. (1986) *Money, Accumulation, and Crisis (New York: Harwood Academic Publishers).*

Furtado, C. (1972) *Analise do 'Modelo' Brasileiro* (Rio de Janeiro: Civilização Brasileira).

Gerschenkron, A. (1951) *A Dollar Index of Soviet Machinery Output, 1927–28 to 1937* (Santa Monica, CA: RAND Corporation).

Gerschenkron, A. (1962) *Economic Backwardness in Historical Perspective* (Cambridge, Mass.: Harvard University Press).

Harris, J. R. and Todaro, M. P. (1970) 'Migration, Unemployment, and Development: A Two-Sector Analysis', *American Economic Review*, vol. 60, pp. 126–42.

Hawkins, D. (1948) 'Some Conditions of Macroeconomic Stability', *Econometrica,* vol. 16, pp. 309–22.

Hirschman, A. O. (1958) *The Strategy of Economic Development* (New Haven, Conn.: Yale University Press).

Hoffman, W. G. (1958) *The Growth of Industrial Economies* (Manchester: Manchester University Press).

Houthakker, H. S. (1976) 'Disproportional Growth and the Intersectoral Distribution of Income', in Cramer, J. S., Heertje, A. and Venekamp, P. (eds) *Relevance and Precision: Essays in Honor of Pieter de Wolff* (Amsterdam: North-Holland).

Jorgenson, D. W. (1961) 'The Development of a Dual Economy', *Economic Journal*, vol. 71, pp. 309–34.

Jorgenson, D. W. (1967) 'Surplus Agricultural Labor and the Development of a Dual Economy', *Oxford Economic Papers*, vol. 19, pp. 288–312.

Kaldor, N. (1955) 'Alternative Theories of Distribution', *Review of Economic Studies*, vol. 23, pp. 83–100.

Kaldor, N. (1976) 'Inflation and Recession in the World Economy', *Economic Journal*, vol. 86, pp. 703–14.

Kalecki, M. (1971) *Selected Essays on the Dynamics of the Capitalist Economy* (New York: Cambridge University Press).

Kalecki, M. (1976) *Essays in Developing Economies* (London: Harvester Press).

Kanbur, S. M. R. and McIntosh, J. (1986) 'Dual Economy Models: Retrospect and Prospect', Department of Economics, University of Essex.

Kaneda, H. (1986) 'Rural Resource Mobility and Intersectoral Balance in Early Modern Growth', Chapter 17 in this volume.

Kuznets, S. S. (1955) 'Economic Growth and Income Inequality', *American Economic Review*, vol. 45, pp. 1–28.

Kuznets, S. S. (1966) *Modern Economic Growth: Rate, Structure, and Spread* (New Haven, Conn.: Yale University Press).

Lele, U. and Mellor, J. W. (1981) 'Technical Change, Distributive Bias and Labor Transfer in a Two- Sector Economy', *Oxford Economic Papers*, vol. 33, pp. 426–41.

Lewis, W. A. (1954) 'Economic Development with Unlimited Supplies of Labor', *Manchester School of Economics and Social Studies*, vol. 22, pp. 139–91.

Lewis, W. A. (1958) 'Unlimited Labor: Further Notes', *Manchester School of Economics and Social Studies*, vol. 26, pp. 1–32.

Lluch, C. (1977) 'Theory of Development in Dual Economies: A Survey', Development Research Center, World Bank.

Londono, J. L. (1985) 'Ahorro y Gasto en una Economia Heterogena: El Rol Macroeconomico del Mercado de Alimentos' (Bogota: FEDESARROLLO).

Lustig, N. (1980) 'Underconsumption in Latin American Economic Thought: Some Considerations', *Review of Radical Political Economics*, vol. 12, pp. 35–43.

Luxemburg, R. (1968) *The Accumulation of Capital* (New York: Monthly Review Press).

Mahalanobis, P. C. (1953) 'Some Observations on the Process of Growth of National Income', *Sankhya*, vol. 12, pp. 307–12.

McCarthy, F. D. and Taylor, L. (1980) 'Macro Food Policy Planning: A General Equilibrium Model for Pakistan', *Review of Economics and Statistics*, vol. 62, pp. 107–21.

McCarthy, F. D., Taylor, L. and Talati, C. H. (1987) 'Trade Patterns in Developing Countries, 1964–82' *Journal of Development Economics*, forthcoming.

Mitra, A. (1977) *Terms of Trade and Class Relations* (London: Frank Cass).

Nayyar, D. (1978) 'Industrial Development in India; Some Reflections on Growth and Stagnation', *Economic and Political Weekly*, vol. 12, pp. 1265–78.

Nurkse, R. (1953) *Problems of Capital Formation in Underdeveloped Countries* (Oxford: Basil Blackwell).

Ocampo, J. A. (1986) 'New Developments in Trade Theory and LDCs', *Journal of Development Economics*, vol. 22, pp. 129–70.

Pasinetti, L. L. (1981) *Structural Change and Economic Growth* (New York: Cambridge University Press).

Preobrazhenski, E. (1965) *The New Economics* (Oxford: Clarendon Press).

Quisumbing, M. A. R. and Taylor, L. (1986) 'Resource Transfers from Agriculture', in *The Balance Between Industry and Agriculture in Economic Development*, vol.3, Sukhamoy Chahravaty (ed.) *Manpower and Transfers* (London: Macmillan, 1988).

Rattso, J. (1986) 'Macroadjustments in a Dual Economy when the Domestic Terms of Trade Are Controlled', Institute of Economics, University of Trondheim.

Roemer, J. (1982) *A General Theory of Exploitation and Class* (Cambridge, Mass.: Harvard University Press).

Rosenstein-Rodan, P. N. (1943) 'Problems of Industrialization in Eastern and South-Eastern Europe', *Economic Journal*, vol. 53, pp. 202–11.

Rosenstein-Rodan, P. N. (1961) 'Notes on the Theory of the Big Push', in Ellis, H. S. and Wallich, H. C. (eds) *Economic Development in Latin America* (New York: St Martin's Press).

Rothschild, E. (1986) 'A Divergence Hypothesis', *Journal of Development Economics*, vol. 23, pp. 205–26.

Rothschild, E. (1987) 'An Extramediate Matrix and other Adjustments' Program in Science, Technology and Society, Massachusette Institute of Technology.

Rybczynski, T. M. (1955) 'Factor Endowment and Relative Commodity Prices', *Economica*, vol. 22, pp. 336–41.

Sah, R. K. and Stiglitz, J. (1984) 'The Economics of Price Scissors', *American Economic Review*, vol. 74, pp. 125–38.

Salter, W. E. G. (1959) 'Internal and External Balance: The Role of Price and Expenditure Effects', *Economic Record*, vol. 35, pp. 226–38.

Schultz, T. W. (1964) *Transforming Traditional Agriculture* (New Haven, Conn.: Yale University Press).

Schultz, T. W. (1978) *Distortions of Agricultural Incentives* (Bloomington, Ind.: Indiana University Press).

Sen, A. K. (1963) 'Neo-Classical and Neo-Keynesian Theories of Distribution', *Economic Record*, vol. 39, pp. 54–64.

Simon, H. (1947) 'Effects of Increased Productivity upon the Ratio of Urban to Rural Population', *Econometrica*, vol. 15, pp. 31–42.

Solow, R. M. (1956) 'A Contribution to the Theory of Economic Growth', *Quarterly Journal of Economics*, vol. 70, pp. 65–94.

Steindl, J. (1952) *Maturity and Stagnation in American Capitalism* (Oxford: Basil Blackwell).

Stolper, W. F. and Samuelson, P. A. (1941) 'Protection and Real Wages', *Review of Economic Studies*, vol. 9, pp. 58–73.

Streeten, P. (1959) 'Unbalanced Growth', *Oxford Economic Papers*, vol. 11, pp. 167–90.

Swan, T. W. (1960) 'Economic Control in a Dependent Economy', *Economic Record*, vol. 36, pp. 51–66.

Sylos-Labini, P. (1984) *The Forces of Economic Growth and Decline* (Cambridge, Mass.: MIT Press).

Syrquin, M. (1986) 'Sector Proportions and Economic Development: The Evidence since 1950', chapter 2 in this volume.

Tavares, M. da C. (1972) *Da Substituição de Importações ao Capitalismo Financeiro* (Rio de Janeiro: Zahar Editores).

Taylor, L. (1979) *Macro Models for Developing Countries* (New York: McGraw-Hill).

Taylor, L. (1983) *Structuralist Macroeconomics* (New York: Basic Books).

Taylor, L. (1985) 'Demand Composition, Income Distribution, and Growth', Department of Economics, Massachusetts Institute of Technology.

Taylor, L. (1986) 'Economic Openness: Problems to Century's End', Department of Economics, Massachusetts Institute of Technology.

Taylor, L. and Bacha, E. L. (1976) 'The Unequalizing Spiral: A First

Growth Model for Belinida', *Quarterly Journal of Economics*, vol. 90, pp. 197–218.

Todaro, M. P. (1969) 'A Model of Labor Migration and Urban Unemployment in Less Developed Countries', *American Economic Review*, vol. 59, pp. 138–48.

Urata, S. (1986) 'Sources of Economic Growth and Structural Change: An International Comparison', chapter 7 of this volume.

Young, A. (1928) 'Increasing Returns and Economic Progress', *Economic Journal*, vol. 38, pp. 527–42.

2 Sector Proportions and Economic Development: The Evidence since 1950

Moshe Syrquin*
BAR-ILAN UNIVERSITY, ISRAEL AND HARVARD INSTITUTE FOR INTERNATIONAL DEVELOPMENT

1. INTRODUCTION

Economic development is a process of transformation as much as, if not more than, a process of expansion. The transformation of the economic system from a low productivity, predominantly rural, agrarian economy to one predominantly urban, industrial, with higher productivity, is not instantaneous. It is a gradual process during which productivity has to increase in most segments of the economy and, at the same time, the centre of gravity shifts from lower to higher productivity units. Disaggregation is therefore essential for the analysis of development.

There is no unique way of defining the disaggregated units. The principal requirement is that the categories should be of relevance for economic analysis and policy. They also have to be identifiable *ex ante* to avoid circular reasoning ('low productivity in the informal sector' where this is defined as comprising units of low productivity).

The next section addresses the question of why disaggregate the economy into sectors or industries. Section 3 reviews the evidence on sector proportions and development in about one hundred countries since 1950, and section 4 presents some attempts to account for the main determinants of long-run structural change.

2. CONCEPTUAL BACKGROUND

In the early literature on development we find many examples of disaggregation: modern-traditional, consumption goods-capital goods, etc. The most prominent approach has probably been the one that segments economic activity into sectors of production or industries. This approach has some similarities with the older stages theories of the historical school (Hoselitz, 1960). Its modern versions go back to Fisher (1939), Clark (1940) and Kuznets (1957)[1] and are now formalised in the United Nations' standard classifications of production and trade data. Production sectors can be differentiated on various criteria:

(a) *Similarity on the production side*. This criterion encompasses techniques of production, inputs used (land, natural resources), factor proportions, etc. Agriculture and mining are to be treated separately because of their dependence on specific inputs. Within manufacturing, sectors differ in their production functions (economies of scale, elasticity of substitution, factor proportions).

(b) *Demand*. One of the best established stylised facts is the evolution of the composition of final demand with the level of development (Engel effects). Systematic differences in income elasticities imply that the composition of demand changes in predictable ways with the level of development. Changes in relative prices can also have a significant effect on demand. Over long periods of time, however, they have not altered the direction of change implicit in income elasticities, at least not for broad categories. The matching of commodities (the subject of demand analysis) and industries is not always trivial or invariant over time. Price effects and changes in technology can vary the sectoral location where a demand is satisfied. Appliances (manufacturing) have substituted for domestic services, while size and differentiation have led a variety of services to be transacted in markets instead of being provided within the firm (accounting).

(c) *Tradability*. Real trade theory has always dealt with individual sectors, rather than with the aggregate economy. Traditionally, sectors were differentiated by specific inputs or factor proportions. A more recent added feature is the tradability of the output.

(d) *Location*. While not always technologically determined, industrial activities tend to be concentrated in urban areas. Since initially the population tends to be predominantly rural, industrialisation is closely linked with migration and urbanisation.

(e) *Policy and perceptions*. In some cases a sectoral division is important because it is perceived to be so and influences policy decision. Some examples are the socialist bias against services and in favour of heavy industry, and the strategy of import-substitution industrialisation.

(f) *Productivity differential*. A key element in the process of transformation, one that adds weight to the sectorisation of the economy, is the pattern of differential productivity growth observed in the long-term experience of industrial and semi-industrial countries. The faster growth of factor productivity in modern-urban-industrial activities, is one of the factors behind the static differential in factor returns across sectors. Since adjustment is not instantaneous, the differential persists for long periods of time, often widening in the initial stages of development. It eventually induces a reallocation of resources which, by improving the efficiency of the economy, becomes itself an important source of growth.

Recognising the analytical value of dividing economic activity into sectors does not mean that in practice the division can always be carried out in a clear-cut way. Part of the process of modern economic growth is an increase in specialisation and a growing differentiation of occupations. In the early stages of development an individual may be engaged in a variety of trades, and it is therefore difficult to assign him or her to any one occupation. This was the central point in Bauer and Yamey's (1951) critique of Clark's primary-secondary-tertiary distinction. The difficulty is an empirical, not a conceptual one, and moreover, one that diminishes in importance as development proceeds.[2]

3. EXPERIENCE SINCE 1950

Empirical studies of sector proportions in modern times go back to Colin Clark's comparisons of structures of employment primarily for developed economies. With the development of national income accounts in the post-war period, it became feasible to do comparative analysis for a large number of less developed countries. In 1957, Kuznets published his famous study on 'Industrial Distribution of National Product and Labor Force'. For this study Kuznets assembled information for fifty-nine countries and established some systematic relations of change by comparing average structures for countries grouped by income level without the use of econometric methods. This task was taken up by Chenery (1960), who estimated the patterns of industrialisation after deriving them as reduced forms from a general Walrasian framework.

Although Kuznets warned that differences are to be expected between cross-country patterns and time series relations within countries, his own research (1957, 1971) showed a high degree of similarity between the two, at least for the structure of production.[3]

The main characteristics of the transformation that emerged from the long-term experience in the advanced countries, and from the cross-country comparisons, were a shift in the production of commodities from primary to manufacturing activities, and a mild increase in the share of services in total output. In the case of employment a shift away from primary was also observed but with a lag, implying an initial drop in relative labour productivity in that sector. Part of the transfer of labour from the primary sector went into industry but, on average, the main beneficiary was the services sector. The repeated confirmation of the above trends in various studies across countries or over long periods of time justify their being regarded as stylised facts of the transformation. Not everyone agrees on this however. Dissenters have questioned (a), the universality of the patterns, or (b), the value of historical patterns for today's developing countries. I will present a brief account of the two arguments and return to them after presenting empirical results for a large number of countries since 1950:

(a) Jameson (1982), among others, has argued that the decline in the share of agriculture in value added is not as universal as the cross-section results seem to indicate. A warning is then added that relying on the cross-country result may impart an anti-agricultural bias to development policy. What is needed, according to this view, is a reorientation of development strategy that will emphasise and give priority to agriculture.

(b) The historical shift of labour from agriculture to industry was achieved under different conditions from those prevailing today. The rates of population growth were significantly lower during the process of industrialisation in advanced countries and, in addition, there were no industrial countries dominating the scene at the time. These facts, it is argued, are behind the failure of industry to absorb labour at rates comparable to the historical ones. The result has been an overexpansion of employment in services, particularly in the low productivity 'informal sector'.

3.1 Patterns of Transformation since 1950

One of the reasons for doing cross-country comparisons in the late 1950s was the absence of time series information in developing countries. Subsequent studies of development patterns made use of the accumulation of temporal data, still in a cross-country framework (Chenery and Taylor, 1968, Chenery and Syrquin, 1975). It has now become feasible to analyse individual time series for a large number of countries. The experience of about one hundred countries during the period 1950–83 was the subject of a recent study (Syrquin and Chenery, 1986). In every country where the time series data were long enough, for each sectoral share in output or employment, a simple regression was run against the log of income per capita:

$$x = a + b \ln y$$

where x stands for a share in GDP or employment and y for per capita income. The estimates of b are measures of structural change with respect to income per capita, but not necessarily with respect to time, save for the case of a constant growth rate of per capita income. The objective of estimating equation (1) is not to test a given theory, nor to determine a direction of causality but, rather, to obtain a measure of association between structure and income growth that is easily comparable for a large number of countries.

Table 2.1 summarises the results of estimating equation (1) for about one hundred countries (as shown in Appendix Table 2A.1). The dependent variables were the shares of agriculture and manufacturing in GDP, in current and constant prices, and the shares of agriculture and industry in employment. The 'industry' category includes, besides manufacturing, also mining, construction, gas, electricity and water. The results are presented as unweighted averages of the individual income slopes (the b's) for groups of countries ranked by level of development, as in the 1986 *World Development Report*. The individual estimates refer to the whole period for which data were available. The periods range from eleven to thirty-four years. For any given period, actual change in a share can approximately be obtained as the product of b times the increase in the log of

Table 2.1 Time-series relations by income groups

	Low Income	*Lower-middle income*	*Upper-middle income*	*Industrial market economies*
n	31	32	21	19
Average growth rate of y	0.8	2.0	3.8	3.2
$n < 0$	8	2	0	0
Average income slopes				
VALUE ADDED				
Current price shares: n	28	30	21	18
Agriculture	– .24	–.17	–.12	–.09
$n > 0$	3	4	0	0
Manufacturing	.06	.05	.03	–.05
$n < 0$	6	5	7	13
Constant price shares: n	18	27	16	16
Agriculture	– .19	–.14	–.11	–.05
$n > 0$	2	4	0	1
Manufacturing	.07	.04	.06	.03
$n < 0$	3	4	1	6
LABOUR FORCE				
n	28	31	20	19
Agriculture	–.10	–.20	–.22	–.18
$n > 0$	6	2	0	0
Industry	.05	.07	.08	.01
$n < 0$	5	4	2	9

Source: Appendix table.

---

y. The first part of the table gives average growth rates of per capita income for the four income groups (growth rates are OLS estimates). The data indicate a clear acceleration of growth with the level of development. Negative estimates for the whole period were found in eight countries in the lowest income group and in two lower-middle income countries.

Turning to the income slopes, the most striking result is the almost universal inverse association with income of the shares of agriculture in income and employment. Of the ninety-seven countries for which long enough time series were available, only in seven did the income coefficient for the share in value-added at current prices come out positive. In three of them (Liberia, Nicaragua, and Zambia), the estimated coefficient did not differ significantly from zero. In another three (Niger, Senegal, and Somalia), per capita income *fell* during the period; hence the positive coefficient signifies that the share of agriculture diminished in spite of the decline in income. The seventh, Burma, is the only true exception to this general phenomenon.

The average income slopes of the share of manufacturing at current prices is positive in developing countries but diminishes with the level of income. There are many more exceptions in this case than in the case of agriculture. In almost one-third of the cases recorded, the estimated slope is negative. It is instructive to identify the main cases with negative income elasticities. Among the very low income countries we find some with negative growth (Niger, Somalia). In oil exporting countries (Algeria, Congo, Egypt, Iraq, Iran),[4] the decline in industry is the result of the oil boom – 'Dutch disease'. In a third group there was a fall in the manufacturing share but from extremely high initial values (Hungary, Israel, Yugoslavia). Finally, in virtually every industrial country, there was a shift from industry to services at some point during the period. For the period as a whole, negative slopes were estimated in thirteen of the eighteen countries defined as industrial and with the required data.

The decline in the share of employment in agriculture follows the one in value added but with a lag. Since initially the share of employment exceeds the one in output, relative labour productivity in agriculture declines. In the upper-middle income group, relative labour productivity in agriculture often improves. It is interesting to note the large size of the income slope of agricultural employment in industrial countries.

Comparing the results for value added in current and constant prices we find that in almost every case the relative prices of agriculture and manufacturing declined in the period. At constant prices, the decline in agriculture's share was smaller and the increase in manufacturing larger than was true of the current price shares. Offsetting these changes in relative prices were the relative increases in the prices of mining and nontradables. A significant part of the shift in industrial countries, from tradables to services, was therefore a price effect.

I now return to the dissenting views mentioned above, about the universality and applicability of stylised facts derived from cross-country comparisons.

The results in Table 2.1 suggest that the association of growth with a reallocation of economic activity away from agriculture is among the most robust of the stylised facts of development. It is in the same league with Engel's law, which is in fact one of its determinants.

Sustained development requires improvements in all fronts so that no sector becomes a bottleneck. It is possible to design an agriculture-based strategy of economic growth (Mellor, 1986), but it will have to be one that recognises that: 'Economic development is a process by which an economy is transformed from one that is dominantly rural and agricultural to one that is dominantly urban, industrial and service in composition' (p. 67). It is more useful to acknowledge, as Mellor does, that there are natural limits to an agriculture-based strategy than to attribute a sectoral bias to an observed universal pattern.

The post-war decline in the share of agriculture in the labour force has been as pronounced as the similar decline that took place in the historical experience of today's advanced countries. But, it is sometimes argued, the migrating workers have nowadays gone mostly into services and not into manufacturing. The figures in Table 2.1 confirm that only between one-third and one-half of the decline in agriculture's share was taken up by industry (which includes mining, construction, and public utilities). Has this pattern really been that different from the experience of the industrial countries? Thanks to the efforts of Maddison (1980), we have information on the sectoral composition of employment in sixteen advanced countries at two distant dates: 1870 and 1950. The averages of the income slopes between the two points are -.27 for agriculture and .09 for industry. These values indicate a faster rate of transformation with respect to income (not necessarily per unit of time, since the growth rate of income per capita was much lower). The income slope for industry is slightly higher than the ones in Table 2.1 for middle-income countries. As a proportion of the corresponding slope for agriculture it does not differ at all. Almost thirty years ago, Kuznets' analysis of long-term trends in the industrial distribution of the labour force led him to a similar conclusion: 'In fact, in most countries the substantial decline in the share of the A sector is compensated by a substantial rise in the share of the S sector – not by a rise in the share of the M sector' (1957, p. 32).

Cross-country patterns are not always a reliable guide to variation over time. In the case of employment structure, however, the correspondence between the two has, on the average, been quite high.

4. SOURCES OF INDUSTRIALISATION

In the recent study on patterns of development cited before (Syrquin and Chenery, 1986), the individual time series for about one hundred countries were pooled to estimate average patterns of change since 1950. We define the transition range as the income interval from $ 300 to $ 4000 (in 1980 US $). Over this range the predicted transformation in the output of commodities is for the share of primary output (agriculture and mining) to decline by 28 percentage points, while the manufacturing share goes up on the average by 12 percentage points. How can we account for such expected changes?

There are two main approaches that try to identify the determinants of structural transformation. The first approach is to build a model (ideally a general equilibrium one) incorporating behavioural, technological, and institutional relations, as well as assumptions about the functioning of markets. The second approach, more limited but simpler to implement, derives the proximate sources of growth and structural change from identity-based decompositions within a general, not fully specified, economic model.

The second approach, applied below, complements the descriptive patterns of the previous section, which can be regarded as reduced-form equations. The decomposition of change into its proximate sources relates the patterns to a more complete structure which might have given rise to them. It is an approach which can be applied to a large number of countries, giving us an order of magnitude of the various effects. Systematic discrepancies help to identify important effects not yet considered. As other partial approaches it is subject to limitations. Markets and prices are not studied directly and therefore the links to policy are not immediate. Associations are identified but causality cannot be determined. The crucial assumption in this type of analysis, which goes back to Kuznets, is that there are some universal factors powerful enough to produce 'common features of significance in the economic growth of nations' (Kuznets, 1959, p. 170).[5]

Price endogenous models deal with some of the deficiencies of statistical analysis. These models usually focus on a specific issue within a country. They rely on the historical experience of an economy, but have to do without the variation afforded by the comparative approach across countries. A computable general equilibrium model is better suited to confront issues of causality and to probe beyond 'proximate' sources of growth by endogenizing effects regarded as exogenous in the statistical approach. The extent to which additional relations can be incorporated into a model is limited by the available data. The main contribution of applied general equilibrium models has been the ability to go beyond partial effects to general equilibrium analysis. To date, however, such models have contributed little to unravel the threads between structural change and growth.

Computable general equilibrium models are most useful for short- or medium-run studies. For long-term analysis their strength – the ability to model market behaviour – becomes a weakness. Long-run transformation involves significant changes in the nature and working of markets and institutions.

The two approaches are best seen as complements rather than rival. The remainder of this section presents results from decomposition analysis of structural change.

4.1 A First Approximation

An identity-based decomposition is a useful first approximation to account for industrialisation over the transition.

A model of industrialisation as a system-wide phenomenon was first presented by Chenery, Shishido, and Watanabe (1962) in a study of the transformation of the Japanese economy. It was subsequently revised and adapted to simulate the transformation over the complete transition range, on the basis of cross-country data (Chenery and Syrquin, 1980, 1986). The approach starts with the material balance equations:

$$X_i = W_i + D_i + T_i \qquad (2)$$
$$V_i = v_i X_i \qquad (3)$$

where the index i refers to a sector, X is gross output, W is intermediate demand, T is net trade (exports minus imports), V is value added and v is the value-added ratio. Expressing all the absolute magnitudes as shares of GDP, which equals $V = \Sigma\ V_i$, and combining (2) and (3) we obtain:

$$V_i V = v_i\,[W_i/V + D_i/V + T_i/V] \qquad (4)$$

Changes in the sectoral shares in value added can be accounted for by changes in the composition of demand (intermediate and final), changes in the composition of trade, and changes in the value-added coefficient as in equation (5):

$$\Delta\ (V_i/V) = \bar{v}_i\ [\underset{(a)}{\Delta\ (W_i\ \ V)} + \underset{(b)}{\Delta\ (D_i/\ V)} + \underset{(c)}{\Delta\ (T_i\ /V)}] + \underset{(d)}{(\overline{V_i/V})\ \Delta\ v_i/\ v_i} \quad (5)$$

(a bar over a variable means that its value is set at the mean of the initial and terminal levels)

All elements needed to calculate equation (5) over the transition are derived from cross-country regressions as follows. For intermediate production (*W* and *V*) I rely on a recent comparative study of inter-industry relations that estimated systematic patterns of change from data on eighty-three input-output matrices (Deutsch and Syrquin, 1986). Final demand and trade come from the study of patterns of development since 1950 (Syrquin and Chenery, 1986). The trade regressions distinguish primary and manufacturing exports and imports. For final demand I assume that food consumption generates demands from the primary sector only, and that the manufacturing sector supplies one half of non-food consumption and investment (the other half represents construction and other nontradables). With these assumptions we can now compute equation (5) for the transition range. The results appear in Table 2.2.

Over the course of the transition there is a significant shift in value added from primary production to manufacturing and nontradables. The average patterns in Table 2.2 show a very close correspondence between the directly estimated shift (the last row), and the one calculated by the right-hand side of equation (5). Changes in domestic demand (Engel effects) account directly for less than one-half of the change in structure, and changes in net trade for about 10 per cent on the average.

The contribution of intermediates has two components. First there is a very significant increase in the demand for manufacturing products to be used as intermediates, and a decline in the relative use of intermediate inputs from the primary sectors. These trends reflect the evolution from a comparatively simple to a more diversified, roundabout system with a higher degree of fabrication and specialisation. The substitution of fabricated materials for natural ones is due to changes in technology and also to changes in relative prices. The second component refers to variations in the ratio of value added to

Table 2.2 Accounting for the transformation: a first approximation

	Impact on the share in value-added of: Primary	Manufacturing
Sources of change		
(a) Intermediate demand	−.06	.18
(b) Final demand		
Food consumption	−.20	
1/2 of nonfood consumption and investment		.08
(c) Net trade	−.05	.05
Changes in gross output $\Delta(X_i/V)$	−.31	.31
Mean value-added ratio (v_i)	.71	.35
(d) Changes in value added ratios		
Δ v_i	−.20	.03
Δ v_i/v_i	−.27	.09
Implied changes in value-added shares	−.30	.13
Directly estimated shares in value added:		
at y = $300	.44	.12
at y = $4000	.16	.24
change	−.28	.12

Source: Syrquin and Chenery, 1986.

gross output in a sector. In agriculture this ratio tends to decline with the rise in income, or equivalently, the use of purchased intermediate inputs per unit of output tends to increase. As shown in Table 2.2, this factor accounts for about one-quarter of the decline in the share of primary in total GDP.

4.2 Growth-accounting on the Demand Side

In an input-output model, the variation in intermediate use can be further attributed to changes in final demand, trade, and input-output coefficients.

Let intermediate demands be related to output by a linear input-output technology (A), and let u_i be the share of total domestic demand of sector i supplied from domestic sources substituting into the material balance equations (2) and solving for X, we have in vector notation:

$$X = R\,(\hat{u}D + E) \tag{6}$$

where $R = [I - \hat{u}A]^{-1}$, E is exports and the 'hat' over u indicates a diagonal matrix.

From this accounting identity we derive a growth accounting system relating output growth to its sources on the demand side,[6] by differentiating (6) with respect to time ($\dot{x} = dx/dt$):

$$\dot{X} = \underset{(a)}{R\hat{u}\dot{D}} + \underset{(b)}{R\dot{E}} + \underset{(c)}{R\dot{\hat{u}}(D + W)} + \underset{(d)}{R\hat{u}\dot{A}X} \tag{7}$$

The four factors measure the total (direct and induced) effects of:

(a) domestic demand expansion (DD),
(b) export expansion (EE),
(c) import substitution (IS), and
(d) changes in input-output coefficients (IO).[7]

The sources in equation (7) were recently computed for a group of semi-industrial countries in the post-war period (Chenery, Robinson and Syrquin, 1986). They were also calculated for a prototype country during the transition period. The focus in the present paper is on changes in sector proportions, whereas equation (7) refers to absolute changes in output. It can easily be adapted for the purpose at hand, by comparing its results to a balanced growth case where all the right-hand elements in the equation expand at the same rate, with unchanged import and input-output structures. Deviations from proportionality or changes in output shares are now related to nonproportional expansion in demand and exports and to changes in import and input-output coefficients.

Table 2.3 gives the sources of structural change in gross output for the transition range, broken into four income intervals.[8] The results from the cross-country model of industrialisation show that the fall in the primary share is mostly due to demand (Engel effects) at low income levels, and to trade effects afterwards. Trade effects contribu-

Table 2.3 Sources of strutural change: demand-side decomposition

Income interval (1980 US $)	*DD*	*EE*	*IS*	*IO*	*Total*	*Change in share in gross output*
	Sources of changes in structure per cent					
	Decline in share of primary					
$300 – $600	67.6	14.5	12.7	5.2	100	–6.8
$600 – $1200	48.8	19.8	22.9	8.5	100	–5.7
$1200 – $2500	29.6	26.1	33.3	11.0	100	–4.9
$2500 – $5000	13.4	29.5	45.9	11.2	100	–3.9
						–21.3
	Increase in share of manufacturing					
$300 – $600	22.5	19.5	30.5	27.5	100	5.3
$600 – $1200	15.3	26.3	29.0	29.4	100	5.5
$1200 – $2500	10.3	31.4	30.7	27.6	100	6.0
$2500 – $5000	6.9	35.7	27.8	29.6	100	5.9
						22.7

Source: Chenery and Syrquin, 1980.

ting to a decline in share mean a lower than proportional expansion in primary exports and import liberalisation (defined as an increase in the sectoral import coefficient). Changes in input-output coefficients account for part of the decline at all income levels. This effect captures the substitution of fabricated materials for natural resources induced by technology and relative prices.

The rise in manufacturing share, which best represents the process of industrialisation, is less due to high income elasticities and more to trade and technology. Import substitution is quite significant at all income levels. A more disaggregated analysis would show early import substitution in consumer goods, shifting to producer and capital goods at higher levels of development. The increase in the overall density of the input-output matrix that accompanies development is especially important in heavy industry (Deutsch and Syrquin, 1986). It shows here as large contributions of the IO term to the rise in the manufacturing share.

4.3 Growth-accounting on the Supply Side

The last section presented sources of growth from the demand side. As with other growth-accounting exercises, it starts from an accounting identity and decomposes growth or structural change into its proximate sources without necessarily implying causality. An alternative growth-accounting approach, this time from the supply side, is the Abramovitz-Solow-Denison decomposition of the sources of growth. It starts from a production function and relies more on economic theory than the demand side accounting, but it obtains an important source of growth-total factor productivity – as a residual. Factor accumulation and productivity growth are related in this section to the change in sector proportions.

Changes in sector proportions in GDP clearly imply differential rates of sectoral growth. During the transition, agricultural output expands at a lower rate than the average economy-wide rate, and manufacturing at a higher than average rate. To what extent are these differentials due to primary inputs and to productivity growth? Some orders of magnitude are presented in Table 2.4, which is based on the long-run model of industrialisation mentioned in the previous section. On the basis of cross-country information on the variation with income of primary input-output ratios and of factor elasticities, it derives a consistent set of sectoral growth-accounting equations of the form:

$$g_{Vi} = \alpha_i g_{Ki} + (1 - \alpha_i) g_{Li} + \lambda_i \qquad (8)$$

where g_x is the growth rate of a variable x, V is value added, K is capital, L is labour, α is the elasticity of output with respect to capital and λ the growth of total factor productivity (TFPG). The weighted growth rates of inputs can be combined into one term (g_F) labelled total input growth.[9]

The results in Table 2.4 show that the decline in the share of agriculture is mostly due to differential productivity growth. Only in the higher income interval, when labour is declining absolutely in agriculture and productivity has risen significantly, is the low growth of total inputs more important in accounting for the decline in the output share.

TFPG in manufacturing is generally above the rate for the economy, but it accounts for only a small part of the faster-than-average growth of the sector's output. Total input growth, primarily capital, is the principal source of the rising share of manufacturing output.

The figures for the whole economy indicate an acceleration of growth as the economy moves over the transition. Such an acceleration was observed in the long-term experience of the industrial

Table 2.4 Growth of output, inputs and productivity (per cent)

Annual growth rates	*Income intervals (1980 US $)* 300–600	600–1200	1200–2500	2500–5000
Value added				
Agriculture	3.9	3.9	3.5	2.7
Manufacturing	5.7	6.8	7.6	7.8
Total	4.8	5.7	6.3	6.6
Total input				
Agriculture	3.8	3.7	2.6	1.2
Manufacturing	4.8	5.2	5.5	5.3
Total	4.1	4.3	4.0	3.7
Total factor productivity				
Agriculture	0.1	0.2	0.9	1.5
Manufacturing	0.9	1.6	2.1	2.5
Total	0.7	1.4	2.3	2.9
Effect of reallocation	0.15	0.3	0.55	0.75

Source: Syrquin, 1986.

countries, and in the more recent experience of semi-industrial economies. The acceleration results from higher investment shares and from an acceleration of aggregate TFPG. This last effect has several components. First, TFPG tends to go up in most productive sectors. Secondly, as the weight of sectors with high TFPG increase (manufacturing), the overall rate goes up as well. Finally, the aggregate rate of TFPG will increase if resources shift from sectors with low (marginal) productivity to high productivity sectors. The cross-country simulation model, from which the figures in Table 2.4 are derived, has some illustrative calculations of this effect of resource reallocation. The last row in the table shows annual rates of growth accounted for by this resource shift. They account for as much as one-quarter of aggregate TFPG and appear to be of importance in the explanation of growth acceleration in middle income countries.

5. CONCLUDING COMMENTS

This paper has presented evidence about the pervasiveness of one element of the structural transformation, namely the change in sector proportions during the process of development. Structural change is intimately related to economic development. That change can retard growth if its pace is too slow or its direction inefficient. It can also contribute to growth if it improves the allocation of resources. Market forces tend to move the economic system toward equilibrium, but they are blunted by inflexibility in the system and high adjustment costs, by shocks from external events and unbalanced productivity growth, and even by government policies.

This paper focused on average patterns and on their proximate determinants. It is important to emphasise that nowhere is it implied that there is a unique path. The timing and sequencing of the stages in the transformation are significantly influenced by structural characteristics (size, resource endowments) and by government policies. Nevertheless, the overall nature of the transformation over long periods of time has had a remarkable degree of uniformity.

Appendix Table 2A.1
Time-series relations by country

Country	*Growth rate of income per capita*	*Value-added shares current prices Agric.*	*Manu.*	*constant prices Agric.*	*Manu.*	*Employment Agric.*	*Ind.*
Low income							
Chad	–2.8	–.19	.01*	–.21	.02	.16	–.08
Ethiopia	1.4	–.50	.17	–.70	.15	–.23	.06
Bangladesh	0.4	–.35	.06			–.07*	.06*
Mali	0.9	–.43	.00*			–.64	.26
Nepal	0.3	–.58*	.02*			–.16	.00*
Zaire	–0.1	–.42	.05*	–.08	.06	–.07*	.03*
Burkina	1.1	–.59	.14			–.39	.31
Burma	1.9	.16	.09	.04*	.00*	–.04	–.00*
Malawi	2.2	–.23	.15			–.12	.04
Uganda	–0.5	–.43	.09	–.08	.10		
Burundi	2.5	–.33	.04*			.08	.03
Niger	–1.5	.64	–.16			.07	–.03
Tanzania	1.9	–.16*	.13	–.30	.08	–.13	.04
Somalia	–0.6	.67	–.10*	.60	–.05*	.09	–.06
India	1.5	–.33	.03*	–.40	.11	–.15	.09
Rwanda	0.9	–.76	.35	–.53*	.33	–.05*	.01*

Appendix Table 2A.1 continued

Country	*Growth rate of income per capita*	*Value-added shares current prices Agric.*	*Value-added shares current prices Manu.*	*Value-added shares constant prices Agric.*	*Value-added shares constant prices Manu.*	*Employment Agric.*	*Employment Ind.*
CAR	0.2	–.19*	.09*	–.31	.02*	–.25	.09
Togo	2.1	–.56	–.02*			–.21	.10
Benin	0.7	–.35	.05*			–.28	.26
China	4.6	–.09	.12	–.32	.18		
Guinea	1.3					–.17	.14
Haiti	0.5					–.02*	.00*
Ghana	–0.8	–.47	.12	–.09*	.02*	.31	–.17
Madagascar	–0.7	–.36	–.02*			.17	–.06
S. Leone	0.8	–.12*	–.01*	–.25	.02*	–.31	.16
Sri Lanka	2.0	–.09	.06	–.15	–.03	–.06	.02
Kenya	2.1	–.08	.05	–.17	.18	–.15	.08
Pakistan	2.8	–.27	.08	–.23	.06	–.07	.03
Sudan	0.6	–.14*	–.02*	–.20	–.01*	–.07*	.03*
Afghanistan	0.2						
Mozambique	–0.3	–.14	.04*	–.06*	.06	–.07*	.04*
Lower-middle income							
Senegal	–0.3	.41	–.19	.25	–.25	.19*	–.15
Liberia	0.7	.09*	.10	.14	.10	–.22	.09
Mauritania	1.8	–.29	.02	–.35	.01*		
Bolivia	1.1	–.17	–.05	–.08	.04	–.26	.13
Yemen AR	4.6			–.36	.03	–.06	.03
Indonesia	3.3	–.40	.05	–.29	.08	–.23	.05
Zambia	2.0	.01*	.09*	–.05	–.04*	–.14	.04
Honduras	0.9	–.52	.19	–.31	.13	–.28	.27
Egypt	3.1	–.11	.15	–.21	–.04	–.11	.21
El Salvador	1.1	–.13	.08	–.20	.12	–.29	.14
Iv. Coast	1.8	–.33	.09	–.32	.07	–.19	.04
Zimbabwe	1.4	–.09	.20	.02*	.06*	–.19	.08
Morocco	1.4	–.11	.02	–.16	.00*	–.20	.14
Papua	1.9	–.39	.10			–.11	.06
Philippines	2.6	–.04*	.10	–.11	.09	–.27	.06
Nigeria	2.2	–.70	.02	–.51	.06	–.28	.14
Cameroon	2.0	–.09	–.00*	–.24	.08	–.11	.05
Thailand	3.6	–.14	.08	–.16	.11	–.09	.05
Nicaragua	1.2	.01*	–.02*	–.09	.02*	–.30	–.01*
Costa Rica	2.5	–.10	.08	–.11	.15	–.34	.07
Peru	1.6	–.30	.08	–.18	.12	–.21	–.00*
Guatemala	2.0					–.22	.10
Congo P.R.	2.8	–.19	–.05	–.17	–.02*	–.35	.17
Turkey	3.1	–.25	.11	–.26	.09	–.33	.06

continued on page 50

Appendix Table 2A.1 continued

Country	*Growth rate of income per capita*	*Value-added shares current prices Agric.*	*Manu.*	*constant prices Agric.*	*Manu.*	*Employment Agric.*	*Ind.*
Tunisia	4.5	–.06	.07	–.08	.08	–.23	.15
Jamaica	2.1	–.13	.04	–.06	.03*	–.23	.03
Dominican R.	2.6	–.11	.03	–.16	.04	–.25	.07
Paraguay	2.3	–.11	.01*	–.12	–.00*	–.11	.01
Ecuador	3.4	–.17	.02*			–.07	–.02
Colombia	2.3	–.20	.04	–.14	.04	–.46	.05
Angola	–1.3	–.04*	.04	.08	.01*	.03*	–.01*
Lebanon	1.1	–.07	.03			–.63	.08
Upper-middle income							
Jordan	6.5	–.06	.03*	–.09	.04	–.18	.15
Syria	3.3	–.10	–.18	–.14	–.16	–.26	.14
Malaysia	4.0	–.12	.16	–.13	.09	–.16	.05
Chile	0.8	–.06	.08*	–.06	.08*	–.21	.01*
Brazil	4.4	–.05	.02	–.08	.03	–.22	.09
Korea	4.9	–.21	.13	–.21	.26	–.28	.15
Argentina	1.2	–.16	.05*	–.15	.05*	–.19	–.13
Panama	3.6	–.11	.03	–.12	.04	–.21	.04
Portugal	4.6	–.14	.07	–.18	.07	–.16	.08
Mexico	3.2	–.11	.02	–.08	.08	–.27	.09
Algeria	2.3	–.28	–.02*			–.64	.19
South Africa	2.3	–.11	.03			–.06	–.01
Uruguay	1.0	–.13	.01*	–.04	.00*	–.30	.10
Yugoslavia	4.7	–.13	–.03	–.10	.10	–.29	.17
Venezuela	2.3	–.02	.11	–.03	.04	–.34	.09
Greece	5.3	–.09	.03	–.11	.08	–.14	.06
Israel	4.3	–.08	–.03			–.08	.02
Iran	5.7	–.19	–.01*			–.12	.08
Iraq	4.3	–.12	–.03			–.12	.08
Taiwan	5.6	–.15	.17	–.14	.19		
Hungary	5.9	–.02	–.09	–.07	.04	–.13	.06
Industrial market economies							
Spain	4.2	–.15	–.03*	–.07	.13	–.28	.12
Ireland	2.8	–.06	.12			–.26	.15
Italy	4.0	–.09	.05	–.07	.11	–.26	.11
New Zealand	1.6	–.26	.06*			–.16	–.00*
Belgium	3.3	–.06	–.07	–.04	.07	–.08	–.09
UK	2.0	–.03	–.22	–.00*	–.10	–.06	–.11
Austria	4.3	–.09	–.09	–.05	.04	–.19	–.00*
Netherlands	3.0	–.09	–.14	–.03	–.01*	–.11	.08

Appendix Table 2A.1 continued

Country	*Growth rate of income per capita*	*Value-added shares current prices Agric.*	*Manu.*	*constant prices Agric.*	*Manu.*	*Employment Agric.*	*Ind.*
Japan	6.4	–.11	.04	–.07	.14	–.18	.07
France	3.7	–.08	–.04	–.06	.08	–.20	.03
Finland	3.7	–.12	–.06	–.12	.08	–.33	.06
Germany	4.0	–.05	–.07	–.02	.03	–.15	.03
Australia	2.3	–.14	–.14	–.05	–.07	–.13	–.10
Denmark	2.9	–.12	–.09	–.07	.06	–.21	.02
Canada	2.7	–.03	–.08	–.04	–.00*	–.17	–.08
Sweden	2.7	–.08	–.14	–.05	–.02*	–.18	–.09
Norway	3.2	–.06	–.07	–.06	–.06	–.21	.00
US	1.9	–.02	–.13	–.03	.02*	–.15	–.08
Switzerland	2.3					–.15	–.01*

Notes: Growth of income per capita (y) from $\ln y = \alpha + \beta t$.
Income slopes for shares from $x = a + b \ln y$.
* t ratio less than 2.

Notes

* I thank Jeff Williamson for his comments at the conference and in writing, and Yosi Deutsch for help and advice on the empirical part.

1. Kuznets's early works already show interest in the differential evolution of commodities and sectors. He analysed their behaviour in terms of logistic and Gompertz curves (1930).
2. A slightly different list of criteria for distinguishing sectors appears in Taylor (1986). The paper also discusses various approaches with implications for sector proportions.
3. Chenery and Taylor (1968) compared their cross-country results with long time-series in nine advanced countries. On average the cross-country regressions accounted for about 80 per cent of the historical decline in the primary share.
4. Libya and Saudi Arabia (not in the table) showed the same pattern.
5. The main universal factors in Kuznets's presentation – labelled by him 'transnational factors' – are the industrial system based on the application of the technological potential afforded by modern science, a community of human wants and aspirations, and the organisation of the world into nation-states (see also Syrquin, 1987).
6. The 'demand' label has become customary for this approach. Trade and comparative advantage are more supply than demand driven.

7. In deriving the equation, use is made of the result that $\dot{R} = -R\dot{A}R$. In discrete fixed weight decompositions, interaction terms arise, unless the weights of both periods are mixed. There are two extreme ways of mixing them, equivalent to Paasche and Laspeyres formulations. In Chenery, Robinson and Syrquin (1986) an average of the two is shown in the tables.
8. In the original study, the intervals were in 1964 US$ and went from $100 to $1500. To convert them into 1980 dollars I used a factor of 3 for the lowest income level, and a slightly higher factor for higher income levels. The differential factor is based on the apparent relative appreciation of real exchange rates in higher income countries (see Syrquin and Chenery (1986); Wood, 1986).
9. As in most growth accounting exercises, the sources are regarded as independent from each other.

References

Bauer, P.T. and Yamey, B.S. (1951) 'Economic Progress and Occupational Distribution'. *Economic Journal*, vol. 61, pp. 741–55.

Chenery, H.B. (1960) 'Patterns of Industrial Growth', *American Economic Review*, vol. 50 (September), pp. 624–54.

Chenery, H.B., Robinson, S., and Syrquin, M. (1986) *Industrialization and Growth: A Comparative Study* (New York: Oxford University Press).

Chenery, H.B., Shishido, S., and Watanabe, T. (1962) 'The Pattern of Japanese Growth, 1914–1954', *Econometrica*, vol. 30, pp. 98–139.

Chenery, H.B. and Syrquin, M. (1975) *Patterns of Development, 1950–1970* (London: Oxford University Press).

Chenery, H.B. and Syrquin, M. (1980) 'A Comparative Analysis of Industrial Growth', in Matthews, R.C.O. (ed.) *Economic Growth and Resources*, vol. 2. *Trends and Factors*, IEA (London: Macmillan; New York: St Martins Press).

Chenery, H.B. and Syrquin, M. (1986) 'Typical Patterns of Transformation', in Chenery, H.B., Robinson, S. and Syrquin, M. *Industrialization and Growth: A Comparative Study* (New York: Oxford University Press).

Chenery, H.B. and Taylor, L. (1968) 'Development Patterns among Countries and over Time'. *Review of Economics and Statistics* vol. 50 (November) pp. 391–416.

Clark, C. (1940) *The Conditions of Economic Progress* (London: Macmillan)

Deutsch, J. and Syrquin, M. (1986) 'Economic Development and the Structure of Production', processed.

Fisher, A.G.B. (1939) 'Production: Primary, Secondary, and Tertiary', *Economic Record* vol. 15, pp. 24–38.

Hoselitz, B.F. (1960) 'Theories of Stages of Economic Growth', in Hoselitz, B.F. (ed.) *Theories of Economic Growth* (New York: The Free Press).

Jameson, K.P. (1982) 'A Critical Examination of "The Patterns of Development"', *Journal of Development Studies*, vol. 18 (July), pp. 431–46.

Kuznets, S. (1930) *Secular Movements in Production and Prices* (Boston: Houghton Mifflin).

Kuznets, S. (1957) 'Quantitative Aspects of the Economic Growth of Nations: II. Industrial Distribution of National Product and Labor Force', *Economic Development and Cultural Change*, vol. 5 (July): supplement.

Kuznets, S. (1959) 'On Comparative Study of Economic Structure and Growth of Nations', in National Bureau of Economic Research *The Comparative Study of Economic Growth and Structure* (New York: NBER).

Kuznets, S. (1971) *Economic Growth of Nations: Total Output and Production Structure* (Cambridge, Mass.: Harvard University Press).

Maddison, A. (1980) 'Economic Growth and Structural Change in the Advanced Countries', in Leveson, I. and Wheeler, J.W. (eds) *Western Economies in Transition* (Colorado: Westview Press).

Mellor, J.W. (1986) 'Agriculture on the Road to Industrialization', in Lewis, J.P. and Kallab, V. (eds) *Development Strategies Reconsidered* (New Brunswick, NJ: Transaction Books for the Overseas Development Council).

Syrquin, M. (1986) 'Productivity Growth and Factor Reallocation'. In Chenery, H.B., Robinson, S. and Syrquin, M. *Industrialization and Growth: A Comparative Study* (New York: Oxford University Press).

Syrquin, M. (1987) 'Patterns of Structural Change', to appear in Chenery, H.B. and Srinivasan, T.N. (eds), *Handbook of Development Economics*, forthcoming, North-Holland.

Syrquin, M. and Chenery, H.B. (1986) 'Patterns of Development: 1950 to 1983'. Economic Analysis and Projections Department, World Bank, processed.

Taylor, L. (1986) 'Theories of Sectoral Balance', Chapter 1 in this volume.

Wood, A. (1986) 'Puzzling Trends in Real Exchange Rates: A Preliminary Analysis', processed at Institute of Development Studies, University of Sussex.

World Bank (1986) World Development Report 1986 (New York: World Bank and Oxford University Press).

3 Structural Change and Economic Growth in Developing Countries

Vadiraj R. Panchamukhi, R.G. Nambiar and R. Mehta

RESEARCH AND INFORMATION SYSTEM FOR THE NON-ALIGNED AND OTHER DEVELOPING COUNTRIES, NEW DELHI

1. INTRODUCTION

The connection between economic structure and the level of development has been one of the most extensively explored themes in both historical and cross-section studies. The most fruitful among these include Clark's (1940) analysis of the changes in the use of labour with rising income, Kuznets's study (1957) comparing elements of national accounts, and the more recent works of Chenery (1960) and Chenery and Syrquin (1975) exploring the patterns of development in the post-war period with more refined and improved methods. The central thesis emerging from these comparative studies is that for countries to grow, there must be 'structural transformation'. This transformation involves shifts in resource allocation from primary to secondary activities, mainly manufacturing. This is borne out by the fact that growth occurs with certain regularities; and the direction of the change is the same, if one looks at one country at different points of time, and at different countries at the same point of time.

In what follows, we consider a sample of ninety-two developing countries. The intertemporal data is drawn for the period of 1970 to 1984.[1] The ninety-two country sample accounts for 97 per cent of the developing world's population, and 94 per cent of its GDP. Using this time series and the cross-section data, we try to describe and analyse the growth and structural change sequences of the 'developing world'. An inquiry along these lines is, in our opinion, very helpful to further our understanding of the development process in general, and in particular the sources of economic growth in this part of the world. Moreover, this might help to capture the common perceptions of structural change and adjustment in developing countries more than in the earlier works (Chenery, and others) where the sample is heavily characterised by a mix of prominent developed and developing countries.

The plan of the rest of the paper is as follows. Section 2 lays the ground work of the study; it traces, using the time-series data of the sample countries, changes in the type-structure of national product, changes in the products shares of different sectors (agriculture, industry and services), instability affecting the time to time movements, and the direction of intersectoral shifts. Section 3 explores the nature of the association between structural changes and growth. Using the pooled data of fifty-four non-oil developing countries over the period 1970–84, we first establish the statistical association between the structural changes and the changes in the level of development; and then study for each country the manner of intersectoral substitution with rises in income levels. Sections 4 and 5 are devoted to investigating the relation between sectoral growth rates and overall growth rates. Section 4 examines specifically the question whether countries having a smooth transition process have higher growth rates than those which exhibit violent fluctuations in the relative sectoral trends; and in section 5 we test two of the theoretical postulates: Kaldor's hypothesis and Verdoorn's law. Finally, in Section 6, we carry out a detailed examination of the determinants of output growth in developing countries by postulating a relationship between the output growth and several macroeconomic variables.

Attention is also drawn to the fact that only aggregate indicators such as relative sectoral shares in product are examined here. Greater penetration in depth can contribute greatly to the elucidation of much that is obscure at present, but for pragmatic reasons, attention is focused only on broad sectoral movements rather than on microeconomic discussions.

2. STRUCTURAL CHANGES

Table 3.1 presents the comparative growth picture of sectoral shares (agriculture, industry and services) in the total product of different countries of our sample. This has been computed by fitting time trends to the individual share series of the sample countries. In order

Table 3.1 Growth of sectoral share in GDP for developing countries, 1970–84

S.no. Growth rate (% per ann.)	*All countries* Agr. (no.)	Ind. (no.)	Ser. (no.)	*Non-oil countries* Agr. (no.)	Ind. (no.)	Ser. (no.)
1. < − 1	12	5	1	11	1	1
	(7.4)	(10.2)	(0.4)	(9.3)	(0.6)	(0.6)
2. −1 to −0.5	19	9	4	15	8	3
	(24.2)	(3)	(0.5)	(26.5)	(4.3)	(0.7)
3. −0.5 to 0.0	28	12	5	25	10	5
	(41.7)	(12.9)	(4.5)	(48.3)	(11.6)	(7.1)
Declining trend	*59*	*26*	*10*	*51*	*19*	*9*
(1–3)	(73.2)	(26.1)	(5.4)	(84.1)	(16.4)	(8.4)
4. 0.0* (no change)	26	25	36	22	22	30
	(20.4)	(39.1)	(22.7)	(14.4)	(51.4)	(13.3)
5. 0.0 to 0.5	5	21	21	2	19	19
	(6.1)	(25.5)	(39.4)	(0.9)	(20.6)	(51.6)
6. 0.5 to 1.0	2	13	20	2	12	15
	(0.4)	(5.4)	(25.9)	(0.6)	(7.7)	(24.1)
7. > 1.0	0	7	5	0	5	4
	(–)	(3.9)	(6.5)	(–)	(3.8)	(2.5)
Increasing trend	*7*	*41*	*46*	*4*	*36*	*37*
(5–7)	(6.5)	(34.8)	(71.9)	(1.5)	(32.2)	(78.3)
Total	92	92	92	77	77	77
	(100.0)	(100.0)	(100.0)	(100.0)	(100.0)	(100.0)

Notes: Growth rates (percentages per annum) are geometric rates, calculated by exponential curve fitting to time series data of share of each sector in total GDP (at 1975 prices), by least squares.

Figures in parentheses represent weights which are on the basis of GDP in US $.

The samples period is not exactly 1970–84 for all countries (see Panchamukhi, Nambiar and Mehta (1986) for the sample period of each country and sources of data).

* Statistically insignificant value of a regression coefficient denotes no change in sectoral share per annum during the sample period.

---

to economise on space, we have summarised these growth results in the form of a frequency distribution where the number of countries are listed against each of growth intervals (see also Figure 3.1).

The broad impression conveyed by the table is the tremendous diversity of experience within the developing world. Thus, in a

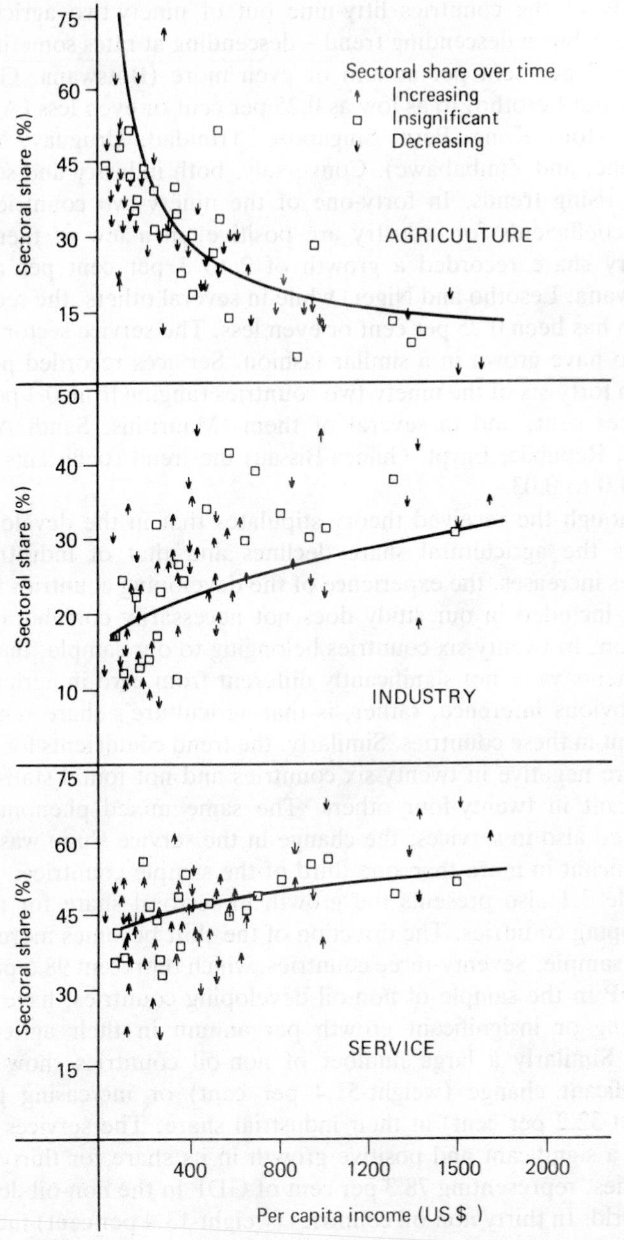

Note: The regression lines are based on the results given in Table 3.4A.

Figure 3.1 Income level and sectoral share

majority of the countries fifty-nine out of ninety-two agricultural share exhibits a descending trend – descending at rates sometimes as high as 2 per cent per annum or even more (Botswana, Guinea-Bissau and Lesotho) to as low as 0.25 per cent or even less (Algeria, Brazil, Hong Kong, Peru, Singapore, Trinidad, Uruguay, Yemen Republic, and Zimbababwe). Conversely, both industry and services reveal rising trends. In forty-one of the ninety-two countries, the trend coefficients for industry are positive; in a few of them, the industry share recorded a growth of 2 to 3 per cent per annum (Botswana, Lesotho and Niger) while in several others, the recorded growth has been 0.25 per cent or even less. The service sector is also seen to have grown in a similar fashion. Services recorded positive rises in forty-six of the ninety-two countries ranging from 0.1 per cent to 3 per cent, and in several of them (Mauritius, Saudi Arabia, Yemen Republic, Egypt, Guinea-Bissau) the trend coefficients varied from 0.0 to 0.03.

Although the received theory stipulates that in the development process the agricultural share declines and that of industry and services increases, the experience of the developing countries for the period included in our study does not necessarily corroborate this situation. In twenty-six countries belonging to our sample, the trend coefficients were not significantly different from zero in agriculture. The obvious inference, rather, is that agriculture's share remained constant in these countries. Similarly, the trend coefficients for industry were negative in twenty-six countries and not found statistically significant in twenty-four others. The same mixed phenomena is observed also in services; the change in the service share was nil or insignificant in more than one-third of the sample countries.

Table 3.1 also presents the growth in sectoral share for non-oil developing countries. The direction of the shift becomes more sharp in this sample. Seventy-three countries, which represent 98.5 per cent of GDP in the sample of non-oil developing countries, have either declining or insignificant growth per annum in their agricultural share. Similarly a large number of non-oil countries show either insignificant change (weight-51.4 per cent) or increasing growth (weight-32.2 per cent) in their industrial share. The services sector shows a significant and positive growth in its share for thirty-seven countries, representing 78.3 per cent of GDP in the non-oil developing world. In thirty non-oil countries (weight-13.4 per cent) the share of the services sector has remained constant during the sample period.

Table 3.2 Shifts in sector share and instability in developing countries, 1970–84

S.No. Growth rate (% p.a.)	Coefficient of instability	Agriculture			Industry			Services		
		0 to .25	*.25 to .5*	*.5 to 1.0*	*0 to .25*	*.25 to .5*	*.5 to 1.0*	*0 to .25*	*.25 to .5*	*.5 to 1.0*
1. < − 1		9	1	2	4	0	1	0	1	0
2. −1 to −0.5		16	3	0	5	4	0	2	2	0
3. −0.5 to 0.0		10	12	6	0	9	3	0	3	2
Declining trend (1–3)		*35*	*16*	*8*	*9*	*13*	*4*	*2*	*6*	*2*
4. 0.0* (no change)		0	1	25	0	0	25	1	1	34
5. 0.0 to 0.5		3	2	0	8	7	6	9	6	6
6. 0.5 to 1.0		2	0	0	9	3	1	14	6	0
7. > 1.0		0	0	0	6	2	1	4	1	0
Increasing trend (5–7)		*5*	*2*	*0*	*23*	*12*	*8*	*27*	*13*	*6*
Total		40	19	33	32	25	37	30	20	42

Notes: Growth rates (percentage per annum) are geometric rates, calculated by exponential curve fitting to time series data of share of each sector in total GDP (at 1975 prices), by least squares.

The coefficient of instability is the ratio of error sum of squares to the total sum of squares of sectoral share around the trend value.

* Statistically insignificant value of a regression coefficient denotes insignificant change in sectoral share during the sample period.

Table 3.2 presents the results on the fluctuations around the trend in sectoral shares and how these fluctuations are related to the trend behaviour of the sectoral share in GDP. Such interrelations affecting the sectoral share and its progression were worked out by using the standard error around the trend as a measure of instability.

The data reported in the table show the lack of any change (or change in the opposite direction) in sectoral share linked to a higher coefficient of instability in a number of countries. To illustrate this – the coefficient of instability in agriculture was above 0.50 in thirty-three countries: and twenty-five of them recorded a lack of any significant change in the share of agriculture. Conversely, fifty-nine countries in the sample recorded declining trends in agriculture's share, and in fifty-one of them the coefficients of instability were less than 0.50.

The direction of structural change – whether the expansion path tilts towards industry or services – is of particular interest to students of economic development. Both Clark and Kuznets showed independently in their works that the expansion path along which the underdeveloped countries of the world progress, is first towards industry and then from industry to services. Clark traced the observation of this relationship back to Sir William Petty and proposed that the shift from agriculture to industry and from industry to services in the course of economic growth be called Petty's Law.

It remains now to elucidate the direction of the structural change. Our approach relies on sectoral product ratios (industry relative to agriculture, services relative to agriculture and services relative to industry) regressed against time trend. The resulting coefficients will tell us about the nature of the shift of one sector relative to the other. These results are arrayed in Table 3.3 according to the magnitudes of their coefficients – decreasing, constant, and increasing, and within each category several sub-intervals are presented; the number shown against each such interval is the number of countries falling in that sub-interval.

A close inspection of these results is interesting. When the significance of the three distributions (agriculture-industry, agriculture-services, and industry-services) in the table is assessed, the tendency for a secular rise in the shares of non-agricultural sectors is hinted. Comparisons also illustrate that shifts from agriculture to services are more frequent than either shift from agriculture to industry, or from industry to services.

Table 3.3 Inter-sectoral shift in sectoral shares (to total GDP), for developing countries, 1970–84

S.No.	*Inter-sector shift per annum (%)*	*From Agr. to Ind. no. of countries*	*weight*	*From Agr. to Ser. no. of countries*	*weight*	*From Ind. to Ser. no. of countries*	*weight*
1.	< −10	5	11.31	0	0	3	0.24
2.	−10 to −2	6	5.31	4	0.78	10	6.79
3.	−2 to 0	3	0.86	3	0.54	1	2.68
	Subtotal (1 to 3)	*14*	*17.48*	*7*	*1.32*	*14*	*9.71*
4.	0	*33*	*23.70*	*35*	*30.13*	*50*	*57.86*
5.	0 to 2	14	16.09	8	2.08	3	9.58
6.	2 to 10	21	26.96	25	37.75	16	14.54
7.	> 10	10	15.77	17	28.72	9	8.31
	Subtotal (5 to 7)	*45*	*58.82*	*50*	*68.55*	*28*	*32.43*
	Total	*92*	*100.00*	*92*	*100.00*	*92*	*100.00*

Notes: The intersectoral shift of a country has been estimated by regressing the share of one sector in total GDP at 1975 prices to another on time-trend during the sample period (mostly 1970–84).

Negative value of a coefficient denotes the shift is in reverse direction, while the zero value (i.e. insignificant coefficient) that there is no intersectoral shift.

Weights are on the basis of GDP in US $.

3. STRUCTURAL CHANGE AND DEVELOPMENT PATTERN

In assessing the significance of our findings reported in section 1, two kinds of questions arise. One is whether the changes observed in the structure of developing countries are in conformity with the prevailing views on development patterns, and the other is whether the deviations are big enough to damage or discredit the existing ideas. We set out next to explore this association between development and structural change.

The prevailing views of the development pattern are based on the universality of such structural transformations (Clark, 1940; Kuznets, 1957, 1971; Chenery, 1960; Chenery and Taylor, 1968; Chenery and Syrquin, 1975; Kader, 1985). The universality of the structural shifts is based on the assumptions that the composition of consumer demand varies in a similar fashion with rising income, the rate of growth of capital accumulation exceeds that of the labour force, and all countries have access to international trade, capital inflow and technology. Then, we are told, that with rising income or any other index of economic development, variables describing production, domestic use, international trade and resource allocation change over time in a predictable manner. These variables usually show a period of fairly rapid changes followed by deceleration and in some cases even a reversal of the direction of change.

We attempt here to seek empirical verification of the model.

For the purpose of this investigation, three development variables – per capita income, population and openness – were chosen,[2] all of them are generally believed to be closely associated with economic development. For measuring openness[3] two alternative indicators have been employed: (i) exports to GDP ratio and (ii), ratio of total trade (exports + imports) to GDP.

The following non-linear specification was used to test the links between the above variables and structural change.[4]

$$\text{In } S = a + b_1 \ln\bar{Y} + b_2 (\ln\bar{Y})^2 + b_3 \ln N + b_4 (\ln N)^2 + b_5 \ln O$$

where:

S = dependent variable (shares of agriculture, industry and services relative to total GDP);
Y = per capita income (at 1975 prices) in US \$);
N = population;
O = degree of openness, i.e., exports to GDP ratio, or proportion of (exports + imports) in total GDP (at 1975 prices).

A number of striking features emerge from the regression results. First, while comparing the results, we find that they are not greatly affected by our alternative measures of openness. Both the estimated relations produced the expected results of declining agricultural share and rising industry and service shares with rising Export-GDP ratio, or rising trade share.

Secondly, the size variable, represented by population, has the expected sign in all the estimated equations, but is found to be statistically insignificant (Table 3.4A). Thirdly, the anticipated relationship between per capita income and the sectoral shares broadly tends to hold good. In all the estimated equations, the b_1 coefficient is statistically significant and has also the expected sign. The expectations that with rising income, the share of agriculture tends to

Table 3.4.A Regression results of production relations for developing countries (dependent variable: ln) (sectoral share in total GDP:S)

Sector	*Constant*	ln $\bar{Y}$	$(\ln Y)^2$	ln N	$(\ln N)^2$	$\ln \frac{(X+M)}{GDP}$	R^{-2}	F	*Sample size*
I. Agr.	–1.2477	–0.5623	–0.0699	–0.0792	–0.0007	–0.2911	0.65	267.18	709
	(5.84)	(33.70)	(34.70)	(1.65)	(0.26)	(10.83)			[52]
II. Ind.	–1.9233	0.3284	0.0430	0.0305	0.0075	0.2844	0.50	140.84	709
	(9.73)	(21.27)	(22.62)	(0.69)	(2.98)	(11.43)			[52]
III. Ser.	–0.5277	0.1177	0.0123	0.0011	–0.0013	0.0533	0.38	87.17	709
	(4.91)	(14.00)	(11.92)	(0.04)	(0.98)	(3.94)			[52]

Notes: The estimation of the regression equations have been carried out by Least Squares using panel data of various non-oil-developing countries over time.

Y = per capita income (in US \$) at 1975 prices; N = population; X, M, GDP and S represent total exports, total imports, gross domestic product and sectoral share, at 1975 prices.

Figures in parentheses represent t-values.

Figures in square brackets, in the last column, represent total number of countries.

Table 3.4.B Comparison of estimates of production structure

S.No.	Sectoral share	Per capita income (US $) 100	250	500	1000	2000	Implicit elasticity
	OUR STUDY						
1.	Agriculture	0.41	0.31	0.24	0.17	0.11	–0.1548
2.	Industry	0.18	0.20	0.23	0.27	0.33	0.1156
3.	Service	0.42	0.45	0.48	0.51	0.56	0.0481
	KADER'S STUDY						
1.	Primary	–	0.55	0.48	0.36	0.29	–0.1151
2.	Industry	–	0.10	0.14	0.19	0.22	0.3000
3.	Service	–	0.35	0.38	0.45	0.49	0.0952

Notes: Kader's results are based on a cross-section of sixty-four countries (equally divided between developed and developing countries) for the year 1976, while our results are based on pooled cross-section and time-series data of fifty-two developing countries over the period of 1970–84, at 1975 prices. The implicit elasticity has been computed over the range US $ 250–1000.

Sources: 1. Table 3.4.A.
2. Kader, 1985, p. 208.

---

decelerate faster (judged from b_2 coefficients) are also met; the coefficient b_2 for agriculture is negative and significant.

Fourthly, rising per capita income is seen as lifting up both the industry share as well as the services share (Figure 3.1). More important than this is the fact that they are lifted up in an accelerated manner as judged from the b_2 coefficients; these coefficients are positive and statistically significant. The expectation that the expansion path could be represented by an inverted U (or S)-shaped curve, is, however, not met. In other words, over this stage in the process of economic development, whether measured within the experience of one developing country or by cross-country comparisons, we are led to the unexpected conclusion that industry and service shares rise and accelerate almost in a parallel fashion.

Before proceeding, it might be relevant to compare these results with those obtained from one of the earlier looks at the development pattern as in Table 3.4B. The comparisons illustrate a point that is perhaps obvious, but worth emphasising. There are strong *a priori* grounds for expecting both industry and service share to rise concurrently with the rise in income. However, the rate at which services tend to replace other sectors in general, and industry in particular, appears to be much larger when judged within the experiences of developing countries than that exhibited in the alternative scenario drawn from a sample of both developed and developing countries (Table 3.4B, last column).

A useful working hypothesis about the reason for these striking differences in the development pattern is that it lies in certain developments in the last fifteen or twenty years involving essential economic and social reforms which have brought the 'visible hand' of modern government and the vast resources it can mobilise, both nationally and internationally, right into the centre of economic affairs. Growth, employment generation, poverty alleviation and creating social infrastructures became the major goals of national economic policies, as indeed they should be. The change led to a period of unprecedented governmental activity.

Among other factors that might have stimulated service growth in developing countries are: (1) The historical role of the urban middle class in wholesale trade and distribution, (2) The operation of the demonstration effect in these countries, creating similar demand patterns to those of high income countries, (3) The comparative advantage of many low income countries in providing final product services such as tourism, (4) The intensification through urbanisation of the demand for a wide variety of services ranging from basic infrastructure to general welfare, to financial and commercial services.

The obvious inference is that as the incomes rose, the demand for services must have risen at the same rate or greater than that for industry.

To probe into this, we make a closer scrutiny of the patterns of intersectoral substitution with respect to income changes. These intersectoral substitution elasticities were derived by using a non-linear equation where the year-to-year changes in inter-sectoral ratios (industry relative to agriculture, etc.) were regressed against per capita income changes and relative prices:[5]

$$\ln S_i/S_j = a + b_1 \ln \bar{Y} + b_2 (\ln \bar{Y})^2 + b_3 (\ln P_i/P_j) + b_4 (\ln P_i/P_j)^2$$

where:

P_i = price index of i-th sector.

Since the observations on prices of each sectors are not available, we have taken the implicit deflators of corresponding sectors. The estimation of the above equation has been carried out for around 80 countries.

Table 3.5 presents the values of the mean elasticity[6] of intersectoral shift with respect to per capita income for eighty-two countries. Our results give a wide range of mean values of elasticity of substitution between industrial and agriculture sectors (*eIA*). In fifty-four countries (out of eighty-one countries) the mean elasticity of substitution between industry and agriculture is positive, i.e., the industrial sector is substituting agriculture. It is clear from Table 3.5A that this shift is quite significant in a number of developing countries to the order of more than 2 per cent per annum. However, the trend of elasticity in individual countries is mixed. Out of eighty-one countries, forty-three have positive value of b_2 which implies that industry is replacing agriculture at an increasing rate with the rise in per capita income. In twenty-five countries, the shift has a declining trend. Only in 3 countries (Dominican Republic, Nigeria and Zambia), is agriculture substituting industry at an increasing rate.

The results of an intersectoral shift between service and agriculture sectors (*eSA*) also show that the service sector is substituting agriculture in more than 64 per cent of (fifty-five out of eighty-six countries) of developing countries that is, those for which the mean elasticity is positive.

The results comparing values of *eIA* and *eSA* are summarised in Table 3.5B. We have grouped the countries into the two categories $eIA > eSA$, or $eIA < eSA$. The data in the table show that in thirty-six out of total eighty-one countries, the growth process involved, on average, industry substituting agriculture. In all these countries the elasticities of substitution between agriculture and industry were higher than the corresponding elasticities between agriculture and services. This is to say, the structural transformation in these countries has taken place in the conventionally expected fashion. Conversely, in thirty-one other countries the development has led more to substitution of agriculture by services than substitution by industry. But a somewhat less usual phenomenon is that in fourteen countries, structural transformation is reversed – agriculture is substituting both industry as well as services, as the level of income rises. As far as the rate of change of elasticity (with respect to change in per

Table 3.5A Intersectoral substitutions, per capita income and relative prices, for developing countries (elasticity of intersectoral shift w.r.t. mean per capita income)

Class no.	*Industry/Agriculture Frequency class (eIA)*	*no. of countries*	*Class no.*	*Industry/Agriculture Frequency class (eSA)*	*no. of countries*
1.	< −5	3	1.	< −5	7
2.	−5 to −1	9	2.	−5 to −1	11
3.	−1 to −.50	3	3.	−1 to 0	10
4.	−.50 to 0	12	4.	0 to 0.5	19
5.	0.0 to 0.5	16	5.	0.5 to 1	6
6.	0.5 to 1.0	8	6.	1 to 2	10
7.	1 to 2	10	7.	2 to 3	9
8.	2 to 5	9	8.	3 to 20	10
9.	> 5	11			
	Total	81			82

Notes: The elasticity of intersectoral shift with respect to per capita income has been estimated for each country by the application of time-series data to the following equation:

$$S_i\, S_j = a + b_1 \ln \bar{Y} + b_2 (\ln \bar{Y})^2 + b_3 (\ln P_i\, P_j) + b_4 (\ln P_i\, P_j)^2$$

where, S_i = Share of i-th sector in total GDP at 1975 prices; Y = per capita income; P_i = Price index of i-th sector.

The mean value of elasticity (e) has been estimated using the mean per capita income during the sample period.

Table 3.5B Comparison of intersectoral elasticities with respect to per capita income between (a) agriculture and industry, and (b) agriculture and services (no. of countries)

eIA > eSA		*eSA > eIA*		
eIA = + *ve* *eSA* = + *ve*	*eIA* = + *ve* *eSA* = − *ve*	*eSA* = + *ve* *eIA* = + *ve*	*eSA* = + *ve* *eIA* = − *ve*	*eSA* = − *ve* *eIA* = − *ve*
24	12	18	13	14

Notes: *eIA* = Elasticity of substitution between agriculture and industry. A positive value of *eIA* shows that industry is substituting agriculture, when *eIA* = 0, there is complementarity between the two sectors, when *eIA* is negative, agriculture substitutes industry.

eSA = Elasticity of substitution between agriculture and services.

capita income), b_2, is concerned, forty-two (out of eighty-one) countries have higher values in industry – agriculture substitution than in services – agriculture substitution. In the remaining thirty-nine countries, the coefficients show the reverse order.

This heterogeneity in response of different sectors in developing countries suggests that the factors that are important in explaining the development path of developing countries differ from those that loom large in explaining the developed countries' patterns, and that there is some evidence that the main forces at work are the differential responses of production and service sectors operating in these countries.

4. STABLE TRANSITION AND GROWTH RATES

Until this stage, we have sought to find the time-path of growth in the developing countries. A question which is perhaps most relevant in the present context and to which the data used here can supply a much firmer answer is whether the balance between agriculture and industry is conducive to rapid economic growth. We probe into this aspect next.

In order to understand this, we first select those countries which follow stable growth in their sectoral share. In other words, we select those countries which have low values of instability coefficient (Table 3.2) in the industrial share as well as the agricultural share. Using the pooled time-series data of these countries, we then estimate the relations between intersectoral share, per capita income and relative prices. Based on the regression results, we next estimate the confidence interval of the mean elasticity, e (= $1.47 + 0.34 \ln \bar{Y}$, where $\bar{Y}$ is mean per capita income) by taking $e \pm 3S$ limits, where S represents standard deviation of mean elasticities. Countries falling below the lower limit were classified as agricultural oriented; countries falling above the upper level were classified as industrialised; and countries falling in the prescribed range of mean were categorised as 'stable growth countries'.

Table 3.6 presents a comparative picture of growth performance in countries under agricultural orientation, stable growth, and industrial orientation. It emerges from the table that a large number of countries which fall under the category of 'stable growth' have scored higher growth rates; but the converse is not true; there are countries

Table 3.6 Growth rate, stable growth and developing countries – frequency distribution

Pattern of stable Growth / *Growth rate*	*< 0*	*0–2*	*0–4*	*4–6*	*6–8*	*8–10*	*Total*
ALL DEVELOPING COUNTRIES				*No. of countries*			
Agr. orientation	2	11	13	10	7	3	46
	(4.35)	(23.92)	(28.26)	(21.74)	(15.21)	(6.52)	(100.00)
Stable growth	1	1	2	3	5	2	14
	(7.14)	(7.14)	(14.29)	(21.43)	(35.72)	(14.28)	(100.00)
Ind. orientation	3	3	8	6	2	3	25
	(12.00)	(12.00)	(32.00)	(24.00)	(8.00)	(12.00)	(100.00)
NON-OIL DEVELOPING COUNTRIES							
Agr. orientation	2	10	12	10	3	1	38
	(5.26)	(26.32)	(31.59)	(26.31)	(7.89)	(2.63)	(100.00)
Stable growth	0	1	2	3	2	2	10
	(0.0)	(10.00)	(20.00)	(30.00)	(20.00)	(20.00)	(100.00)
Ind. orientation	2	3	8	5	2	3	23
	(8.69)	(13.05)	(34.60)	(21.74)	(8.69)	(13.05)	(100.00)

Notes: The selection of countries among the three groups, (1) agriculture orientation, (2) stable growth, and (3) industrial orientation, has been done on the basis of the three ranges of elasticity of intersectoral (industry *v*. agriculture) share with respect to per capita income. The range of *eIA* for stable growth has been computed by regressing the intersectoral share of industry to agriculture with respect to per capita income and corresponding relative prices using pooled data for a group of countries over time. The range is Mean Elasticity ± 3 S.D. of mean elasticity. The group of the countries has been selected on the basis of coefficient of instability in agricultural share and industrial share.

Figures in parentheses represent the percentage distribution.

with a high growth rate, which do not fall in to the category of 'stable growth'.

5. SECTORAL GROWTH AND OVERALL GROWTH

Having enumerated and discussed the nature and direction of the shift in the structure of national product that have occurred in these countries in the course of the past one and a half decades, we may now ask (i) what is the relationship between sectoral growth in output and overall growth? and (ii) what is the relationship between growth and productivity? We shall dwell upon these two questions in this section.

5.1 Industry as an Engine of Growth

The driving force behind economic growth, according to Kaldor, is growth in the manufacturing sector. 'Fast rates of growth', Kaldor (1967, p. 7) writes, 'are almost invariably associated with the fast rate of growth of the secondary sector, mainly manufacturers'. In his examination of the question why Britain's rate of economic growth was so much lower than that of other industrial countries, Professor Kaldor discovered a close positive association between growth in manufacturing output and the overall growth of the economy. This association, according to him, lay in the fact that productivity expands faster in the manufacturing sector than in others because the former operates under increasing returns to scale; the resulting expansion in productivity helps to expand output in other sectors. The increasing capital intensity in agricultural production, for instance, cannot be sustained without expansion in manufacturing activity. In fact it is advocated that the industrial sector is the engine of growth.

That the manufacturing sector plays a key role in the process of development is a view widely accepted among economists. Its importance for growth in other sectors has been indicated by concepts such as backward and forward linkages, and the leading sector thesis. But more important than this, the inference to be drawn from Kaldor's analysis is that the manufacturing sector continues to play a key role in countries even after the goals of modernisation and industrialisation are achieved.

It is a moot question to ask – is industry the engine of growth for the developing countries of recent vintage? Kaldor (1967), UN (1970) and Stoneman (1979), have considered the alternate specifications for testing the strength of causal linkages between sectoral growth and overall growth rates.

We shall indicate briefly what impressions we have derived from testing this hypothesis with the data for developing countries for the period 1970–84. The regression results obtained from a ninety-one-country sample are reported in Table 3.7A. The detailed analysis of these results does not seem to corroborate Kaldor's thesis. In equation (1) where the GDP growth is regressed with the growth rate in industry's value added, the results indicate the existence of a strong correlation between the industrial growth and the overall growth rate. But equation (2), where we have sought to capture the relationship between service growth and the overall growth, tells almost a similar story of positive association. In the face of the likelihood that both types of growth linkages are present, it must be admitted then that forming clear expectations about the applicability of this hypothesis to developing countries is not wise. Since the regression results of (1) have positive constant term and slope coefficient significantly less than unity, we anticipated positive association between overall growth rate, and the excess of the growth of industrial output over that of non-industrial sectors. Our expectations along these lines, are, however, not met as seen from the results in equation (5).

The table also presents the relationship between the growth rate of GDP and the growth of agriculture. Equation (3) shows that there is no significant relationship between the two, and if at all, it is in the wrong direction. This obviously is a reflection of the concurrent occurrence of a falling agricultural and increasing non-agricultural sector.

5.2 Productivity Growth and Overall Growth

Many studies have been conducted in the literature to examine the relationship between productivity growth at sectoral level and overall growth in GDP. Certain apparent empirical regularities are also found to exist in these relationships. In particular, it is observed, notably by Verdoorn, that there exists a strong positive relationship between productivity growth in the manufacturing sector and the overall growth of the economy.

In his famous article Verdoorn (1949) presents empirical evidence on the constancy of the ratio between productivity growth and output growth in industry in the long run. From the examination of a historical series on production for industry as a whole and for individual sectors of industry in a number of countries, he concludes that the average value of the elasticity of labour productivity with respect to output is around 0.5, the lower and upper limits being 0.41 and 0.57 respectively.

Empirical testings of Verdoorn's law have been made by Kaldor (1967, 1975), UN (1970), Cripps and Tarling (1973), Kendrick (1974), Rowthorn (1975), Choi (1983) etc. using the data of the developed countries.[7] Here we test the law using the cross-section data of fifty-one developing countries for the period 1975–80. Table 3.7B presents the results relating to the two specifications – Kaldor's and Rowthorn's[8] – for three sectors: agriculture, industry, and services. Our main findings are the following:

1. The results support Verdoorn's law in the industrial sector; there is a positive significant relationship between productivity growth and output growth. The estimated value for the industrial sector is not statistically different from the value obtained by Verdoorn (i.e., 0.5). A somewhat similar pattern emerges in the results of the services sector also, but the value of the coefficient is higher than in the industrial sector. Though there is high correlation between productivity growth and output growth in the agricultural sector also, the regression coefficient is not statistically different from unity, and the trend of productivity growth is negative. It indicates that Verdoorn's law does not hold well and that factors other than those assumed in the law are at work. Kaldor (1967) and UN (1970) also obtain similar results for non-manufacturing sectors.
2. The results also support Kaldor's observation that there should be a positive correlation between the output growth and employment growth, i.e., expansion of the output requires a simultaneous expansion of employment. However, the coefficient is significant for the industrial sector only. So Kaldor's proposition is valid in the industrial sector as far as the developing countries are concerned.
3. The regression results obtained through the application of Rowthorn's specification do not support the positive association between productivity growth and employment growth. In all the three sectors of economy the relationship is negative and statistically significant. Rowthorn obtained positive association between

Table 3.7A Regression result relating to Kaldor's hypothesis: industry as an engine of growth

Eqn no.	*Regression equation*	$\bar{R}^2$	F	*Sample size*
(1)	$gGDP = 1.8110 + 0.4666gI$ (8.13) (13.85)	0.68	191.89	91
(2)	$gGDP = 1.4369 + 0.4818gS$ (4.00) (8.58)	0.45	73.68	91
(3)	$gGDP = 3.9700 - 0.0337gA$ (0.24)	–0.01	0.06	91
(4)	$gGDP = 1.6524 + 0.5289gNI$ (6.33)	0.35	40.09	73
(5)	$gGDP = 3.3746 + 0.0152(gI \text{ to } gNI)$ (0.51)	–0.01	0.26	73

gGDP = Growth rate of gross domestic product, during 1970–84 (annual average)

gI = Growth rate of gross value added in industrial sector, during 1970–84 (annual average)

gA = Growth rate of gross value added in agriculture sector, during 1970–84 (annual average)

gS = Growth rate of gross value added in service sector, during 1970–84 (annual average)

gNI = Growth rate of gross value added in non-industrial sector, during 1970–84 (annual average)

The estimation has been carried out for cross-country data by least squares.

Figures in parentheses represent *t*-value.

---

productivity growth and employment growth. Commenting on Rowthorn's specification, Kaldor suggests that a statistically significant and positive relationship between productivity growth and employment growth may not necessarily support Verdoorn's law. It has been suggested by Kaldor (1975) and Choi (1983) that the negative sign may be due to the fact, 'that technological change is endogenous or that growth of employment is endogeneously determined or both' (cf. Choi, 1983).

In short, the results tend to support Verdoorn's law, particularly in the industrial sector – productivity growth in the developing countries can be associated with output growth. The law does not hold well for non-industrial sectors even in developing countries. Our results are thus in conformity with those obtained by Kaldor and UN. The value

Table 3.7B Regression results relating to Verdoorn's law, for developing countries

	KALDOR'S SPECIFICATION $gP = a + b.gY$				$gE = a + b.gY$				ROWTHORN'S SPECIFICATION $gP = a + b.gE$			
	a	b	R^{-2}	F	a	b	R^{-2}	F	a	b	R^{-2}	F
Agriculture	–1.1627 (1.86)	0.9485 (6.26)	0.43	39.22	1.2860 (2.02)	0.0457 (0.30)	–0.02	0.09	2.4153 (4.71)	–0.9419 (7.16)	0.50	51.29
Industry	–4.4418 (6.15)	0.6543 (7.15)	0.50	51.65	4.8065 (5.90)	0.3060 (2.97)	0.13	8.81	1.6867 (1.37)	–0.4242 (2.73)	0.11	7.45
Services	–4.0510 (6.07)	0.8001 (8.50)	0.59	72.32	4.2998 (6.07)	0.1619 (1.62)	0.03	2.64	3.8529 (3.34)	–0.6407 (3.46)	0.18	11.98

Notes: gP = Growth rate of productivity, annual average (at 1975 prices), during 1975–80.
gY = Growth rate of gross value added, annual average (at 1975 prices), during 1975–80.
gE = Growth rate of employment, annual average, during 1975–80.

The growth rates are in terms of percentage and represent compound rates.

The results are based on cross-section data for fifty-one developing countries.

Figures in parentheses represent *t*-values.

of coefficient (0.65) lies in the range obtained by Verdoorn and others.

6. SOURCES OF GROWTH

In appraising structural change in developing countries, we can hardly neglect factors determining growth. We take this up in the present section.

There are two possible ways of approaching the problem. The first is to use the concept of an aggregate production by which we try to 'account' for economic growth. Denison (1962, 1974), for example, accounts for economic growth through the growth of labour and capital inputs, with the unexplained residual assumed to represent 'technological growth' or 'productivity growth'. In Kendrick (1961), both tangible and intangible goods are measured in an attempt to reduce this residual. Though this approach is very useful to the understanding of growth, the concept of a production function is not very helpful if it is not a stable function or if there are very large unexplained shifts in it. It is of very little help to measure these changes if we do not know what they are.

The second approach (Robinson, 1971; Kormendi and Mequire, 1985, etc.) rests on a relationship between the growth rate of real product and macroeconomic variables. According to this approach, changes in the growth rate of real product are attributable to changes in several economic variables such as the level of development, population, sectoral distribution of output, inflation rate, monetary variables, to name a few.

Using this latter approach, we tested the association between growth rate and some macroeconomic variables *á la* Kormendi and Mequire (1985) using cross-country data of seventy-three developing countries.

The growth rate of developing countries is explained by the following regression equation:[9]

$$gGDP = a + b_1 \bar{Y}_0 + b_2 gP + b_3 INS + b_4 DIST + b_5 gINF + b_6 INV + b_7 GC + b_8 EX$$

where:

$gGDP$ = growth rate of real GDP;

$\bar{Y}_0$ = per capita income (in US \$), at 1975 prices, in the initial year, i.e., 1970;

gP = growth rate of population;

INS = coefficient of instability (i.e., ratio of error sum of squares to total sum of squares) around the growth trend;

$DIST$ = Distribution of gross domestic product in terms of sectoral output (i.e., ratio of gross value added in non-agriculture to agriculture, i.e., non-agr./agr.);

$gINF$ = growth rate of inflation;

INV = ratio of gross domestic investment to GDP (at 1975 prices), during 1970–84;

GC = ratio of government consumption to GDP (at 1975 prices); in alternative I, we have taken mean value and in alternative II, the annual average growth rate (gGC);

EX = ratio of exports to GDP (at 1975 prices); in alternative I, we have taken mean value and in alternative II, the annual average growth rate of the ratio (gEX) is taken.

The growth rates (i.e., *gGDP*, *gP*, *gGC* *gEX*, and *gINF*) have been computed as average annual geometric rates using exponential curve fitting; *DIST*, *INV*, *GC* and *EX* represent the simple averages during the period 1970–84.

Table 3.8 presents the regression results relating to the determinants of the growth, using cross-section data (a) seventy-three developing countries and (b), sixty non-oil developing countries.[10] Our results show that the overall fit is satisfactory in terms of coefficient of determination. Some of the broad findings of our results are summarised below:

1. The regression coefficient of $\bar{Y}_0$ has the expected negative sign, i.e., the higher the level of initial per capita income the lower is the growth rate of real output.[11] The coefficient is significant in most of the fits. Its value shows that with one unit increase in the initial level of per capita income, the output growth rate declines by around 0.001 per cent.
2. The statistical insignificant value of b_2 (Table 3.8) shows that the effect of population growth on the growth of real output is not significant.
3. The hypothesis proposed by Black (1979) concerning risk-return

trade-off stands rejected. The regression coefficient of output instability (*INS*) is negative and significant in all the fits. It shows that countries which had lower degree of fluctuations around the trend, also experienced larger overall growth rates.[12]

4. The impact of distribution of output (by type of industries) on growth rate is supported by our results. The coefficient of non-agr./agr. is positive and significant. It shows that with 1 per cent shift in the output mix from agriculture to non-agriculture, the growth rate increases by 0.09 per cent, on an average, in the developing countries.
5. Inflation rate has a negative relation with the GDP growth rate for developing countries as shown by the estimated value of b_5. It supports the hypothesis explored by Stockman (1981) and Kormendi and Mequire (1985) that higher inflation reduces economic activity.
6. The expected relation of positive association between investment rate and GDP growth rate is corroborated by the significant value of b_6. In non-oil developing countries, the impact of investment rate is more prominent than in oil-exporting countries.
7. The 'supply side' hypothesis concerning the effects of government expenditure is not supported by both the alternatives considered in this study. It is assumed in this hypothesis that the taxes used to support government expenditure generally distort the optimum resource allocation and hence the growth of output. The sign of the regression coefficient is negative, as expected, but is insignificant in both the alternatives.[13]
8. Finally, the results do not support the hypothesis that the high growth rate of GDP is closely associated with openness. The coefficient of *EX* is insignificant in both the alternatives. If we take the Alternative II (i.e., growth rate of export to GDP ratio) for non-oil developing countries, the coefficient is positive and significant at 5 per cent level.

CONCLUDING OBSERVATIONS

In this paper we have tried to examine the validity or otherwise of some of the received hypotheses on structural changes and the determinants of growth for the developing countries in the recent periods. We summarise below our main findings:

Table 3.8 Determinant of growth in developing countries: regression results

S.no.	Dep. var.	Const.	Y_0	gP	INS	Non-agr./agr.	gINF	INV	GC	EX	gGC	gEX	R^{-2}	F	Sample size
ALL COUNTRIES															
1.	gGDP =	4.0852	–0.0010	0.3552	–5.9379	0.0323	–0.0239	6.6529	–4.4099	1.8142			0.69	20.65	73
		(4.55)	(3.67)	(1.61)	(7.93)	(1.97)	(2.07)	(2.44)	(1.01)	(1.28)					
2.	gGDP =	4.0815	–0.0015	0.1714	–5.5628	0.1685	–0.0283	6.7554			0.0150	0.0421	0.70	18.08	60
		(4.03)	(4.01)	(0.80)	(7.37)	(2.42)	(2.65)	(2.14)			(0.21)	(0.71)			
3.	gGDP =	4.0299	–0.0010	0.2367	–5.8449	0.0350	–0.0263	7.4987					0.68	27.02	73
		(4.76)	(3.69)	(1.15)	(8.66)	(2.15)	(2.35)	(3.20)							
NON-OIL DEVELOPING COUNTRIES															
1.	gGDP =	4.7984	–0.0017	0.0777	–5.1133	0.0931	–0.0252	9.5467	–6.4939	–0.5822			0.65	14.89	60
		(4.27)	(2.28)	(0.29)	(6.14)	(2.14)	(2.04)	(2.49)	(1.26)	(0.31)					
2.	gGDP =	4.5470	–0.0019	–0.0138	–5.5690	0.0948	–0.0214	7.1200					0.66	19.71	60
		(4.13)	(2.52)	(0.05)	(8.24)	(2.46)	(1.92)	(2.38)							
3.	gGDP =	4.7043	–0.0015	–0.0158	–4.9724	–0.0313	–0.0211	6.8162			0.0088	0.1462	0.69	15.05	52
		(4.38)	(1.70)	(0.07)	(6.76)	(0.21)	(1.95)	(2.22)			(0.13)	(2.23)			

Notes: *gGDP* = growth rate of GDP at constant prices, 1970–84, (annual average).
Y_0 = per capita income (in US \$) (at 1975 *prices*), in 1970.
gP = growth rate of population, 1970–84, (annual average).
INS = coefficient of instability (i.e., ratio of error sum of squares to total sum of squares around growth trend).
GC = mean ratio of government consumption to GDP (at 1975 prices), during 1970–84.
EX = mean ratio of exports to GDP (at 1975 prices), during 1970–84.
gGC = growth rate of government consumption to GDP ratio (at 1975 prices), during 1970–84.
gEX = growth rate of exports to GDP (at 1975 prices), during 1970–84.
gINF = growth rate of inflation (per annum), during 1970–84, (annual average).
INV = mean ratio of gross domestic investment to GDP (at 1975 prices), during 1970–84.
Non-Agr./agr. = mean ratio of gross value added in non-agriculture to agriculture (at 1975 prices), during 1970–84.

The estimation has been carried out by least squares using cross-country data.

Figures in parentheses represent *t*-values.

1. One of the main points emerging from our examination of developing countries data for the last one and half decades is that there have been shifts in the structure of national product from agriculture to non-agriculture, and there appears to be little doubt about the existence of this movement. However, there is tremendous diversity of experience as far as the direction of the shift is concerned in replacing agriculture by industry or service sector in the different countries. In the face of this, it must be admitted that framing clear expectations about the universality of development process as perceived by Clark, Kuznets, Chenery and others is a rather more hazardous business than commonly envisaged in the literature.
2. Kaldor's thesis that industry is the engine of growth based on the observations of British and other developed countries[14] also stands discredited when tested against the experiences of developing countries. We detect, in the course of our investigation, a strong linkage between the growth of the service sector and overall growth in GDP of these countries. This, of course, is not to deny the role of manufacturing in the process of development. Though it is crucial, the historical fact is that manufacturing activity in developing countries has been kept in a state of suppression due to sluggish restructuring of production and trade at the global level.
3. The empirical testing of Verdoorn's law of linkages between productivity growth and overall growth of GDP seems to confirm the stability of the elasticity of productivity growth with respect to output growth and vice versa. It is interesting to note that this elasticity coefficient is higher for the service sector than for the industrial sector. The empirical analysis in this part requires a good deal of improvement since the concept of, and data on, employment in the different sectors are subject to a number of limitations.
4. Our elaborate empirical testing on the determinants of overall growth, based on the cross-section data for developing countries, brings out a number of interesting and policy-relevant results. We mention here four of them: (i) the higher the instability in growth, i.e., the higher the fluctuations around the trend, the lower is the overall growth of GDP, (ii) the higher the inflation rate, the lower is the overall growth of GDP, (iii) the shift from agriculture to non-agricultural sector does not have very large growth-inducing effects on the overall growth of the economy, (iv) it is not univer-

sal that larger export orientation necessarily implies higher growth via the efficiency expansion route.

In conclusion, we must add the remark that our results are of an exploratory nature but they are good enough, we believe, to question some of the received theories and hypotheses in the field of structural changes and growth. We hope that further analysis of the recent experiences of the developing countries will provide clues for formulating theories more relevant for their development process.

Notes

1. The sample period is not same for all ninety-two countries (see Panchamukhi, Nambiar and Mehta (1986) for the coverage of sample period and sources of data). The testing of various hypotheses is carried out using the data of these countries. However, the sample size was not uniform in all the cases due to lack of data.
2. The choice of the explanatory variables is based on the experience of similar studies (see Chenery, 1960; UN, 1963; Chenery and Syrquin, 1975, among others).
3. A number of studies have shown the importance of the degree of openness in the analysis of structural change (see Chenery and Taylor, 1968, among others).
4. The double log quadratic functional form is often used in the literature and follows number of properties (see Chenery and Taylor, 1968; Gemmell, 1982; Ballance *et al.*, 1982; Kader, 1985). The rationale for introducing the non-linear form in the equation is that it allows for the decline in elasticity with rising income or population. This formula avoids the necessity of subdividing the sample by income level and size.
5. It may be noted that in most of the earlier studies, the relative prices are not included in the specifications of the model. The coefficients b_1 and b_2 will show the degree of intersector substitution process and the rate at which this process continues (see Panchamukhi, and Margreiter, 1982, for details).
6. The mean elasticity of intersectoral substitution [i.e., $b_1 + 2b_2 \ln\bar{Y}$] have been estimated by using the mean value of per capita income ($\bar{Y}$) during the sample period.
7. See Choi (1983) for a review of the literature on Verdoorn's law.
8. Kaldor emphasised that to test Verdoorn's law the relationships between (i) productivity growth (gP) and output growth (gY), i.e. $gP = a + b.gP$, and (ii) between employment growth (gE) and output growth (gY), i.e. $gE = a + b.gY$, should be examined. On the other hand, Rowthorn argues that there is no gain in estimating two equations as the Verdoorn's law can be tested by establishing the relationship between productivity growth (gP) and employment growth (gE), i.e., $gP = a +$

$b.gE$. Since $gP = gY - gE$, the three equations used by Kaldor and Rowthorn are not independent of each other. In 1975 Kaldor emphasised the estimation of two equations, because he believed that the significance of association between gY and gE will tell whether Verdoorn's law asserts something significant about the real economies or is simply a statistical illusion.

9. The cited variables have been included in the model to test various hypotheses related to the growth rate of GDP. See Barro (1984), Fisher (1979), Stockman (1981), Buiter (1977), Kormendi (1983) and Kormendi and Mequire (1985), Panchamukhi *et al.* (1986) etc. for rationale of different variables.
10. The choice of countries is based on the availability of data.
11. It may be due to technology diffusion as well as the diminishing returns for a given technology (Barro, 1984).
12. To explore the association between growth rate and the dispersion in the trend of output growth, we regressed the growth rate of output on the coefficient of instability (measured as ratio of error sum of squares to total sum of squares around output trend), for (a) total GDP, (b) agriculture sector, (c) industrial sector, (d) service sector, using cross-country data. Our results show that there is a negative and highly significant relationship between growth and rate and output instability.
13. It should be noted that the variable used, in this exercise, includes government consumption expenditure, and other expenditures like investment could not be included for lack of data. In a number of developing countries, the share of public investment is quite substantial. It is feasible that if the appropriate variable of government expenditure is taken, the value of coefficient may change.
14. Kaldor (1967), UN (1970), Stoneman (1979).

References

Ballance, R., Ansari, J. and Singer, H.W. (1982) *The International Economy and Industrial Development* (Osmun: Allanhelt).

Barro, R.J. (1984) *Macroeconomics* (New York: Wiley.)

Black, F. (1979) 'Business Cycles General Equilibrium', unpublished MIT working paper (Cambridge, Mass: MIT).

Buiter, W.H. (1977) 'Crowding out and the Effectiveness of Fiscal Policy', *Journal of Public Economics*, vol. 7, pp. 309–28.

Chenery, H.B. (1960) 'Patterns of Industrial Growth' *American Economic Review*, vol 50, pp. 624–54.

Chenery, H.B. and Syrquin, M. (1975) *Patterns of Development: 1950–1970* (London: Oxford University Press).

Chenery, H.B. and Taylor, L. (1968) 'Development Patterns among Countries and Over Time' *Review of Economics and Statistics*, vol. 50, pp. 391–416.

Choi, K. (1983) *Theories of Comparative Economic Growth* (Ames: Iowa State University Press).

Clark, C. (1940) *The Conditions of Economic Progress* (London: Macmillan).

Cripps, T.F. and Tarling, R.J. (1973) *Growth in Advanced Capitalist Economies 1950–70* (Cambridge University Press).

Denison, E.F. (1962) *The Sources of Economic Growth in the US and the Alternatives Before US* (New York: Comm. of Economic Development).

Denison, E.F. (1974) *Accounting for Slower Economic Growth: The United States in the 1970's* Washington, DC: The Brookings Institution).

Fisher, S. (1979) 'Anticipations and the Non-neutrality of Money', *Journal of Political Economy* vol. 87, pp. 228–52.

Gemmell, N. (1982) 'Economic Development and Structural Change: The Role of Service Sector', *Journal of Development Studies*, vol. 19, pp. 37–66.

Jameson, K.P. (1982) 'A Critical Examination of "The Patterns of Development"', *Journal of Development Studies*, vol. 18, pp. 431–46.

Kader, A.A. (1985) 'Development Patterns among Countries Re-examined', *The Developing Economies*, vol. 23, pp. 199–220.

Kaldor, N. (1966) *Causes of Slow Rate of Growth of the United Kingdom: An Inaugural Lecture* (Cambridge University Press).

Kaldor, N. (1967) *Strategic Factors in Economic Development* (Ithaca: Cornell University Press).

Kaldor, N. (1975) 'Economic Growth and the Verdoorn Law – A Comment on Mr. Rowthorn's Article', *Economic Journal*, vol. 85, pp. 891–96.

Kendrick, J.W. (1961) *Productivity Trends in the United States* (New York: NBER).

Kendrick, J.W. (1974) *Postwar Productivity Trends in the United States, 1948–69* (New York: NBER).

Kormendi, R.C. (1983) 'Government Spending, Government Debt and Private Sector Behaviour', *American Economic Review*, vol. 73, pp. 994–1010.

Kormendi, R.C. and Mequire, P.G. (1985) 'Macroeconomic Determinants of Growth: Cross-Country Evidence', *Journal of Monetary Economics*, vol. 16, pp. 141–63.

Kuznets, S. (1957) 'Quantitative Aspects of Economic Growth of Nations. II. Industrial Distribution of National Product and Labour Force', *Economic Development and Cultural Change*, vol. 5, supp.

Kuznets, S. (1971) *Economic Growth of Nations: Total Output and Productions Structure* (Cambridge University Press).

Lewis, W.A. (1954) 'Economic Development with Unlimited Supplies of Labor', *The Manchester School* vol. 22, pp. 139–91.

Lewis, W.A. (1980) 'The Slowing Down of the Engine of Growth', *American Economic Review*, vol. 70, pp. 555–64.

Panchamukhi, V.R. and Margreiter, G. (1982) 'Transition Processes in Economic Development – Some New Dimensions', mimeo, UNIDO.

Panchamukhi, V.R. Mehta, R. and Tadas, G.A. (1987) *Savings, Investment and Trade in the Third World: A Macroeconomic Analysis* (New Delhi: Research and Information System for the Non-aligned and Other Developing countries).

Panchamukhi, V.R. Nambiar, R.G. and Mehta, R. (1986) '*Structural Change and Economic Growth in Developing Countries*' Discussion Paper (New Delhi: Research and Information System for The Non-aligned and Other Developing Countries).

Robinson, S. (1971) 'Sources of Growth in Less Developed Countries: A Cross-Section Study', *Quarterly Journal of Economics*, vol. 95, pp. 391–408.

Rowthorn, R.E. (1975) 'What Remains of Kaldor's Law?', *Economic Journal*, vol. 85, pp. 897–901.

Stockman, A.C. (1981) 'Anticipated Inflation and the Capital Stock in a Cash-in-Advance Economy', *Journal of Monetary Economics*, vol. 8, pp. 383–93.

Stoneman, P. (1979) 'Kaldor's Law and British Economic Growth: 1800–1970', *Applied Economics*, vol. 11, pp. 309–20.

UN (1963) *A Study of Industrial Growth*, (New York: Dept. of Economics and Social Affairs).

UN (1970) *Economic Survey of Europe 1969* (New York: UN).

Verdoorn, P.J. (1949) 'Factors that Determine the Growth of Labour Productivity', *L'Industria*, vol. 1, pp. 3–11.

4 The Constraints on Industrialisation: Some Lessons from the First Industrial Revolution

Jeffrey G. Williamson
HARVARD UNIVERSITY

1. INTRODUCTION

It is easy to be misled by history. Development economists and economic historians have both spent a great of energy identifying the stylised facts and 'patterns' of development. Indeed, in uncritical moments we tend to gauge an economy's performance by its ability to replicate or even exceed those stylised patterns. This is an ineffective use of history since it fails to exploit the information available. What we *really* want to know is how economies perform in response to technological events and changing world market conditions. What we *really* want to assess is the ability of an economy to reach its full potential, that is, to come close to optimal growth and industrialisation.

This paper offers one such effort. Section 2 illustrates that Britain – the first industrial nation – was an outlier by the standards of estimated 'patterns' of development. She committed a far larger share of her resources to industry than did the Third World in the 1950s and 1960s. She also had lower rates of capital and skill accumulation. Section 3 shows that British factor markets failed, and section 4 assesses the impact of that failure by implementing a computable general equilibrium model. It appears that factor market failure seriously constrained British industrialisation.

2. THE DEVELOPMENT TRANSITION: THE CONTEMPORARY THIRD WORLD, NINETEENTH CENTURY EUROPE AND BRITAIN COMPARED

2.1 The patterns of development

Using a large cross-section drawn from the 1950–70 decades, Chenery and Syrquin (1975) have estimated the structural features of the industrial revolution. They regressed various structural attributes on two explanatory variables, income per capita and population size. Table 4.1 reports predicted values generated by their estimated non-linear regressions when population size is fixed at 10 million and when income per capita is allowed to vary between $300 and $900 (in 1970 US dollars). According to Kravis *et al.* (1978, Table 4, col. 5), this range captures the contemporary growth experience from agrarian underdevelopment to newly-industrialised country. According to Crafts (1985, Table 3.2, p. 54) it also includes a good share of the full range of Britain's eighteenth and nineteenth century industialisation experience:

$300	*$550*	*$900*
India	Egypt	Turkey
Indonesia	Honduras	Brazil
***	***	***
Britain c.1700	Britain c.1830	Britain c.1870

By 1890 Britain was well out of the $300–$900 range, and Belgium, Denmark and Germany had joined her by 1910.

Relying on the recent work of Crafts (1985, ch. 3), Table 4.1 makes it possible to compare the development transition in the contemporary Third World (1950–70) with that of Europe in the nineteenth century (1830–1900). The predicted structural features across the same range of per capita incomes suggests roughly similar 'patterns'. Birth and death rates fall – tracing out the demographic transition, the urban share rises as agriculture undergoes a relative demise, urban-based industrial and service activities rise in relative importance, the (labour) productivity gap between agriculture and the rest of the economy becomes more pronounced, the investment share in GNP rises as does the school enrolment rate. All of these trends are, of course, well known stylised facts of development.

Given that nineteenth century Europe developed in a different

Table 4.1 The development transition: nineteenth century Europe (1830–1900) and the contemporary Third World (1950–70) compared (Third World predicted values in parentheses)

Variable	Income per capita, 1970 (US dollars) $300	$550	$900
Crude birth-rate	38.8	34.0	30.0
	(44.8)	(38.8)	(32.6)
Crude death-rate	28.9	23.7	19.5
	(19.0)	(14.1)	(10.9)
Crude rate of natural increase	9.9	10.3	10.5
	(25.8)	(24.7)	(21.7)
Urbanisation (%)	13.0	30.5	44.7
	(20.5)	(34.0)	(46.2)
Share of labour force in (%):			
Agriculture	75.4	55.9	40.1
	(66.7)	(57.3)	(46.6)
Industry	10.1	24.6	36.4
	(8.9)	(15.3)	(21.9)
Services	14.5	19.5	23.5
	(24.4)	(27.4)	(31.5)
Share of income from (%):			
Agriculture	54.2	38.0	24.9
	(46.3)	(34.6)	(24.9)
Industry	18.1	24.8	30.3
	(14.5)	(20.5)	(26.2)
Services	27.7	37.2	44.8
	(39.2)	(44.9)	(48.9)
Implied productivity gap	.72	.68	.62
Agriculture v. economy-wide	(.69)	(.60)	(.53)
Investment share (%)	10.5	14.2	17.2
	(15.4)	(18.3)	(20.8)
School enrolment rate (%)	17.4	36.0	51.2
	(35.4)	(52.2)	(66.3)

Sources: The Third World predicted values are from Crafts (1985, Table 3.1, pp. 50–51), based on Chenery and Syrquin (1975, Table 3, pp. 20–21), where simulated values are for 10 million population size, and the income per capita figures are based on Kravis *et al.* (1978, Table 4, col. 5). The European predicted values are from Crafts (1985, Table 3.3, p. 55), based on a pooled regression, European countries 1830–1900, using the same Chenery-Syrquin estimation equations, and where predicted values control for 10 million population size. The enrolment rate is the per cent of those aged 15–19 in school. In the case of nineteenth century Europe, the labour force refers to males only.

price, demographic and technological environment, pursued different policies, and had quite different institutional arrangements influencing factor market behaviour, and given that the Chenery-Syrquin regressions ignore such influences, it is remarkable how close the estimated nineteenth century Europe patterns of development are to those for the contemporary Third World. Yet Table 4.1 also highlights some very important differences. First, the European population burden was far lower. Secondly, the level of urbanisation was also lower, encouraging the belief that the contemporary Third World is 'over-urbanised' (e.g., Hoselitz, 1955, 1957). Thirdly, the agricultural productivity gap was smaller, perhaps implying that resource allocative policies have penalised agriculture more in the contemporary Third World than in nineteenth century Europe. Fourthly, the city-based service sector was far smaller in nineteenth century Europe, reflecting a larger public sector and social overhead commitment in the contemporary Third World.

Fifthly, the contemporary Third World has allocated far more resources to accumulation. This can be seen in the predicted investment shares, but it is even more apparent in the school enrolment rate. While the low European investment rates can be explained in part by the lower rates of population and labour force growth, the lower school enrolment rates cannot. These more modest European commitments to accumulation help account for the fact that the contemporary Third World has enjoyed far more rapid rates of development: while income per capita in the low and middle income countries combined grew at 2.4 per cent per annum in the 1960s and 1970s (International Bank for Reconstruction and Development, 1980, Table 1, pp. 110–11), it grew at only 0.7 and 1.3 per cent per annum in Europe during the first and second half of the nineteenth century (McGreevey, 1985).

Finally, given income levels, note that nineteenth century Europe had a more 'modern' industrial structure than the contemporary Third World, and underwent a more dramatic change in input and output mix over the same range of per capita incomes. Over the \$300 and \$900 per capita income range, the agricultural employment share declined by 35.3 per cent (from 75.4 to 40.1 per cent) in nineteenth century Europe, while it declined by only 20.1 per cent (from 66.7 to 46.6 per cent) in the contemporary Third World. More dramatic European structural change is also apparent in the industrialisation measures: the industrial employment share rose by 26.3 per cent in nineteenth century Europe, while it rose by only 13 per cent in the contemporary Third World. The phrase 'more dramatic' must be used with caution, of course, since the Third World has undergone per capita income growth rates twice that of nineteenth century Europe.

The key message emerging from Table 4.1 is that there is no unique path to industrialisation revealed by the Chenery-Syrquin regressions. On the contrary, the evidence suggests that comparative advantage, policy and institutions matter, and that income per capita is a very imperfect indicator of such forces. Table 4.1 implies that the Chenery-Syrquin regressions need to be augmented by information on factor endowment, the world market environment and policy regimes, as well as on those institutional arrangements which have influenced resource allocation.

Table 4.2 repeats the 'patterns' exercise, but here it is British experience that is compared with estimated European norm. Given the per capita income estimates for Britain at each of the five benchmark dates between 1760 and 1890, Crafts is able to predict the structural features of the British economy had it followed the normal nineteenth century European pattern. While the British industrial revolution traces out roughly the same patterns as does nineteenth century European history, there are two important differences worth noting. First, the rate of structural change was more dramatic in Britain than would have been true had Britain followed the European norm. Secondly, the investment share and the school enrolment rate were far lower. This finding is consistent with other evidence which documents very low rates of capital deepening and skill formation in Britain (Williamson, 1984; 1985a, ch. 7). It follows that Britain departed from Third World patterns even more dramatically than did the rest of nineteenth century Europe.

There is an anomaly emerging from Table 4.2 well worth pondering. How can we explain the coincidence of relatively rapid structural change with low rates of accumulation in Britain? And what were the labour market, migration and inequality implications of this kind of growth regime?

2.2 Migration, Labour Markets and the Development Transition in Britain

Britain's growth regime of rapid structural change during her development transition could not have been facilitated without a significant rearrangement of economic resources, labour in particular, and

Table 4.2 The development transition: nineteenth century Europe and Britain compared, 1760–1890 (European predicted values in parentheses)

Variables	*1760*	*1800*	*1840*	*1870*	*1890*
Income per capita, 1970 US dollars	399	427	567	904	1130
Crude birth-rate	33.9	37.7	35.9	35.2	30.2
	(36.5)	(36.0)	(33.7)	(30.0)	(28.2)
Crude death-rate	28.7	27.1	22.2	22.9	19.5
	(26.4)	(25.9)	(23.4)	(19.4)	(17.5)
Urbanisation (%)	na	33.9	48.3	65.2	74.5
	na	(23.2)	(31.4)	(44.8)	(51.3)
Share of labour force in (%):					
Agriculture	52.8	40.8	28.6	20.4	14.7
	(66.2)	(64.0)	(54.9)	(40.0)	(32.8)
Industry	23.8	29.5	47.3	49.2	51.1
	(16.9)	(18.6)	(25.3)	(36.5)	(41.9)
Services	23.4	29.7	24.1	30.4	34.2
	(16.9)	(17.4)	(19.8)	(23.5)	(25.3)
Share of income from (%):					
Agriculture	37.5	36.1	24.9	18.8	13.4
	(46.6)	(44.8)	(37.2)	(24.8)	(18.9)
Industry	20.0	19.8	31.5	33.5	33.6
	(21.3)	(22.0)	(25.2)	(30.3)	(32.8)
Services	42.5	44.1	43.6	47.7	53.0
	(32.1)	(33.2)	(37.6)	(44.9)	(48.3)
Implied productivity gap:	.71	.88	.87	.92	.91
Agriculture v. Economy wide	(.70)	(.70)	(.68)	(.62)	(.58)
Investment share (%)	6.0	7.9	10.5	8.5	7.3
	(12.2)	(12.6)	(14.4)	(17.2)	(18.6)
School enrolment rate (%)	na	na	na	16.8	38.5
	na	na	na	(51.4)	(58.2)

Sources: Crafts (1985, Table 3.6, pp. 62–63). The figures in parentheses are European predicted values based on the Chenery-Syrquin regressions underlying Table 2, using Britain's income per capita and population size. The remaining figures are actual British experience. Labour force refers to males only.

---

especially in an environment of relatively slow accumulation and labour force growth. We have already seen in Table 4.1 and 4.2 that the level and rate of urbanisation in Britain was quite impressive by the standards of nineteenth century Europe. How do Britain's rates of city growth and rural emigration compare with the contemporary Third World?

Todaro (1984, p. 13) has recently reported the rate of city growth in the Third World in the 1960s and 1970s to have been 4.32 per cent per annum. While these city growth rates are double the English rates in Table 4.3, we must remember that everything was growing more rapidly in the Third World, including overall population (2.3 versus 1.2 per cent per annum). In any case, the English city immigration rates were not so much lower in the three decades 1776–1806, ranging between 1.10 and 1.91 per cent per annum, than they were in the Third World, averaging 1.79 per cent per annum. Thus, while England's growth was more gradualist, she recorded city immigration rates on par with the exploding urban Third World. Furthermore, with the exception of the war-induced good times for English agriculture between 1801 and 1806, rural emigration took place at every point over the century 1776–1871. Indeed, the emigration rate about doubled over the nineteenth century, and between the 1830s and the 1860s it compares favourable with the Third World. While the rates estimated for the Third World in the 1960s and 1970s range between 0.97 and 1.37 (Preston, 1979, p. 197; Kelley and Williamson, 1984, p. 93), they range between 1.01 and 2.10 in Britain from the 1830s onward.

Judged by the standards of the contemporary Third World, therefore, rural-urban labour migration in nineteenth century Britain seems to have been quite impressive. Indeed, how else could British labour markets have accommodated the relatively rapid transformation from rural agriculture to urban industry?

All was not optimal in British labour markets, however, but rather there is abundant evidence that labour markets 'failed' even by the standards of the contemporary Third World. As Table 4.4 shows, real wage gaps between city and countryside (for relatively homogeneous unskilled labour) were quite pronounced, and apparently even bigger than in the Third World. Labour market failure of this sort clearly implies that labour costs were too high in the city, that non-agricultural employment was choked off, and that the rate of industrialisation was less than optimal, even in Britain where the rate of structural change was unusually rapid.[1]

2.3 Labour Markets, Inequality and the Kuznets Curve

We have already seen that Britain had far lower school enrolment rates than was achieved both in the rest of nineteenth century Europe and in the contemporary Third World. We have also seen that

Table 4.3 City growth and the labour transfer in England, 1776–1871

Years	Urban Share (%)	Annual Percentage Rate City growth	City immigration	Rural emigration	Percent of city growth due to immigration
1776–1781	25.9	2.08	1.26	0.86	59.5
1781–1786	27.5	1.81	1.62	0.50	89.0
1786–1791	29.1	2.20	1.37	0.56	61.1
1791–1796	30.6	2.17	1.20	0.79	53.7
1796–1801	32.2	2.08	1.10	0.83	51.9
1801–1806	33.8	2.15	1.91	–0.18	88.2
1806–1811	35.2	2.07	0.59	1.07	27.5
1811–1816	36.6	2.40	1.37	0.59	55.6
1816–1821	38.3	2.39	1.06	0.87	42.8
1821–1826	40.0	2.61	1.12	1.19	41.4
1826–1831	42.2	2.33	1.06	1.14	44.0
1831–1836	44.3	2.08	1.04	1.01	48.7
1836–1841	46.3	2.04	0.83	1.20	39.5
1841–1846	48.3	2.41	1.23	1.57	49.7
1846–1851	51.2	2.05	0.97	1.73	45.9
1851–1856	54.0	2.06	0.77	1.54	36.4
1856–1861	56.4	2.08	0.60	1.60	27.9
1861–1866	58.7	2.35	1.06	2.10	43.7
1866–1871	62.0	2.29	1.15	2.05	48.6

Source: Williamson (1985b), Tables 4 and 5.

Table 4.4 Urban-rural wage gaps for unskilled labour: Britain (1830s) and the Third World (1960s and 1970s) compared

Observation	Wage gap (%)	Source
A. *Nominal wages*		
South of England	106.2	Williamson, 1986, Table 2.
North of England	36.3	*Ibid.*
All England	73.2	*Ibid.*
Third World	41.4	Williamson, 1986, Table 4, based on Squire, 1981, Table 30, p. 102.
B. *Real wages*		
All England	52.1	Williamson, 1986, Table 10, excluding disamenities premium for city life.
Third World	35.2	Nominal, deflated by cost-of-living differentials in Peru (Thomas, 1980, p. 89) and Ghana (Knight, 1972, p. 209).

Britain's commitment to industrialisation was more pronounced. To the extent that urban activities were far more skill-intensive, these two events implied a growth regime of skill scarcity and earnings inequality.

There is abundant evidence confirming these predictions. Inequality was on the rise in Britain during her period of dramatic industrialisation up to the 1860s, and earnings inequality was a key component of that experience (Williamson, 1985a). Indeed, the rural-urban wage gap and the wage premium for skilled labour were the main ingredients driving that regime of earnings inequality on the upswing of the British Kuznets curve.

3. WHAT DRIVES INDUSTRIALISATION?

3.1 In the Long Run

The Chenery-Syrquin estimated 'patterns' are certainly useful descriptive devices, but they tell us very little about the sources of industrialisation. Since the regression is a reduced form equation derived from an unspecified model of growth and structural change, we have no way of knowing which correlates of income per capita are driving the observed industrialisation patterns. Are they supply-side forces? Demand-side forces? And what about the link between the current structure of the economy and subsequent growth?

From Adam Smith to W. Arthur Lewis, there has been no shortage of models which have been used to theorise about the determinants of industrialisation, although efforts to give empirical content to those theories have been scarce. The principal hypotheses include Engel effects, capital deepening, Malthusian pressure on the land, favourable world market conditions and unbalanced technological advance. The quantitative importance of each of these forces need not be the same for each country and for each epoch, thus making it possible to observe considerable variance in historical paths to industrialisation. Since Engel effects reflect a passive response to income per capita improvements and since Malthusian forces reflect a passive response to the demographic transition – itself driven by the industrial revolution, since capital deepening reflects an accumulation response to buoyant investment demands embodied in industrial revolutionary forces already set in motion – it seems sensible to focus on technological events as the driving force.

Unbalanced productivity advance is an attribute of all industrial revolutions. The most rapid rates of total factor productivity growth are always focused on modern industrial and capital-cum-skill intensive service sectors. In the simplest two-sector model, technological imbalance of this sort raises factor productivities in the favoured modern sector, encouraging resource transfers as well as an aggregate supply response of those factors which are used most intensively in the technologically favoured sector. Labour migrates to the modern sector where labour scarcities have been created, capital chases after labour seeking high returns, savings and accumulation are stimulated economy-wide as the more capital-intensive sector expands, and investment requirements are augmented, skill formation is encouraged by bottlenecks created in modern skill-intensive activities, city growth takes place and industrialisation unfolds.

These are the underlying long-run determinants of industrialisation and the development transition. Short run constraints on the transition can, however, have an important impact, and to the extent that these are country-specific, considerable variance in the observed 'patterns' of development is guaranteed.

3.2 Factor Market Failure and Constraints on Industrialisation

Short run constraints on industrialisation can take two forms.

First, demand conditions matter. Although they are stressed at length in the development literature, it is not Engel effects and income elasticities of demand which matter most. Rather, it is price elasticities of demand which serve to condition the industrialisation response to unbalanced productivity advance. After all, if the modern sector is faced by relatively inelastic product demand, then technological improvements on the supply side will serve to lower price rather than raise output. Under such circumstances, the increase in the marginal physical productivity of factors in the technologically favoured sector will be offset by output price decline, the rise in the *value* of marginal products will be forestalled, immigration to the favoured sector will be choked off, accumulation responses will be suppressed, and the transformation of the economy retarded. Why might demand-side problems differ across industrial revolutions? One likely reason would lie with world market conditions and the export orientation of the economy. Those economies which best satisfy the small country assumptions of commodity price exogeneity are most likely to exhibit more dramatic rates of industrialisation given the same unbalanced productivity shock.

Secondly, factor markets matter. If agricultural labour responds only sluggishly to excess demands created by unbalanced productivity growth favouring the city-based modern sectors, the cost of labour will rise in the city, and the industrialisation effort will be choked off. We have already seen, in fact, that British labour markets *did* fail since wage gaps between city and countryside had become very pronounced by the 1830s, even more pronounced than in the contemporary Third World. Can this labour market failure help to account for some of Britain's peculiar development experience? If agricultural capital and investible surplus respond only sluggishly to city-based excess demands created by unbalanced productivity growth favouring modern sector, industry will be starved for capital and the industrialisation effort will be forestalled. Capital market failure has long been a staple in the literature on the British industrial revolution, and it has an equally long tradition in contemporary development economics. The view that British capital markets were sharply segmented was so common in the early nineteenth century that the classical economists took it for granted that industrial profits and their investment were the central determinants of accumulation and thus industrialisation. And there is certainly evidence which supports the view of capital market failure. By the mid-nineteenth century, rates of return on industrial capital were close to 2.5 times that of agriculture (Williamson, 1986).

3.3 Linking Factor Market Failure to Industrialisation in the Long Run

Did factor market failure have an important quantitative impact during the first industrial revolution? Lurking behind that question is an implicit counterfactor: if there had been an optimal allocation of labour and capital in the 1830s, would national income have been raised significantly? Would industrialisation have taken place much more rapidly?

I propose to offer a test. Figure 4.1 motivates the empirical exercise where labour markets are used to illustrate the point. This familiar diagram shows employment distribution between agriculture and industry in the presence of wage gaps. In the presence of labour market failure and the wage gaps, national income is simply the sum

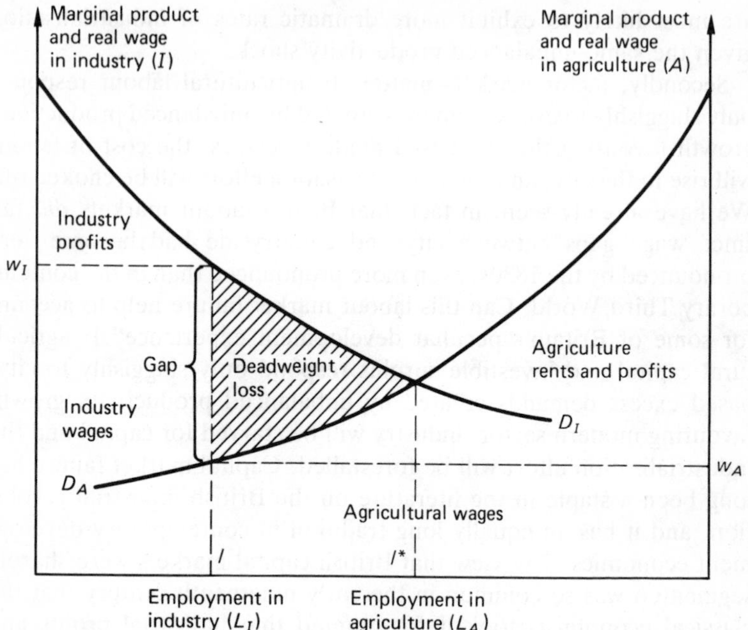

Figure 4.1 'Partial' general equilibrium analysis of wage gaps in two sectors

---

of the area under each of the two derived labour demand curves, up to employment at l. If the wage gaps are eliminated, migrants will have left agriculture for industrial employment, and the labour force will be optimally distributed at l^*. National income has now been increased by the shaded area, or by the elimination of the deadweight loss, sometimes called the 'Harberger Triangle'. Development economists have always made much of the gains from such reallocations from backward agriculture to modern industry. We shall see below that such allocative gains would have been significant to Britain too. More important to the issue at hand, however, is that labour market failure would suggest that British industrialisation was severely constrained to the extent that the employment distribution gap l-l^* turns out to be large.

Labour market failure could have suppressed industrialisation in another way, through the distributional impact documented in Figure 4.2. With the disappearance of wage gaps, and with the emigration of

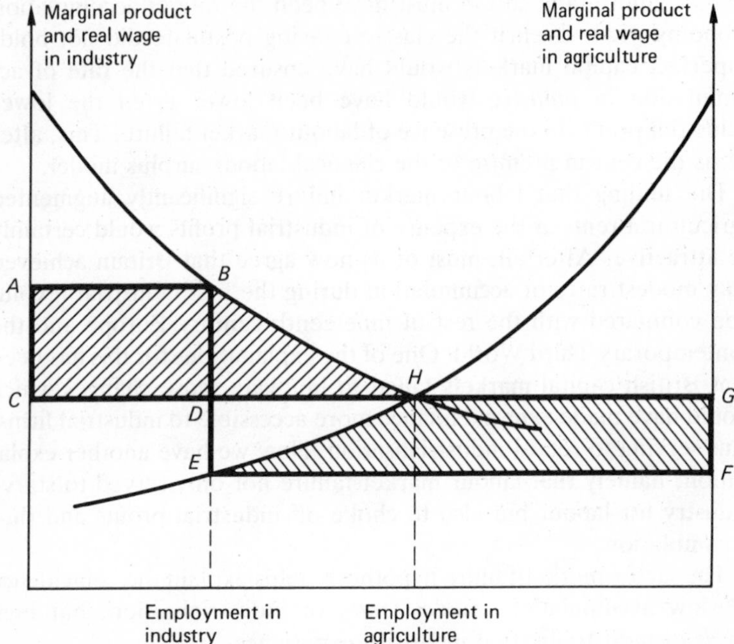

Figure 4.2 'Partial' general equilibrium analysis of wage gaps: who gains and who loses?

labour from agriculture, wages would have risen in agriculture and fallen in industry. Given no change in capital, land or technology in either of the two sectors, it follows that rents (plus profits) in agriculture would have diminished, and profits in industry would have increased. Profits in industry would have risen by ABHC, an increase in their producers' surplus. Rents (or, more likely, farmers' profits) in agriculture would have fallen by EHGF, a decrease in their producers' surplus. It is not clear, however, how common labour would have fared. Labourers would have gained in agriculture but lost in industry for an ambiguous net change of DGFE minus ABDC. As we shall see below, these distributional influences were far greater than the deadweight losses, and it may have mattered. To the extent that the classical saving postulate held, industrial capital accumulation was seriously choked off by labour market failure. That is, if the reinvestment rate out of industrial profits was far higher than out of rents, then aggregate savings were lower in the presence of labour market failure, and so too must have been the rate of accumulation economy-wide. Even if the classical saving postulate did not hold, imperfect capital markets would have ensured that the rate of accumulation in *industry* would have been lower given the lower industrial profits in the presence of labour market failure. This, after all, is the central premise of the classical labour surplus model.

The finding that labour market failure significantly augmented agricultural rents at the expense of industrial profits would certainly be attractive. After all, most of us now agree that Britain achieved only modest rates of accumulation during the first industrial revolution compared with the rest of nineteenth century Europe and the contemporary Third World. One of the explanations for this has been that British capital markets failed to innovate those changes which would have made external finance more accessible to industrial firms, thus serving to retard industrialisation. Now we have another explanation, namely that labour market failure not only served to starve industry for labour but also to choke off industrial profits and thus accumulation.

The factor-market-failure hypothesis helps explain the coincidence of slow accumulation, modest *rates* of industrialisation, but high *levels* of industrialisation in British experience.

4. CONSTRAINTS ON INDUSTRIALISATION: COULD BRITAIN HAVE DONE BETTER?

4.1 Computable General Equilibrium (CGE) Modelling for Nineteenth Century Britain

I propose to measure these factor-market-failure constraints on British industrialisation by appealing to a simple computable general equilibrium model.

A multisector and multifactor CGE model is especially useful for long-run analysis where neoclassical assumptions do the least damage to the macroeconomic facts, and where industrialisation and distributional issues are the main focus. It has become a familiar tool in modern public finance (Shoven and Whalley, 1984) where tax incidence and other distribution issues have been addressed. CGEs have been used at least as often in the development literature to assess industrialisation, trade, urbanisation, and distribution issues (Dervis *et al.*, 1982; Kelley and Williamson, 1984). CGE models have an equally long tradition in economic history, although the cliometrician usually deals with more parsimonious models dictated by more severe data constraints (James, 1984). More to the point, CGEs have been used to quantify the sources of British industrialisation during the first industrial revolution (Williamson, 1985a) One such model was estimated with data drawn from the early nineteenth century, after which it was asked to predict British trends between 1821 and 1861. Since the model reproduced the British development transition extremely well (Williamson, 1985a, ch. 9), I have been encouraged to use it for various counterfactual exercises.

The model used here contains four sectors consisting of agriculture, industry, services and mining. The economy is a net importer of agricultural products and raw materials, and a net exporter of manufactures. The prices of the three tradables – intermediate raw materials, agricultural products and manufactures – are taken to be set by world market condtions and commercial policy, while the non-tradables – services and the products of the mines – are determined endogenously by domestic demand and supply. There are four social classes who supply inputs to the production process: landlords supply agricultural land, capitalists and farmers supply capital, common labour supplies unskilled labour, and skilled labour supplies human capital. Factor market failure is introduced into the model by allowing for wage and rate of return gaps between agriculture and city activities.

The model was designed to deal with comparative static issues only, but it is ideal for making exactly the kinds of distributional and resource allocative assessments posed above. It can be used to assess the impact of factor market failure on sectoral employment (were there too many farmers?), on national income (how big were the deadweight losses?), on industrialisation (was industrial output constrained?), and on distribution (were profits, and thus accumulation, too low?).

4.2 British Factor Market Failure Assessed

Table 4.5 assesses the impact of both labour and factor market failure in the nineteenth century British economy. Here, the modelled economy is allowed to seek a new equilibrium in response to an environment in which wage gaps are absent. Capital is permitted to migrate as well, but with a slight twist. Where labour market failure is eliminated in column (1), capital can migrate, but only enough to

Table 4.5 The impact of factor market failure

	(1) *Eliminating the wage gap and labour market failure*	*(2)* *Eliminating the rate of return gap and capital market failure*	*(3)* *Eliminating both gaps and labour plus capital market failure (≠(1) + (2)),*
A. *Agriculture*			
Percent rise in agriculture output (constant prices)	−20.4	−56.1	−69.6
Emigrants from agriculture as a percent of agricultural labour force	−39.0	−51.4	−57.9
B. *Non-agriculture*			
Percent rise in non-agricultural output (constant prices)	+13.4	+35.8	+39.8
Percent rise in non-agricultural employment (unskilled labour only)	+15.5	+20.4	+23.0
Percent rise in non-agricultural profits (nominal)	+11.2	+64.6	+78.1
C. *Manufacturing*			
Percent rise in manufacturing output (constant prices)	+24.1	+63.6	+73.4
Percent rise in manufacturing employment (unskilled labour only)	+25.3	+40.6	+47.2
Percent rise in manufacturing profits (nominal)	+23.6	+114.7	+141.8
D. *Economy-wide increase in real GNP*	+3.3	+8.2	+7.0

Notes and Source: Williamson, 1986, Table 13. Column (3) does not add the entries in (1) and (2), but rather reports the impact of the joint change.

maintain the initial rate of return differentials. Where capital market failure is eliminated in column (2), capital is allowed to migrate until rate of return differentials disappear. 'Eliminating labour market failure' means that the 52.1 per cent urban-rural wage for the 1830s (documented in Table 4.4) is removed, while 'eliminating capital market failure' means that the ratio of industrial to agricultural rates of return is reduced to 1 from the 2.4 figure which prevailed in the late 1840s.

To begin with, the results listed in each column of Table 4.5 should satisfy economic intuition. The elimination of the wage gap tends to create labour scarcity in agriculture, output contracts, and employment contracts even more, since the relative scarcity of labour rises there. Symmetrically, output expands in non-agriculture, and employment expands even more since the relative scarcity of labour falls there. The impact on output in the two sectors exceeds that attributable to employment changes alone since capital is allowed to chase after labour and the higher returns. Profits rise in non-agriculture, but by less than output. The explanation is a bit more subtle. Since the capital-labour ratio tends to fall in non-agriculture as the relative scarcity of labour falls, one would have thought that the rate of return to capital there would rise, and thus that profits would rise by even more given that the capital stock has now been augmented. But the rise in profits is somewhat less than expected since the price of the non-tradables (services and mining) declines. In any case, profits do rise with the elimination of wage gaps, primarily because, given the small country assumption, the price of industrial output remains stable. Finally, real GNP increases with the elimination of the deadweight losses associated with labour market distortions.

Now consider the magnitudes.

First, the deadweight losses are significant. Had those wage gaps been eliminated in the 1830s, GNP would have been augmented by 3.3 per cent. These deadweight losses associated with labour market failure are hardly trivial, to be sure, but they are considerably smaller than those associated with capital market failure (8.2 per cent). The combined influence of labour and capital market failure was about 7 per cent of GNP.[2]

Secondly, Table 4.5 suggests that non-agricultural employment in general, and manufacturing employment in particular, must have been seriously choked off by factor market failure. While Britain's industrialisation performance over the first half of the nineteenth century was certainly impressive by the standards of the time, it would have been far more impressive in the absence of those factor market distortions. Indeed, non-agricultural employment would have been 23 per cent higher in the early 1840s. If that increase is stretched over the two decades 1821–41, it implies that non-agricultural employment growth would have been about 3 per cent per annum, not the 2 per cent per annum actually achieved (Deane and Cole, 1962, p. 143). Manufacturing output would have been 73.4 per cent higher. And if that increase is stretched over the period 1815 to 1841, it implies that industrial output growth would have been about 5.2 per cent per annum, not the 3.1 per cent per annum actually achieved (Harley, 1982, p. 276). Thus, factor market distortions seriously inhibited nineteenth century British industrialisation, although capital market failure was the more serious constraint.

Thirdly, these distortions had important distributional implications. Profits in non-agriculture would have been much higher (by 78.1 per cent), and they would have been increased even more in manufacturing (by 141.8 per cent, or, as it turns out, more than double the actual profits which accrued to capitalists there). If the reinvestment rate out of non-agricultural profits was relatively high, as most of us believe, then the elimination of factor market distortions would have resulted in a significant rise in saving and accumulation, especially so in manufacturing. The elimination of the capital market distortions would also have made accessible to all potential savers the high rates of return in industry, and if savings was responsive to the rate of return, then economy-wide accumulation would have obtained an even bigger boost.

5. SOME MORALS FROM HISTORY

We tend to view Third World policy as containing an urban bias. Capital markets favour modern sectors. The terms of trade are twisted against agriculture. Exchange rates and tariffs are manipulated to protect urban industry. Government final demand is oriented toward city services. The list seems endless. While such policy distortions certainly deserve our critical attention, we should not forget that private factor markets may often deflect the developing economy from an optimal industialisation path.

Nineteenth century Britain offers a superb example. Here we have an economy with modest goverment intervention and where private factor markets were highly imperfect – so imperfect, in fact, that British industrialisation was sharply constrained. That fact may seem anomalous given that Britain committed a far larger share of her resources to industry than has the contemporary Third World at the same levels of per capita income. Technological leadership and profound changes in comparative advantage no doubt explain Britain's unusually large commitment to industrialisation, but factor market failure suggests that she was nowhere near an optimal industrialisation path.

This conclusion is based on the empirical application of a computable general equilibrium model to nineteenth century Britain. I see no reason why such applications could not be expanded to augment our comparative quantitative knowledge of the determinants of industrialisation, past and present. Chenery and Syrquin 'patterns' offer the descriptive information to identify the many paths to industrialisation which economic history has generated. The time is now ripe to identify the sources of that industrialisation experience and its optimality. I hope this paper is a start.

Notes

1. It should be pointed out that the presence of wage gaps do not necessarily imply support for the Todaro (1969) thesis. There is no evidence that British unemployment rates were high and rising in British cities at this time. Furthermore, there is no evidence to support the view that urban immigrants had higher unemployment rates or even lower annual earnings than non-immigrants (Williamson, forthcoming).
2. Note that the effects of labour market and capital market failure in Table 4.5 are *not* additive, but, in the case of deadweight losses, offsetting. Albert Fishlow and Paul David (1961) alerted us to that possibility some time ago.

References

Chenery, H. B. and Syrquin, M. (1975) *Patterns of Development, 1950–1970* (London: Oxford University Press).

Crafts, N. (1985) *British Economic Growth During the Industrial Revolution* (Oxford: The Clarendon Press).

Deane, P. and Cole, W. A. (1962) *British Economic Growth 1688–1959 (Cambridge: Cambridge University Press).*

Dervis, K., Melo, J. de and Robinson, S. (1982) *General Equilibrium Models for Development Policy* (Cambridge: Cambridge University Press).

Fishlow, A. and David, P. A. (1961) 'Optimal Resource Allocation in an Imperfect Market Setting', *Journal of Political Economy* vol. 69, no. 6 (December), pp. 529–46.

Harley, C. K. (1982) 'British Industrialization Before 1841: Evidence of Slower Growth During the Industrial Revolution,' *Journal of Economic History* vol. 42, no. 2 (June), pp. 267–89.

Hoselitz, B. F. (1955) 'Generative and Parasitic Cities', *Economic Development and Cultural Change* vol. 3 (April), pp. 278–94.

Hoselitz, B. F. (1957) 'Urbanization and Economic Growth in Asia', *Economic Development and Cultural Change* vol. 5 (October), pp. 42–54.

International Bank for Reconstruction and Development (World Bank) (1980) *World Development Report, 1980* (Washington, DC: IBRD).

James, J. A. (1984) 'The Use of General Equilibrium Analysis in Economic History', *Explorations in Economic History*, vol. 21 (July), pp. 231–53.

Kelley, A. C. and Williamson, J. G. (1984) *What Drives Third World City Growth*? (Princeton, N. J.: Princeton University Press).

Knight, J. B. (1972) 'Rural-Urban Income Comparisons and Migration in Ghana', *Bulletin of Oxford University Institute of Economics and Statistics*, vol. 34, no. 2 (May), pp. 199–228.

Kravis, I. B., Heston, A. W. and Summers, R. (1978) 'Real GDP Per Capita for More Than One Hundred Countries', *Economic Journal*, vol. 88 (June), pp. 215–42.

McGreevey, W. P. (1985) 'Economic Aspects of Historical Demographic Change', *World Bank Staff Working Papers*, no. 685 (Washington DC: The World Bank) September.

Preston, S. H. (1979) 'Urban Growth in Developing Countries: A Demographic Reappraisal', *Population and Development Review*, vol. 5 (June), pp. 195–215.

Shoven, J. B. and Whalley, J., (1984) 'Applied General-Equilibrium Models of Taxation and International Trade: An Introduction and Survey', *Journal of Economic Literature*, vol. 22 (September), pp. 1007–51.

Squire, L. (1981) *Employment Policy in Development Countries* (Oxford: Oxford University Press).

Thomas, V. (1980) 'Spatial Differences in the Cost of Living', *Journal of Urban Economics*, vol. 8, pp. 108–22.

Todaro, M. P. (1969) 'A Model of Labor Migration and Urban Unemployment in Less Developed Countries', *American Economic Review*, vol. 59, no. 1 (March), pp. 138–48.

Todaro, M. P. (1984) 'Urbanization in Developing Nations: Trends, Prospects, and Policies', in Ghosh, P. K. (ed.) *Urban Development in the Third World* (Westport, Conn.: Greenwood).

Williamson, J. G. (1984) 'Why Was British Growth So Slow During the Industrial Revolution?', *Journal of Economic History* vol. 44, no. 3 (September), pp. 687–712.

Williamson, J. G. (1985a) *Did British Capitalism Breed Inequality*? (London: Allen S Unwin).

Williamson, J. G. (1985b) 'The Urban Transition in England, 1776–1871: A Demographic Reconstruction', HIER Discussion Paper no. 1146, Harvard

University (April). Presented to the PAA Meetings, Boston (28–30 March 1985).

Williamson, J. G. (1986) 'Did British Labor Markets Fail During the Industrial Revolution?', HIER Discussion Paper no. 1209, Harvard University (March).

Williamson, J. G. (forthcoming) 'Migrant Earnings in Britain's Cities in 1851: Testing Competing Views of Urban Labor Market Absorption', *Journal of European Economic History*.

5 The Contribution of Agriculture to Economic Growth: Some Empirical Evidence

Erh-Cheng Hwa
THE WORLD BANK*

1. INTRODUCTION

Despite the dominant position occupied by the agricultural sector in a traditional economy, many parts of the developing world have continuously denied agricultural and rural development adequate attention. This bias has often led to stagnant agriculture that, in turn, has resulted in large shortfalls in domestic food production, balance of payment crises, and political instability.[1] For a primitive agrarian economy, it is doubtful whether industrialisation can succeed without the prior or concurrent emergence of a productive agricultural sector.

This paper undertakes a statistical analysis of the significance of the contribution of agriculture to economic growth by the use of cross-section data. The major finding of the paper is that agricultural growth, while strongly linked to industrial growth over the development process, contributes to overall economic growth through its favourable impact on total factor productivity. In fostering productivity, the role of agriculture seems to be no less important than that of export performance. This empirical evidence reinforces the argument that agriculture and rural development should be given priority and be properly supported in an overall development strategy.

Section 2 of this paper analyses the relationship between agriculture and industry during the development process. Section 3 examines the contribution of agriculture to economic growth in the framework of a production function. The summary and conclusions are given in the last section.

2. THE RELATIONSHIP BETWEEN AGRICULTURE AND INDUSTRIAL GROWTH

The transition from a traditional agrarian to an industrialised economy is a dynamic process that inevitably involves complex interactions among many economic as well as social factors. The role of agriculture in the transition varies from country to country, conditioned by factor endowment, institutional arrangements, cultural background, historical factors, policy choices, etc. Nevertheless, drawing from the experience of both developing and developed countries, one can identify several important roles that agriculture plays in the transition process (e.g., Johnston and Mellor, 1961; Johnston and Kilby, 1975):

1. Agriculture generates markets for industrial products, especially light industrial ones that have ready markets in the agricultural sector;
2. it provides food and agricultural raw materials for industrial processing;
3. it builds adequate food supplies that are a crucial factor in sustaining price stability;
4. it provides exports to earn foreign exchange;
5. it supplies the non-agricultural sector with capital and labour;
6. in the case of a market-oriented economy, it eases the processs of industrialisation through the gradual accumulation of entrepreneurship and marketing capabilities in the agricultural sector.

In sum, agriculture supports industrialisation by providing a source of labour, capital and raw materials to other sectors, and by generating demand for industrial products.

At the same time, the relationship between agriculture and industry is one of interdependence and complementarity. For example, while providing inputs to industry, agriculture receives from it modern farm inputs, advanced technologies, and consumption goods to increase its productivity. The statistical significance of this relationship may be tested by the following non-linear model of the Chenery-Şyrquin (1975) type, which relates the rate of industrial growth ($\dot{I}$) to per capita income (YN) and the rate of agricultural growth ($\dot{A}$):

$$\dot{I} = f\,[\dot{A},\, \ln YN,\, (\ln YN)^2] + u, \tag{1}$$

where u is a randomly distributed error term.

The model is derived in the following way. First, assume that the rates of growth of industry and agriculture are both non-linear functions of per capita income variables:

$$\dot{I} = \alpha_I \ln YN + \beta_I (\ln YN)^2 + \varepsilon_I \tag{2}$$

and

$$\dot{A} = \alpha_A \ln YN + \beta_A (\ln YN)^2 + \varepsilon_A, \tag{3}$$

where ε_I and ε_A are random errors.

These are simplified reduced form models for the determination of industrial and agricultural growth. The per capita income variables are used as summary measures of the stage of economic development, in addition to being measures of final demand. Chenery and Taylor (1968) have used similar models to study patterns of economic development. Alternatively, these regressions can be thought of as establishing 'norms' for the rates of growth of industry and agriculture with reference to the stage of economic development, as measured by the level of per capita income. It is possible to test the hypothesis of whether countries with higher industrial growth in relation to 'normal' industrial growth are also those with higher agricultural growth with reference to its norm. This test can be made by regressing the residuals in (2) on the residuals in (3):

$$\varepsilon_I = \gamma \varepsilon_A + u, \gamma > 0, \tag{4}$$

where u is a randomly distributed error term.

Substituting (2) and (3) into (4) and rearranging the terms yields:

$$\dot{I} = \gamma \dot{A} + (\alpha_I - \gamma \alpha_A) \ln YN - (\gamma \beta_A - \beta_I) (\ln YN)^2 + u. \tag{5}$$

This equation is the explicit form of (1). It also implies that the disparity between industrial and agricultural growth, $\dot{I} - \dot{A}$, is a second order non-linear function of per capita income.

The estimation of (5) is conducted using two cross-country samples: one consists of sixty-three countries for the decade of the 1960s and the other has eighty- seven countries for the decade of the 1970s.

Table 5.1 Estimated regression coefficients for annual industrial growth $\dot{I}$

Equation	*Period*	ln*YN*	(ln*YN*)^2	$\dot{A}$	*Constant*	R^2
		Independent variables				
1	1960–70	9.277*	–0.658*	–	–24.331*	0.12
		(2.7)	(2.8)		(2.1)	
2	1970–79	13.068**	–0.904**	–	–40.594**	0.13
		(3.4)	(3.4)		(3.1)	
3	1960–70	6.548	–0.458	0.491*	–16.730**	0.18
		(1.8)	(1.8)	(2.1)	(2.1)	
4	1970–79	9.477**	–0.649*	0.722**	–29.873*	0.28
		(2.6)	(2.6)	(4.1)	(2.4)	

Notes: The numbers under the respective coefficients in parentheses are *t*-statistics. The coefficients with a significance level above 5% are indicated by * and those above 1% by **.

The notations have the following meanings:

$\dot{I}$ = The average annual rate of growth of industry, comprised of mining, manufacturing, construction, and electricity, water, and gas.

YN = GNP per capita in 1970 and 1979 US dollars respectively, for the 1960–70 sample and the 1970–79 sample.

$\dot{A}$ = The average annual rate of growth of agriculture.

Both samples include developing as well as developed countries. The data sources are described in the Appendix; the estimation results based upon the ordinary least square (OLS) method are contained in Table 5.1.

First, note that the regressions with the per capita income variables alone (equations 1 and 2) depict a parabolic curve, a result that indicates that at a relative low level of income, industrial growth will increase as per capita income increases, and that when per capita income reaches a certain level, the rate of industrial growth will reach a maximum and then taper off. This outcome confirms the hypothesis as formulated in (2).[2]

Equations (3) and (4) show that the growth rate of agriculture is a statistically significant variable in explaining industrial growth and that it has raised the R^2s significantly for both samples, especially for 1970–79. Although the statistical significance of the per capita income variables were reduced, the results unambiguously confirm the hypothesis that countries with the above-the-norm performance in industry are also those associated with the above-the-norm performance in

Table 5.2 Disparity between the rates of industrial and agricultural growth: $\dot{I} - \dot{A}$

Per capita income (1979 US \$)	*Growth rates (per cent)* Agriculture $\dot{A}$	Industry $\dot{I}$	Disparity between agriculture & industry $\dot{I} - \dot{A}$
100	0.6	0.4	–0.2
200	1.6	3.3	1.7
250	1.8	4.0	2.2
500	2.4	5.7	3.3
1 000	2.7*	6.6	3.9
1 500	2.7	6.6*	4.0*
2 500	2.5	6.3	3.9
3 000	2.4	6.1	3.7
4 000	2.1	5.6	3.5
8 000	1.4	3.9	2.5
10 000	1.0	3.1	2.1

Note: Indicates the peak value. The calculation is based on equation (4) in Table 5.1 for which the growth rates of agriculture at different per capita income levels are computed by the regression obtained for 1970–79: $-14.942 + 4.995 \ln YN - 0.354 (\ln YN)^2$; (2.0) (2.3) (2.4)
$R^2 = 0.06$. In this regression, the numbers underneath the respective coefficients in parentheses are *t*-statistics.

agriculture over the development process that is manifested through a continuous rise in per capita income.[3] Based on the 1970–79 cross-section data of eighty-seven countries, this association is illustrated in Table 5.2 for a spectrum of per capita income levels. As the table shows, the expected values of industrial and agricultural growth rates, as well as their disparities, exhibit a quick rise over the low- and middle-income range (US\$100 to US\$1500) but only slowly tapers off after the middle income range (US\$1500). While both agriculture and industry are accelerating their respective growth rates in the low-middle income range, the former reaches its peak growth rate at a per capita income level preceding the latter. This result seems to suggest that the development of agriculture precedes that of industry.

3. AGRICULTURE AND OVERALL ECONOMIC GROWTH – A PRODUCTION FUNCTION APPROACH

The empirical evidence presented in the previous section suggests that a significant linkage exists between agriculture and industrial developments during the development process. In this section, I examine the contribution of agriculture to economic growth from the perspective of an aggregate production function. The hypothesis formulated and tested here is that agriculture contributes to economic growth through its impact on the rate of increase in total factor productivity; namely, agricultural growth shifts the aggregate production function upward.

There are at least two reasons why agricultural performance is closely related to the overall productivity of an economy. First, industrialisation and the accompanying urbanisation generate a growing need for food and raw materials that can only be met with adequate agricultural supplies. A poorly performing agricultural sector often results in terms of trade that are against industry, in a loss of foreign exchange, or in an inadequate demand for industrial output that restricts industrial expansion. These conditions hamper the transfer of resources from agriculture to industry or make it very costly. On the other hand, rapid agricultural growth makes feasible more per capita domestic consumption, higher exports of agricultural products, and greater absorption of the agricultural labour force by the industrial sector. Therefore, it enhances the resource transfer from agriculture to industry. As Robinson (1971) has shown, transfers of capital and labour from a lower productivity sector such as agriculture, to a higher productivity sector such as industry, is by itself a distinctive source of economic growth.

Secondly, high agricultural growth largely reflects high average labour productivity in agriculture that, in turn, is supported by the high quality of human resources and physical capital inputs into the rural sector. For instance, comparing the variation in agricultural labour productivity between developing and developed countries, Hayami and Ruttan (1970) found that about two-thirds of it could be accounted for by the difference in technology, as embodied in fixed working capital and in human capital, broadly conceived to include the education, skills, and knowledge capacity present in a country's population. A more recent study by Kawagoe, Hayami, and Ruttan (1985) estimated that only a quarter of the productivity differences could be accounted by conventional inputs (land and livestock).

These considerations argue for the inclusion of the rate of agricultural growth as an additional variable in an aggregate production function to measure both the effect of the efficiency of resource transfers between agriculture and industry, and of the impact of agricultural productivity on aggregate economic growth.

3.1 Other Variables Affecting Productivity Change

Other variables that could influence productivity changes should also be considered in order to minimise misspecification of the production function to be estimated below. Here, I consider two variables – the rate of export growth and the rate of inflation. Other variables affecting changes in total factor productivity are summarised in the error term in the production function.

The contribution of exports to economic growth may be rationalised on such grounds as: (1) exports provide for economy-of-scale operations; and (2) exports enlarge the level of competition of domestic industries. These factors promote efficiency and therefore raise economic growth. A positive correlation between export growth rates and GNP growth rates for developing countries has been found in a number of studies, i.e., Michalopoulos and Jay (1973), Michaely (1977), and Krueger (1978). Recently, in the context of an aggregate production function, Balassa (1978) and Feder (1983) have found that the growth in exports contributes significantly to inter-country differences in the rates of economic growth. These studies differed from earlier ones in that, besides exports, they simultaneously considered the contributions made by factor inputs, i.e., capital and labour. They were, therefore, able to estimate the effects of exports on productivity growth.

The rate of inflation, in contrast, affects economic efficiency negatively for at least the following reasons (Tsiang, 1983). First, by increasing the variance of relative prices, inflation may make efficient planning of production difficult, and productivity suffers as a result. Secondly, inflation accompanied by government controls leads to very low real interest rates. The situation in turn leads to a false sense of the cheapness of capital and to its unproductive use. Below-equilibrium interest rates accompanied by high inflation could turn the flow of savings away from financial institutions and toward unproductive investments such as the hoarding of precious metals and investments in housing. Thirdly, under the fixed exchange rate regime widely adopted by developing countries, inflation overvalues the domestic currency and leads to a loss in international competitiveness. The consequences are stagnant exports and balance-of-payments disequilibria, conditions that often result in a more restrictive trade regime and loss of overall efficiency.

3.2 The Model

To test the hypothesis formulated above, I use the following Cobb-Douglas production function:

$$Y = C\, K^{\alpha} L^{\beta} e^{\log R} \tag{6}$$

where Y = gross domestic product;
C = a scale parameter;
K = capital stock;
L = the labour force; and
$e^{\log R}$ = the rate of technological change over time, which is taken to be synonymous with productivity change.

Rewriting the variables in (6) in terms of the rate of change over time yields:

$$\dot{Y} = \alpha\dot{K} + \beta L + \dot{R}. \tag{7}$$

In the literature of production function analysis, the productivity change ($\dot{R}$) in production function (7) is frequently treated as a 'residual', and the production function is estimated accordingly. The argument presented here assumes that the rate of productivity change is positively influenced by both the rates of agricultural growth ($\dot{A}$) and export growth ($\dot{X}$) but is negatively related to the rate of inflation ($\dot{P}$):

$$(\dot{R}) = a + \gamma\dot{A} + \theta X + \eta P + \varepsilon,\ \gamma, \theta, > 0,\ \eta < 0, \tag{8}$$

where a is a constant term and ε is a residual, assumed to be randomly distributed. Combining (7) and (8) yields:

$$\dot{Y} = a + \alpha\dot{K} + \beta L + \gamma A + \theta\dot{X}, + \eta\dot{P}\varepsilon \tag{9}$$

where $\dot{Y}$ = the average annual rate of growth of GDP;

$\dot{K}$ = the average annual rate of growth capital, proxied by the average investment rate;[4]
$\dot{L}$ = the average annual rate of growth of the labour force;
$\dot{X}$ = the average annual rate of growth of exports;
$\dot{P}$ = the average annual rate of inflation.

Before turning to the empirical estimates, several remarks are in order. First, although (9) includes the export and agricultural growth rates as regressors, it is essentially a production function and is not a national income accounting identity. Secondly, by including the inflation and agricultural growth rates as independent variables, the equation is an extension of the Balassa (1978) model, which includes only export growth as the shifting variable in a production function such as (6). Finally, the production function is assumed to be homogenous across countries.

The production hypothesis (9) is tested with cross-country samples from two periods – 1960–70 and 1970–79 by the ordinary least square method. For the first period, the sample includes fifty-seven countries, of which forty-two are developing. For the second period, because of the greater availability of country data, the sample is increased to eighty-two countries, of which sixty-nine are developing. For both time periods, empirical results are obtained for both the developing country and the whole sample.

3.3 1960–70 Period

First, with respect to the results from the developing country sample, note that the inter-country variations in capital and labour contribute to slightly less than one-third of the inter-country variations in total output (Table 5.3, equation 1). When the rate of export growth ($\dot{X}$) and inflation ($\dot{P}$) are added to the regression, the explained variations in total output rise from one-third to about one-half (equation 2) and labour elasticity falls significantly from 1.6 to 1.0. Export growth and labour force growth are significant at above the 5 per cent level, but capital growth and inflation are not.

When the rate of agricultural growth is introduced as an additional shift variable in the production function, the explanatory power of the function is increased significantly (equation 3). The estimated elasticity of agricultural growth is about 0.6. However, the result also indicates the possibility of a strong collinear relationship between

Table 5.3 Estimated regression coefficients for annual growth in GDP, 1960–70

Equation	*Country*	*Independent variables* $\dot{K}$	$\dot{L}$	$\dot{X}$	$\dot{P}$	$\dot{A}$	*Constant*	R^2
1	*Developing*	0.074 (1.6)	1.593** (3.1)				0.608 (0.5)	0.32
2		0.065 (1.6)	1.030** (2.8)	0.139** (3.4)	–0.039 (1.4)		1.302 (1.1)	0.49
3		0.059 (1.8)	0.302 (0.8)	0.132** (3.9)	–0.036 (1.5)	0.622** (4.2)	0.896 (1.0)	0.66
4	All	0.104** (2.7)	1.162** (4.3)					0.29
5		0.072** (2.2)	0.912** (3.9)	0.158** (4.7)	–0.040 (1.7)			0.51
6		0.076* (2.7)	0.300** (1.3)	0.148** (5.1)	–0.041 (2.0)	0.555** (4.7)	0.782 (1.0)	0.66

Note: This table is based on equation (9) in the text. The numbers underneath the respective coefficients in parentheses are *t*-statistics. The coefficients with a significance level above 5% are indicated by * and above 1% by **.

agricultural growth and labour force growth, since the coefficient of labour force growth is significantly reduced from 1.0 to 0.3, while the coefficients of all other variables are not significantly altered. The result could suggest that in developing countries, agriculture was very labour-intensive during the decade of 1960s, and the agricultural labour force constitutes a predominant part of the total labour force.

Because of the collinear relationship, the contribution of agricultural growth in the production function can be overestimated. I therefore perform an alternative test. This time I test whether the part of economic growth that cannot be explained by growth in factor inputs (capital and labour) and in 'other' factors influencing the productivity change (export growth and inflation) is significantly related to the rate of agricultural growth – that is, I regress the unexplained residuals of the production function on agricultural growth as follows:

$$\dot{Y} - \hat{\alpha}\dot{Y} - \hat{\beta}\dot{L} - \hat{\theta}\dot{X} - \hat{\eta}\dot{P} = \gamma\dot{A} + \varepsilon, \qquad (10)$$

where $\hat{\alpha}$, $\hat{\beta}$, $\hat{\theta}$ and $\hat{\eta}$ are the OLS estimates of α, β, θ, and η, respectively, when agricultural growth ($\dot{A}$) is omitted from (9). This test is equivalent to that based upon (9) if all the regressors in (9) were othogonal to each other, i.e., statistically independent. However, the test appears to be more robust because, unlike total economic growth itself, the residuals of the production function, as indicated by the left-hand side of (10), need not bear any systematic relationship with agricultural growth *a priori*.

The result of this test shows that the marginal contribution of agriculture to economic growth remains significant at the 1 per cent level and that it accounts for an additional 18 per cent of the increase in total factor productivity for developing countries (Table 5.4, equation 1). However, the estimated elasticity of 0.40 is significantly smaller than that estimated by the multiple regression (9).

The estimated coefficients obtained for the all-country sample that includes developed countries are very similar to those obtained for the developing country sample, but they are more efficient, as evidenced by the higher *t*-value (Table 5.3, equations 4–6). The elasticities of export growth are higher, and of agricultural growth lower, an outcome suggesting that, in high-income countries, exports contribute more to economic growth, while agriculture less. Thus, agricultural growth accounted for an additional 16 per cent of the increase in economic growth that is not explained by the variables

Table 5.4 Estimated regression coefficients for annual unexplained growth in GDP

Equation	*Period*	*Country*	$\dot{A}$	*Constant*	R^2	*No. of countries*
1	1960–1970	*Developing*	0.402** (3.0)	–1.223 (2.4)	0.18	42
2		*All*	0.322** (3.0)	–0.880 (2.6)	0.16	57
3	1970–1979	*Developing*	0.348** (4.1)	–0.693 (2.2)	0.20	69
4		*All*	0.353** (4.6)	–0.693 (2.9)	0.21	82

Note: This table is based on equation (10) in the text. The numbers underneath the respective coefficients in parentheses are *t*-statistics. The coefficients with a significance level above 5% are indicated by * and above 1% by **.

---

already included in the model for the all-country sample, compared with the 18 per cent obtained for the developing country sample (Table 5.4, equation 2).

3.4 1970–79 Period

The empirical estimates obtained for the period 1970–79 support the general patterns observed for the earlier period 1960–70 (Table 5.5). However, the estimates for this period are statistically much more efficient compared with the previous sample – all the explanatory variables in alternative models are statistically significant either at the 1 per cent of the 5 per cent level. For instance, the inflation rate is found to be significant at above the 5 per cent significance level, even though its negative impact on economic growth in absolute terms becomes smaller. The estimated coefficients of export growth are nearly twice as large as those obtained for the 1960–70 sample, a result indicating that the marginal contribution of exports to economic growth appears to be greater in the 1970s. The difference seems to lie in the behaviour of developing countries, as, contrary to the 1960–70 results, the impact of exports on economic growth is greater for developing countries than for developed countries. (For instance, compare equations 3 and 6, Table 5.5). There seems to be a shift in development strategy toward export-orientation in the developing world in this period. The contribution of agriculture to the increase in productivity is again found to be highly significant and accounts for 20 per cent of the 'unexplained' variations in economic growth (Table 5.4, equations 3 and 4), which is somewhat higher than the 16–18 per cent estimated for the 1960–70 period. However, for the developing country sample, a 1 per cent increase in agricultural growth is estimated to increase productivity by 0.35 per cent during the 1970s, as compared with 0.40 per cent during the 1960s. The decline in the role played by agriculture in economic growth is consistent with the increased degree of industrialisation in developing countries.

However, the difference in regression results associated with the change in the sample period does not appear to be statistically significant, as the F tests, based on pooled regressions of both time periods for both the developing country sample and all country sample (Table 5.6), fail to reject the null hypothesis that the two samples in either time period are homogenous.[5]

The voluminous empirical literature on production functions (for both developed and developing countries) seems to indicate that the Cobb-Douglas production function with constant-return-to-scale provides a very useful reference, if not always the true underlying function, for studying the input-output relationship. In terms of the empirical estimates of the developing country sample in Table 5.6, constant-return-to-scale implies a capital-output ratio of approximately 1.7.[6] The corresponding estimates for a selected number of countries and periods are: China, 1.5 (1981); India, 2.5 (1978–79); Korea, 1.6 (1968); Japan, 3.5 (1965); and UK, 4.5 (1970) (*China: Long Term Issues and Options, 1985*). Although international comparison of capital-output ratios is difficult because of the crudeness of capital stock estimates, our estimate does not seem to be implausible for the developing countries. In other words, the estimated production function seems to be consistent with a Cobb-Douglas production function with constant-return-to- scale. Furthermore, the empirical evidence seems to confirm the homogeneity assumption of the production function.[7]

In broad terms, the results based upon both periods validate the neoclassical production hypothesis formulated in (9). The intercountry differences in economic growth rates can be explained, not only by the differences in factor endowments (capital and labour inputs), but more important, by the different performance of total factor productivity. The latter, in turn, is favourably affected by

Table 5.5 Estimated regression coefficients for annual growth in GDP, 1970–1979

Equations	*Country*	*Independent variables* $\dot{K}$	$\dot{L}$	$\dot{X}$	$\dot{P}$	$\dot{A}$	*Constant*	R^2
1	*Developing*	0.178**	0.937*				–1.645	0.24
		(3.8)	(2.2)				(1.2)	
2		0.097*	0.745*	0.281**	–0.022*		0.247	0.50
		(2.3)	(2.1)	(5.7)	(2.4)		(0.2)	
3		0.098**	0.625*	0.235**	–0.022*	0.380**	–0.139	0.61
		(2.7)	(2.0)	(5.1)	(2.6)	(4.2)	(0.1)	
4	*All*	0.174**	0.967**				–1.647	0.25
		(4.2)	(3.1)				(1.4)	
5		0.094*	1.041**	0.263**	–0.019*		–0.554	0.49
		(2.5)	(3.9)	(5.8)	(2.1)		(0.5)	
6		0.096**	0.834**	0.219**	–0.020*	0.380**	–0.697	0.61
		(2.9)	(3.4)	(5.3)	(2.5)	(4.6)	(0.8)	

Note: This table is based on equation (9) in the text. The numbers underneath the respective coefficients in parentheses are *t*-statistics. The coefficients with a significance level above 5% are indicated by * and above 1% by a **.

Table 5.6 Pooled regressions

Country	Independent variables						
	$\dot{K}$	$\dot{L}$	$\dot{X}$	$\dot{P}$	$\dot{A}$	Constant	R^2
Developing	0.095**	0.835**	0.219**	–0.020**	0.381**	–0.695	0.60
	(2.9)	(3.5)	(5.4)	(2.5)	(4.7)	(0.8)	
All	0.096**	0.604**	0.188**	–0.021**	0.421**	–0.094	0.61
	(3.9)	(2.7)	(6.5)	(3.0)	(5.9)	(0.1)	

Note: The numbers underneath the respective coefficients in parentheses are *t*-statistics. All coefficients are significant above 1% level except constants, and indicated by **.

---

export and agriculture performances but negatively influenced by the inflation rate.[8]

3.5 Sources of Economic Growth Decomposition

The regression estimates for the 1970–79 period are more stable, efficient, and plausible than those for the earlier period. The estimated equation for the developing country sample for this period is then used for the 'sources' of economic growth analysis discussed below.

Using the estimated regressions on economic growth, the average rate of economic growth during a period can be decomposed according to the 'sources' of economic growth. For the developing countries sample of the 1970–79 period, during which the average rate of growth was 4.4 per cent, the sources-of-growth decomposition indicates that capital and labour jointly contributed about 81 per cent of the growth in total output, with the former contributing 48 per cent and the latter 33 per cent. The remaining 19 per cent is explained by the productivity change, which in turn is determined by agricultural growth, export growth, and inflation (Table 5.7).

It is worth noting that, for the developing countries, the contribution of agricultural growth to productivity growth (0.8 per cent) was greater than that of export growth (0.6 per cent). Agriculture's contributing share in total productivity growth (17 per cent) was thus also greater than that of export growth (14 per cent). The inflation rate, on the other hand, reduced productivity growth by an average of 4 per cent a year.

Table 5.7 Sources of economic growth, 1970–79 (developing country sample)

	Factor contribution		*Average growth*
	Growth rate (%)[a]	*Share (%)*	*rate (%)*
Factors of production:	3.5	81	
of which:			
Capital	2.1	48	21.6[b]
Labour	1.4	33	2.3
Productivity change:	0.9	19	
of which:			
Export	0.6	14	2.7
Inflation	–0.4	–9	19.1
Agriculture	0.8	17	2.0
Other[c]	–0.1	–3	
Total (GDP)	4.4	100	

Note: Calculation is based on equation 3 in Table 5.5.
[a] The numbers in this column are obtained by multiplying the estimated elasticities by the average rates of growth of factors of production.
[b] Average investment rate.
[c] Reflecting the constant term.

The historical growth-accounting analysis suggests that agriculture could be as dynamic as exports in generating economic growth in developing countries. Based on the orders of magnitude estimated, the role of agriculture seems to be no less important than that of exports in fostering productivity, and perhaps even more so.

The policy implication seems to be that in formulating a development strategy, agricultural development ought to be given due attention. This strategy frequently means the adoption of appropriate exchange rate and pricing policies, adequate agricultural investments in infrastructure, research and extension facilities, and suitable training and health services for the agricultural labour force. Moreover, the shift in development strategy in favour of agriculture that brings out significant advances in productivity growth can often be carried out without additional physical investment. It involves institutional reforms such as those performed under land reform programmes in several East Asian newly-industrialising countries in the early stages of their development. China's recent economic reform, which began with the agricultural sector and subsequently brought about spectacular growth in agricultural output, also involved a fundamental reorganisation of the basic mode of agricultural production.

4. SUMMARY AND CONCLUSIONS

Using cross-section data of the 1960s and the 1970s, this paper has shown that the inter-country variation in industrial growth is significantly associated with the inter-country variation in agricultural growth over the development process. It has further demonstrated that agricultural growth induces productivity increases and, therefore, facilitates overall economic growth. Moreover, the paper estimated that the role of agriculture seems to be no less important than that of exports in fostering productivity. As argued in the paper, this result may stem from the fact that rapid agricultural growth raises the efficiency of resource transfers (capital and labour) between the agricultural and non-agricultural sectors, an improvement that results in an increase in overall productivity. Moreover, rapid agricultural growth itself may reflect high agricultural productivity. The paper also confirms earlier findings on the positive contribution of export growth to an increase in productivity. In addition, it shows that inflation has the opposite effect on productivity.

The implication of the main thrust of this paper for the long-term development strategy of low-income developing countries is clear: an appropriate agricultural development strategy should be formulated to accelerate agricultural growth rates. While agriculture is inevitably a 'declining industry' from the point of view of its share in total output over the long term, it should not be excessively squeezed for resources to support non-agricultural activities and therefore 'abandoned' prematurely.

While cross-section analysis serves to uncover the common variables that determine the complex growth processes underlying individual countries and to summarise their 'average' or 'normal' experience and thus is a point of reference for analysing individual country performance, it should be supplemented by detailed country studies to examine country-specific factors. Given the often unique conditions surrounding developing country agriculture, an examination of the contribution of agriculture to economic growth through the use of cross-section data, as was done in this paper, should clearly not be exempt from these remarks. In this context, it is worth noting that, based upon a general equilibrium analysis of the economy of the Republic of Korea, Adelman (1983) finds that its economic development could have fared even better under what she calls an 'agricultural-demand-led industrialisation program' than under a strategy based purely on export promotion.

Appendix: Data Sources

The data used in the paper are obtained mostly from the world development indicators annex of the *World Development Report 1981* (World Bank, 1981c) (1979 GNP per capita in dollars (YN) and the average annual rate of inflation ($\dot{P}$) are from Table 1 of the annex, pp. 134–35; the average annual growth rate of GDP ($\dot{Y}$), agriculture ($\dot{A}$), and industry ($\dot{I}$) are from Table 2, pp. 136–37; and the average annual growth rate of merchandise exports is from Table 8, pp. 148–49). The Taiwan data, however, are obtained from corresponding tables in the *World Development Report 1980* (World Bank, 1981b). The average investment rate (K) for the period and 1970 per capita GNP are obtained from the Economic and Social Data Division of the World Bank. The average annual growth rate of the labour force (L) is derived from the Social Indicators Data Sheets for individual countries compiled by the Economic and Social Data Division as of May 1982.

Notes

* The World Bank does not accept responsibility for the views expressed herein which are those of the author and should not be attributed to the World Bank or to its affiliated organisations. The findings, interpretations, and conclusions are the results of research supported by the Bank; they do not necessarily represent official policy of the Bank. The designations employed, the presentation of material, and any maps used in this document are solely for the convenience of the reader and do not imply the expression of any opinion whatsoever on the part of the World Bank or its affiliates concerning the legal status of any country, territory, city, area, or of its authorities, or concerning the delimitation of its boundaries, or national affiliation.

1. For evidence, see World Bank (1981a).
2. The explanatory power of these regressions is notoriously low, simply suggesting that per capita income variables alone are not sufficient to explain inter-country differentials in industrial growth rates. These results, however, should not be a source of alarm because our major purpose here is to establish the statistical association between industrial

growth rates and per capita income levels, rather than to provide a full model for ascertaining why industrial growth rates differ among countries. This argument also applies to the rest of the regressions reported in Table 5.1.

3. Although the empirical evidence by itself does not necessarily imply that agricultural growth 'causes' industrial growth in the Granger (1969) – Sims (1972) sense, we, nevertheless, have reason to believe, as argued earlier, that agricultural growth does 'push' industrial growth – and it seems that this effect is what the empirical evidence reported above attempts to demonstrate. On the other hand, as also argued, industrial growth could also 'pull' agricultural growth over the course of development and, to the extent that it is true, the estimates are subject to the classical simultaneous equation bias, the correction for which requires the specification of a complete model of agricultural and industrial growth that is beyond the scope of the present paper.
4. Since $\dot{K} = \frac{Y}{K}\frac{I}{Y}$, the use of the investment rate to replace $\dot{K}$ assumes that the average capital-output ratio (K/Y) is a constant across countries.
5. The test statistic is constructed as follows: $F\ (K_C,\ N - K_u) = \dfrac{(SSR_C - SSR_u)/K_C}{SSR_u/(N - K_u)}$ where SSR_C and SSR_u are, respectively the constrained and unconstrained residual sum of squares; K_C and K_u are, respectively, the number of regressors in the constrained and unconstrained regressions; and N is the number of countries in the constrained regression. The constrained regressions are pooled regressions given in Table 5.6. The corresponding unconstrained regressions are given in Table 5.3 and Table 5.5.
6. $\dfrac{1 - 0.835}{0.095} = 1.7$, since $\dot{K}$ is approximated by $\dfrac{I}{Y}$ (see note 4).
7. The model could suffer from heteroskedasticity either because the production function is not homogenous across countries as assumed, or because significant explanatory variables may have been omitted which contain distinctive variances. Whatever may be the cause, it is well known that in the presence of heteroskedasticity in the disturbances of an otherwise properly specified linear model leads to consistent but inefficient estimates, and inconsistent co-variance matrix estimates. As a result, faulty inferences will be drawn when testing a hypothesis in the presence of heteroskedasticity. To test whether heteroskedasticity is present in the regression model and to obtain consistent co-variance matrix in the presence of heteroskedasticity, I employ the White test (White, 1980). The test statistic nR^2 is distributed as χ^2 with $K\ (K + 1)/2$ degrees of freedom; where n, K and R^2 are, respectively, the number of observations, the number of regressors and the squared multiple correlation coefficient from the regression of squared estimated residuals $\varepsilon^2{}_{in}$ on all second order product and cross-products of the original regressors X_{ij}, X_{iK}:

$$\varepsilon^2_{in} = \alpha_o + \sum_{j=1}^{K} \sum_{K=j}^{K} \alpha_s X_{ij} X_{iK},$$

where α's are parameters to be estimated by OLS.

The test is performed for the regressions run for both samples. For the 1960–70 period, the test is applied to equations 3 and 6 (Table 5.3) and for the 1970–79 period to equations 3 and 6 (Table 5.5). The test statistics corresponding to 21 degrees of freedom ($K=5$) and 5 per cent probability of type *I* error is 32.7. For the 1960–70 period, the computed test statistics are, respectively, 18.6 and 16.2 for the developing country sample and all country sample – both are smaller than 32.7 – so the null hypothesis of no heteroskedasticity cannot be rejected. However, for the 1970–79 period, the test statistics are, respectively, 56.3 and 66.3 for the developing sample and all country sample. Hence the null hypothesis of no heteroskedasticity should be rejected. In this case, a consistent co-variance matrix of the disturbance term should be obtained to avoid faulty inferences. The method proposed by White does not require any particular hypothesis regarding the co-variance structure for the purpose of obtaining consistent estimates. By employing the White estimator, the revised *t* values of regression parameters corresponding to the consistent standard errors for both samples are obtained. For the developing country sample, they show that export growth and inflation are significant at 1 per cent significance level. Capital accumulation is significant at 5 per cent level. However, labour force growth and agricultural growth can only pass the 8 per cent significance test. For the larger, all sample country sample, all variables pass the 1 per cent significance test except agricultural growth, whose probability of type *I* error is 6 per cent. The correction for heteroskedasticity thus weakens somewhat the statistical significance of agricultural growth. (I am indebted to Mr Lant Pritchett of MIT for the computer programme that calculates the consistent co-variance matrix of the regression parameters.)

8. Genberg and Swoboda (1986) also report that either the inflation rate or the variation of inflation is detrimental to economic growth.

References

Adelman, I. (1983) 'Beyond Export-Led Growth', Division of Agricultural Sciences, University of California (December).

Balassa, B. (1978) 'Exports and Economic Growth: Further Evidence', *Journal of Development Economics*, vol. 5, no. 2 (June) pp. 181–9.

Chenery, H. and Syrquin, M. (1975) *Patterns of Developments 1950–1970* (London: Oxford University Press).

Chenery, H. and Taylor, L. (1968) 'Development Patterns: Among Countries and Over Time', *Review of Economics and Statistics*, vol. 50 (November) pp. 391–416.

China: Long Term Issues and Options, Annex 5: China's Economic Structure in International Perspective (1985) (Baltimore, Maryland: The Johns Hopkins University Press) Table 3.15, p. 38.

Feder, G. (1983) 'On Exports and Economic Growth', *Journal of Development Economics*, vol. 12, no. 1, pp. 59–73.

Genberg, H. and Swoboda, K. (1986) 'The Medium-Term Relationship Between Performance Indicators and Policy: A Cross-Section Approach',

Division Working Paper no. 1986–2, Country Analysis and Projections Division, The World Bank, Washington, DC (May).

Granger, C.W.J. (1969) 'Investigating Causal Relations by Econometric Models and Cross-spectral Methods', *Econometrica*, vol. 37, pp. 424–38.

Hayami, Y. and Ruttan, V. M. (1970) 'Agricultural Productivity Differences Among Countries', *American Economic Review*, vol. 60, pp. 895–911.

Johnston, F. and Kilby, P. (1975) *Agriculture and Structural Transformation: Strategies for Late Developing Countries* (New York: Oxford University Press).

Johnston, B. F. and Mellor, J. W. (1961) 'The Role of Agriculture in Economic Development', *American Economic Review*, vol. 51, no. 4 (September) pp. 566–93.

Kawagoe, T., Hayami, Y. and Ruttan, V. M. (1985) 'The Inter-country Agricultural Production Function and Productivity Differences Among Countries', *Journal of Development Economics*, vol. 19 (September–October) pp. 113–32.

Krueger, O. (1978) *Foreign Trade Regimes and Economic Development: Liberalization Attempts and Consequences* (Cambridge, Mass.: Ballinger Press).

Michaely, M. (1977) 'Exports and Growth: An Empirical Investigation', *Journal of Development Economics*, vol. 4, no. 1 (March) pp. 49–53.

Michalopoulos, C. and Jay, K. (1973) '*Growth of Exports and Income in the Developing World: A Neoclassical View*', AID Discussion Paper no. 28, Washington DC (November).

Robinson, S. (1971) 'Sources of Growth in Less Developed Countries', *Quarterly Journal of Economics*, vol. 85, no. 3 (August) pp. 391–408.

Sims, C. A. (1972) 'Money, Income and Causality', *American Economic Review*, vol. 62, pp. 540–52.

Tsiang, S. C. (1983) 'Theories of Inflation in Less Developed Countries', A paper prepared for the conference on Inflation in East Asian Countries, Taipei, Taiwan (20–22 May).

White, H. (1980) 'A Heteroskedasticity-Consistent Co-variance Matrix Estimator and a Direct Test for Heteroskedasticity', *Econometrica*, vol. 48 (May) pp. 817–35.

World Bank (1981a) *Accelerated Development in Sub-Saharan Africa: An Agenda for Action* (Washington DC: World Bank).

World Bank (1981b) *World Development Report 1980* (Washington, DC: World Bank).

World Bank (1981c) *World Development Report 1981* (Washington DC: World Bank).

Part II

Sector Proportions and Economic Development: Country Studies

6 Extensions of the Constant-Market-Shares Analysis with an Application to Long-term Export Data of Developing Countries

C.J. Jepma
UNIVERSITY OF GRÖNINGEN

1. INTRODUCTION

The aggregate of exports of individual developing countries is generally considered a crucial variable in development processes. Hence, a wave of studies appeared from the 1950s on, trying to explain differences in export performance between countries, and to relate exports to other economic variables (Michalopoulos and Jay, 1973; Chenery and Syrquin, 1975; Michaely, 1977; Krueger, 1978; Balassa, 1978 and 1979; Chenery, 1980; Lewis, 1980; Jepma, 1986, to give just some examples). Most of the studies explaining differences in export performance employed macroeconomic trade data. They started from the assumption that export performance should be judged on the basis of the deviation between actual export performance and the 'normal' performance, the latter being defined as the change in total exports which would have materialised had export growth equalled the corresponding rate of a reference group. The latter usually consists of either the group of competing countries or the world total. The deviation was subsequently explained by relative competitiveness.

However, by employing such an assumption no attention was given to the impact of the particular characteristics of the composition of the exports of the country considered on its 'normal' export performance (for a related discussion, see Lewis, 1980; Riedel, 1983). This is incorrect because one should not expect that the aggregate exports of a country which mainly exports commodities which are confronted with a stagnating international demand, or whose exports are mainly oriented to stagnant markets, will develop 'normally' at the same rate as other countries' total exports. Consequently, what was totally attributed to changes in the competitiveness of the country, in reality also had to be explained by the composition of the exports, i.e., the commodity breakdown and the regional distribution of the flows.

In the present paper it will be empirically demonstrated that the use of the assumption mentioned is, indeed, not justified, particularly if one is dealing with developing countries' exports. The reason is that the composition of the exports of developing countries fundamentally differs on the average from that of the more developed countries, so that aggregation biases can cause the assumption to be unrealistic. Moreover, it will be demonstrated that the differences in relative export performance between less developed countries can also be largely attributed to differences in the composition of their exports. This may help explain why the differences in export performance between what have become known as the newly-industrialising countries (NICs) on the one hand, and several of the less developed countries on the other, have been so enormous during the post-war period. The conclusion is that any comparative analysis of the exports of developing countries which does not deal with the impact which differences in the composition of the exports can have on total exports, can be misleading.

In the following the impact of the specific characteristics of developing countries' export composition on their total exports will be empirically assessed. This will be done by applying decomposition analyses to determine what part of the change of a particular aggregate trade flow during an interval can be attributed to its composition (commodity breakdown and composition of the regions of destination) on the one hand, and competitiveness (price and non-price) of the export products on the other. The decomposition technique employed is a variant of the so-called constant-market-shares (cms) analysis. The analyses have been applied to long-run disaggregated (value) data of fifty-nine individual developing countries' exports to the OECD area, and in addition to several different specifications of the trade data. A kind of sensitivity analysis was thereby carried out to investigate the stability of the final results of the analysis.

Because the decomposition technique differs in some respects from the traditional specification of the basic identity, some attention will first be devoted to the technique itself. The results are subsequently presented in two steps: first they are presented for the total of the fifty-nine developing countries, so that their export performance can be compared with that of the rest of the world; thereafter the results are aggregated for subgroups of developing countries, enabling us to compare the export performance among groups of developing countries.

2. CMS ANALYSIS AND ITS BASIC PHILOSOPHY

The cms analysis came increasingly into vogue after its first application by Tyszynski (1951) to data concerning international trade flows at the beginning of the 1950s. The basic philosophy of the method is that, if the relative competitiveness of a trade flow of a particular exporter remains the same, its market share will remain constant. If that idea is applied to all subflows at the lowest level of aggregation distinguished in the analysis (whereby trade flows are usually disaggregated according to both the commodity and the region of destination), one finds the 'normal' change (i.e., the change if relative competitiveness *vis-à-vis* all subflows remains constant) of total exports of an exporter by summation. Due to differences in the composition of the exports between the country under consideration and the reference group, the 'normal' growth rate of the total exports of the country can differ from the corresponding rate of the reference group. The actual and 'normal' growth rate of total exports of the country by definition only differ due to changes in relative competitiveness. Thus, the actual change in the total exports of the country differs from the actual change in total exports of the reference group due to differences in the composition of the exports on the one hand, and changes in competitiveness on the other. All these aspects can be quantified by applying the decomposition which splits the change in the trade flow during an interval into a number of components, each having an economic interpretation of its own.

The basic identity on which the method was based was either:

$$\Delta q = \underset{(a)}{S_0\,\Delta Q} + \underset{(b)}{(\sum_i S_{i0}\Delta Q_i - S_0\,\Delta Q)} + \underset{(c)}{(\sum_i\sum_j S_{ij0}\,\Delta Q_{ij} - \sum_i \Delta S_{i0}\,\Delta Q_i)} + \underset{(d)}{\sum_i\sum_j \Delta\, S_{ij}Q_{ij0}} + \underset{(e)}{\sum_i\sum_j \Delta S_{ij}\,\Delta Q_{ij}} \qquad (1)$$

or

$$\Delta q = \underset{\text{(a)}}{S_0 \, \Delta Q} + \underset{\text{(b}')}{(\sum_j S_{j0} \Delta Q_j \; - S_0 \, \Delta Q)} + \underset{\text{(c}')}{(\sum_i \sum_j S_{ij0} \, \Delta Q_{ij} \; - \sum_j S_{j0} \, \Delta Q_i)} + \underset{\text{(d)}}{\sum_i \sum_j \Delta S_{ij} \, Q_{ij0}} + \underset{\text{(e)}}{\sum_i \sum_j \Delta S_{ij} \, \Delta Q_{ij}} \qquad (2)$$

where:

Δ = change per period;
q = the exports of the country considered, to be called country 'A';
Q = the exports of the competing countries considered, to be called 'standard';
$S = q/Q$;

The subscript '0' refers to the export flow in the base period; 'i' refers to the product group considered, and 'j' to the particular region of destination.

In (1) and (2), (a) expresses the reference level for the increase in A's exports if the export structure of A were to be equal to that of the standard, and A's market share would have also stayed equal; (c) in (1) and (b′) in (2) represent the market effect, i.e., the impact of the typical regional distribution of the exports of the country considered, whereas (b) in (1) and (c′) in (2) express the commodity effect, i.e., the impact of the specific commodity composition of the country's exports; (d) expresses the size of a residual which can be attributed to changes in the competitiveness of the country considered; (e) represents the second-order effect. As has been pointed out by Richardson (1970), (e) has an interpretation of its own because it indicates to what extent the increases (decreases) in the market shares of the subflows distinguished in total exports are concentrated on the trade flows for which international demand has a high (low) increase.

Furthermore, with respect to the interpretation of the factors, it can be added: (a) + (b) + (c) represent the change in A's exports if the market shares of all its disaggregated trade flows distinguished in the analysis are kept constant; it therefore logically follows that the difference between (a) on the one hand and (a) + (b) + (c) on the other is due to the fact that the commodity composition and the regional distribution of the country considered differs from the standard. The implication is that if a country has a rather typical export composition compared with the standard – and developing countries generally do, compared with the rest of the world (see Chenery and Syrquin, 1975) – there is a fair chance that its aggregate exports will show a different growth rate than the total exports of the standard even if relative competitiveness does not change.

As has been pointed out by Richardson (1971a, 1971b) and some others, the traditional specification of the method, although still commonly used, has some shortcomings:

1. The size of the commodity and market effect is influenced by the order of their specification (compare the difference between (1) and (2)).
2. There is no yardstick for dealing with the issue whether base-year (like in 1 and 2) or end-year values for S should be used in the identity, and also is it unclear what the length of the interval should be.
3. The specification pays no attention to the impact of changes in the composition of the exports of the country considered during the interval (see also Rothschild, 1975).
4. Attention is devoted only to the total of the effects, so that one gets no insight into the contribution of the separate commodities or regions of destination on the effects derived from the decomposition. Besides, the technique has been criticised for its sensitivity to changes in the specification of the data (see Richardson, 1971b), and for its lack of theoretical foundation (see Jepma, 1986; Richardson, 1970). In several ways we have tried to deal with these shortcomings (for an elaboration, see Jepma, 1986, pp. 20–50).

First, a somewhat more extended decomposition has been developed in such a way that the order problem is 'solved', and some indication can be given of the impact of changes in the export structure on relative export performance. The decomposition now becomes (in symbols):

$$\Delta q = \underset{(a)}{S_0 \Delta Q} + \underset{(b)}{(\sum_i \sum_j S_{ij0} \Delta Q_{ij} - \sum_i S_{i0} \Delta Q_i)} + \underset{(c)}{(\sum_i \sum_j S_{ij0} \Delta Q_{ij} - \sum_j S_{j0} \Delta Q_j)} +$$

$$+ \{ (\sum_i S_{i0}\Delta Q_i - S_0\Delta Q) - (\sum_i \sum_j S_{ij0}\Delta Q_{ij} - \underset{\text{(d)}}{}$$

$$\sum_j S_{j0}\Delta Q_j)\} +$$

$$\underset{\text{(e)}}{\Delta SQ} + \underset{\text{(f)}}{(\sum_i \sum_j S_{ij0}Q_{ij0} - \Delta SQ)} +$$

$$\underset{\text{(g)}}{(\frac{Q_1}{Q_0} - 1) \sum_i \sum_j \Delta S_{ij}Q_{ij0}} +$$

$$\underset{\text{(h)}}{\{\sum_i \sum_j \Delta S_{ij}\Delta Q_{ij} - (\frac{Q_1}{Q_0} - 1) \sum_i \sum_j \Delta S_{ij}Q_{ij0}\}} \quad (3)$$

with:

(a) = the reference level for the increase in *A*'s exports if the export structure of *A* equalled that of the standard;

(b) = the market effect;

(c) = the commodity effect;

(d) = the structural interaction effect, i.e., the impact of the interaction of the specific commodity and market composition of *A* on its export performance. This effect indicates to what extent exporters selling products confronted with a rapid increasing world demand, also sell them at the relatively fast growing markets of destination.
The size of (a) – (d) in (3) corresponds with (a) – (c) in (1) and (2);

(e) = the pure residual, i.e., the impact of the general change in *A*'s competitiveness under the condition of an unchanged structure of *A*'s exports;

(f) = will be called the static structural residual because it represents the impact of the interaction between the changes in *A*'s export structure during the interval, and the sizes of the corresponding international demand directed at *A*'s exports. In contrast with (e) it gives some indication of the effects of changes in *A*'s export structure on its residual.
The size of (e) + (f) in (3) corresponds with (d) in (1) and (2);

(g) = the effect expressing what the change in *A*'s exports is, only in so far as the changes in the market shares interact with the general change in the level of world demand (the pure second-order effect), whereas

(h) = indicates how changes in *A*'s market shares interact with changes in the structure of world demand. The latter effect will be called the dynamic structural residual.

The size of (g) + (h) in (3) corresponds with (e) in (1) and (2).

Secondly, the technique has been applied within the framework of a sensitivity analysis.

Thirdly, to avoid the major drawback of using fixed (trade share) weights, namely that no due attention is given to the fact that the trade structure of any country – and most developing countries in particular – shows some typical pattern on the long run, the decomposition was repeatedly applied on all yearly intervals; the effects over the total interval consist of the summation of the yearly effects.

Fourthly, a method was developed to further decompose the effects derived from the decomposition (see Jepma, 1986, pp. 42–50).

3. THE DATA AND THE SENSITIVITY OF THE CMS RESULTS FOR DATA SPECIFICATIONS

In the following the impact of the several characteristics of aggregate exports on export performance will be empirically assessed by applying decomposition analyses to disaggregated (value) data of fifty-nine developing countries' exports to the OECD area. To prevent us from using the less reliable export data of developing countries, the data consist of the value of imports, disaggregated into the eleven most important regions of destination in the OECD, and ten product categories according to the one-digit SITC-code. The exports of this sample amounted to more than three-quarters of the total export value of the Third World (defined as the group of non-OECD countries, excluding South Africa and the Comecon) on average during the period under consideration, 1966–80. The results can therefore be considered representative for the Third World.

The technique was also applied to some fifteen different specifications of the trade data. They differed from each other in the combination of choices made with respect to the following aspects: the way in which an average was calculated with respect to the trade data, the commodity classes which were taken into consideration, the composition of the reference group, and the way in which the trade data were corrected for incidental non-average data and for great fluctuations (for details on the smoothing procedure developed, see Jepma, 1986, pp. 158–60). With these variants several tests were carried out to investigate the sensitivity of the results of the cms analysis to differences in (combinations of) these four aspects. The analysis can therefore be regarded more or less as a complement to the major 'sensitivity' analysis carried out formerly by Richardson (1971b) in which he looked at the sensitivity to other aspects of the data. Our 'sensitivity' analysis was mainly carried out by applying a regression analysis to the results of combinations of variants which only differed with respect to one of the four aspects mentioned above.

It turned out that the length of the interval for determining the averages of the data (as well as the way of smoothing the data, as we developed it) did not have a substantial influence on the results, generally speaking; the values of the adjusted correlation coefficients for all decomposition effects were larger than 90 per cent. One may therefore conclude that the results are insensitive for these changes in the specification of the basic data. The sensitivity of the results to not incorporating SITC 3 (mineral fuels) into the data turned out to be substantial. This more or less confirmed what could be expected. Finally, the sensitivity to the choice of a standard turned out to be the most significant one in our research.

4. THE RESULTS OF THE DECOMPOSITION ANALYSIS

4.1 Total Developing Countries

After carrying out these sensitivity tests we decided to single out four variants, out of the many variants we used and implemented, which could be regarded as the most relevant ones for the present issue and which differed from each other mainly with respect to what was considered to be the most important in this respect from the sensitivity analysis: the size of the reference group and whether or not SITC 3 was incorporated into the data. The results of these variants which used the total world as a standard (α and ß in Table 6.1 for the variant without, and with, SITC 3, respectively) are summarised below by presenting the effects (a) to (h) of the decomposition, as defined above in (3), in percentages of the actual changes in the

Table 6.1 The results of the cms analysis on trade data of 59 developing countries for the 1966–80 period, using the rest of the world as standard

effect \ Variant		α (all commodity categories excl. SITC 3)		β (all commodity categories)	
a		90		74	
b		−1		1	
c	structural	−10	−12	2	2
d	effect	−1		−1	
e		9		23	
f	residue	10	21	−1	24
g		2		3	
h		0		−1	

Source: OECD, *Trade by Commodities* (Paris: OECD), Series C. Imports, vol. 1966–80.

exports after having aggregated them for all the countries under consideration and for the whole interval.[1]

Table 6.1 clearly illustrates some of the main findings of the analysis:

- the export performance of the sample considered – consisting of most of the less-developed and semi-industrialised countries – developed favourably during the interval compared to that of the total world. If one leaves SITC 3 out of consideration, the increase in exports turned out to be about 11 per cent [(100 – 90)/90] higher than what could be expected on the basis of the general increase in world trade, whereas the corresponding figure for even all trade amounted to 35 per cent [(100 – 74)/74]. The latter may, of course, be attributed to the fact that a considerable number of oil exporters are included in the sample;
- a major distinction between the variant with and without SITC 3 is that from the latter variant it is clear that a major difference between the developing countries and the rest of the world lies in the negative and relatively substantial commodity effect. For the sample this means that, even if the market shares are retained in all trade flows under examination, the increase in their exports will still fall short of the increase that would have materialised if these countries would just have retained their share in overall world

trade by about 10 per cent. The negative impact of the commodity breakdown on the export performance of the sample did not show up, however, in the variant including SITC 3 because the size of the commodity effect (c) here is positive. This implies that the importance of the sample as a supplier of energy in the world is such that the impact of the relatively large share of SITC 3 in the exports on the commodity effect more or less compensates for the negative commodity effect for the rest of the commodity categories;

- generally speaking, the regional spread of the exports and the interaction between commodity and regional distribution (effects b and d) have no substantial influence on the overall export performance of the developing countries. This seems plausible enough, because the sample of countries is so large that one would not expect the regional distribution of the exports of the whole group to differ too much from the corresponding figures for the reference group (and the same holds for interaction of regional and commodity breakdown);
- broadly speaking, the size of the residual effects, which indicate what the impact of competitiveness in the broad sense of the word is on export performance (defined as (e) up to and including (h)), is comparable in both variants. In both cases this residual is very substantial and makes a large contribution to the explanation of the relatively favourable export performance of the developing countries. In all the variants the second-order effect (g + h) is slightly positive, while the static residual (e + f) accounts for the bulk of the size of the residual;
- another major distinction between both variants is the extent to which the increases (decreases) in the market shares are biased towards relatively large (small) trade flows. This factor is very substantial only in the trade data excluding SITC 3, probably due to the fact that that competitiveness of developing countries is such that they can penetrate more easily in a particular trade flow if this is substantial enough to make further entrance of suppliers from developing countries possible.

4.2 Subgroups of Developing Countries

In the actual calculations the cms effects were in fact computed per year and per country, and were even further decomposed subsequently making it possible to draw all kinds of conclusions on all kinds of levels of aggregation and detail.[2]

In this paper we will confine ourselves in this respect to making a distinction between the cms results for the subgroup of eleven countries (within the sample of fifty-one countries for which the 1976–80 data were complete), which are generally regarded as the successful newly-industrialising, export-oriented economies (NICs),[3] (at least for the variants ß and δ using data including SITC 3), and the subgroup of eleven countries that can be considered oil-exporting countries (OILEX),[4] and for the rest of the sample (less-developed, LESSDC). Once again the results were aggregated over the countries and over the years in order to be able to draw general conclusions. The results are reported for all four variants in Table 6.2.

The main conclusion that can be drawn from a comparision of the results of the cms analysis for the NICs and the less-developed countries is that the tremendous difference in export success between the two groups can be attributed to three main factors, the effects of which reinforce the impact on the difference, namely (1) the commodity effect (c), (2) the static residual (e + f), and (3) the dynamic structural residual (h). These are dealt with below.

(1) – Whereas in α the commodity effect is only slightly negative for the NICs, it turned out to be substantially negative for the less-developed countries. Apparently the commodity breakdown of the latter group – which of course cannot be seen separately from their stage of development – has a very strong unfavourable influence on their export performance, although during the development process countries generally are developing in such a way that this negative bias disappears.

(2) – The difference between the NICs and less-developed countries export performance is further reinforced by the fact that the NICs succeeded in substantially increasing their market share, while the less-developed countries could not even keep their shares constant on average.

(3) – Finally, further decomposition showed another factor that may have contributed to the relatively poor export performance of the less-developed countries as compared with the NICs. Although the second-order effect generally turned out to be relatively small, an exception has to be made for effect (h) for the less-developed countries in ε. This effect turned out to be -11 per cent (while the corresponding effect for the NICs was -3 per cent). The latter implies that (in contrast with the NICs) the less-developed countries generally lost their market shares in those export flows which developed relatively favourably. This effect, which may indicate a lack of flexibility in the export performance of the less-developed countries,

Table 6.2 The results of the cms analysis on trade data of subgroups of the 51 developing countries on which the analysis was applied for the 1966–80 period

Variant / effect		α NICs		α LESSDC		β NICs		β OILEX		β LESSDC		δ NICs		δ OILEX		δ LESSDC		ε NICs		ε LESSDC	
a	structural effect	71		132		75		61		172		100		81		181		76		135	
b		–3		3		–2		2		5		3		6		10		3		1	
c		–3	–6	–24	–23	–10	–10	15	15	–28	–26	–9	–4	21	22	–53	–52	20	22	–24	–25
d		0		–2		2		–2		–3		2		–5		–9		–1		–2	
e	residue	26		–28		23		34		–37		4		14		–65		22		–31	
f		6	35	21	–8	10	37	–11	24	23	–16	2	5	–13	–2	44	–29	–17	2	32	–10
g		3		–1		4		3		–2		1		1		–3		0		0	
h		0		0		0		–2		0		–2		–4		–5		–3		–11	

Note: δ and ε differ from ß and α respectively only in the choice of the standard. This is total developing countries in δ and ε instead of total world in α and ß; in ε SITC 3 is excluded, whereas δ includes SITC 3.

Source: see Table 6.1.

also had a substantial negative impact on the export performance of the less-developed countries.

One of the factors that had a strong positive impact on the export performance of the less-developed countries was the fact that the increases (decreases) in market shares were generally concentrated in the relatively large (small) trade flows. It can be doubted, however, whether this should be considered a sign of an advantageous competitive position. Although no further research has been carried out here in this respect, it may be hypothesised that the less-developed countries have in fact, such a weak position as suppliers in the international market that they succeed in penetrating a particular market only if there is enough scope for them to be accepted. The fact that the size of this effect is much smaller or even negative for the NICs may also fit into this hypothesis.

5. SUMMARY AND CONCLUSIONS

In the post-war period a continuous debate has taken place on the participation of developing countries in international trade. In this debate almost no systematic attention has been given to the impact of the composition of the exports of the developing countries on their relative export performance. This can be considered a shortcoming, especially when the export performance of countries with different patterns of specialisation (e.g., agricultural versus industrial exporters) are compared. This paper tries to fill this gap by applying a set of decomposition analyses on the disaggregated yearly export data of fifty-nine developing countries in the 1966–80 period. To this end a new variant of the constant-market-shares approach has been developed, which can be considered an extension of the version usually applied as it solves some of the well-known drawbacks of the traditional method. The main conclusion is that substantial differences in the relative export performance between categories of developing countries can to a great extent be attributed to differences in the commodity composition of the exports and to the speed of adjustment of the exporting economy to changes in the pattern of international demand. They therefore should not be attributed wrongly to differences in competitiveness.

Notes

1. Presenting the aggregated results of γ and δ is pointless, because in these variants the exports of the sample comprise about 80 per cent of the exports of the reference group.
2. The author will supply all the disaggregated results on request.
3. This subgroup comprises the countries Turkey, Israel, Lebanon, Mexico, Brazil, India, Hong Kong, Taiwan, South Korea, Singapore and Malaysia.
4. This subgroup consists of the countries Algeria, Congo Brazzaville, Ecuador, Gabon, Iran, Iraq, Nigeria, Syria, Saudi Arabia, Tobago, Trinidad, Venezuela and Indonesia.

References

Balassa, B. (1978) 'Export Incentives and Export Performance in Developing Countries: A Comparative Analysis'. *Weltwirtschaftliches Archiv*, vol. 114, pp. 24–61.

Balassa, B (1979) 'Export Composition and Export Performance in the Industrial Countries 1953–71', *Review of Economics and Statistics*, vol. 61, pp. 604–7.

Chenery, H. B. (1980) 'Interactions between Industrialization and Exports', *American Economic Review*, Papers and and Proceedings of the American Economic Association, vol. 70, pp. 281–7.

Chenery, H. B. and Syrquin, M. (1975) *Patterns of Development 1950–1970* (London: Oxford University Press).

Jepma, C. J. (1986) *Extensions and Application Possibilities of the Constant Market Shares Analysis; the Case of the Developing Countries' Exports* (Gröningen: University Press).

Krueger, A. O. (1978) *Foreign Trade Regimes and Economic Development: Liberalization Attempts and Consequences* (New York: National Bureau for Economic Research).

Lewis, W. A. (1980) 'The Slowing Down of the Engine of Growth', *American Economic Review*, vol. 70, pp. 555–64.

Michaely, M. (1977) 'Exports and Growth; An Empirical Investigation', *Journal of Development Economics*, vol. 4, no. 1 (March) pp. 49–53.

Michalopoulos, C. and Jay, K. (1973) 'Growth of Exports and Income in the Developing World; a Neo-classical View', AID Discussion Paper no. 28, Washington, DC (November).

Richardson, J. D. (1970) *Constant-Market-Shares Analysis of Export Growth* (Michigan: unpublished PhD thesis, University of Michigan, USA).

Richardson, J. D. (1971a) 'Constant Market Shares Analysis of Export Growth', *Journal of International Economics*, vol. 1, pp. 227–39.

Richardson, J. D. (1971b) 'Some Sensitivity Tests for a "Constant Market-Shares" Analysis of Export Growth', *Review of Economics and Statistics*, vol. 53, pp. 300–4.

Riedel, J. (1983) 'Trade as the Engine of Growth in Developing Countries:

A Reappraisal', *World Bank Staff Working Papers*, no. 555, Washington, DC.

Rothschild, K. W. (1975) 'Export Structure, Export Flexibility and Competitiveness', *Weltwirtschaftliches Archiv*, vol. 111, pp. 222–42.

Tyszynski, H. (1951) 'World Trade in Manufactures Commodities, 1899–1950', *The Manchester School*, vol. 19, pp. 272–304.

7 Sources of Economic Growth and Structural Change: An International Comparison

Shujiro Urata
WASEDA UNIVERSITY, TOKYO AND WORLD BANK*

1. INTRODUCTION

Economic development and structural change result from the interaction of supply and demand factors. Supply factors include the accumulation and efficient use of factors of production such as labour and capital, whereas demand factors include changes in patterns of intermediate and final demand. Although economists realise the need for a simultaneous examination of both supply and demand factors in understanding the mechanisms of development and structural change, no studies have dealt with this issue satisfactorily. Instead, economists have tended to examine either supply or demand factors in isolation. In this paper we will follow the conventional approach, and analyse the effect of demand factors on economic growth and structural change.[1]

Following a methodology applied by Chenery (1960), we decompose the changes in output growth into changes in domestic demand, export demand, import substitution, and intermediate demand within an input-output framework. The approach is similar to growth accounting exercises that decompose observed changes into their sources, but without necessarily revealing the causal links. This methodology is very useful for identifying the effects of economic policies on economic growth and structural change, since the pattern of individual components of demand reflects economic policies.

In the analysis, five large economies, China, India, the Soviet Union, Indonesia, and Brazil, are added to the sample of nine relatively small economies, Japan, Mexico, Turkey, Korea, Yugoslavia, Colombia, Taiwan, Norway, and Israel, which have already been studied extensively under a World Bank research project.[2] One objective of the paper, therefore, is to examine if there are any discernible patterns specific to large economies, regarding the relationship between economic development and the changes in demand factors. Earlier studies such as Chenery and Syrquin (1975) have found that the size of the economy influences the development pattern of the economy significantly. Moreover, this paper attempts to relate the differences in development pattern among the sample economies to different trade strategies pursued in these economies.

The paper is organised as follows – section 2 describes the methodology. Section 3 discusses some characteristics of the economies under analysis, and section 4 discusses the results of an analysis of sources of output growth and structural change. Finally, section 5 presents some concluding remarks.

2. METHODOLOGY[3]

The principal feature of an analysis of demand-side sources of growth is that a material balance equation is used as a basis for the decomposition of output growth. Equation (1) describes the material balance equation:

$$X_i^1 = W_i^1 + D_i^1 + E_i^1 - M_i^1 \tag{1}$$

where:

X = domestic output;
W = intermediate goods consumption;
D = domestic final demand;
E = exports;
M = imports.

Subscript indicates sector: superscript, period.

Denoting the change in a variable by $\triangle$ ($\triangle X_i = X_i^2 - X_i^1$), we can describe the change in output of sector i between periods 1 and 2 by the changes in four demand components for the ith sector as equation (2):

$$\triangle X_i = \triangle W_i + \triangle D_i + \triangle E_i - \triangle M_i \tag{2}$$

Two points are worth noting about equation (2). First, it does not explicitly incorporate inter-industry commodity flows. Secondly, it assumes that domestic goods and imports are perfect substitutes. Although these assumptions facilicate interpretation of the equation, we shall modify them to make them more realistic.

To appreciate fully the workings of inter-industry transactions in the economy (the core of input-output analysis) we replace W_i in equation (1) by $\Sigma_j a_{ij} X_j$: a_{ij} is an input-output coefficient defined as X_{ij}/X_j where X_{ij} indicates a purchase from the ith sector to the jth sector. An assumption of perfect substitutability between domestic goods and imports may apply to some primary goods such as rice and fish, but it does not apply to most manufactured products. To incorporate an assumption of imperfect substitutability between domestic goods and imports, we drop imports (M_i) from equation (1), and introduce the domestic supply ratio (u_i) defined as $[(X_i - E_i)/(W_i + D_i)]$. The higher the domestic supply ratio, the lower the dependence on imports.

Incorporating these two assumptions, we rewrite equation (1) as equation (3):

$$X_i^1 = u_i^1 (D_i^1 + \Sigma_j a_{ij}^1 X_i^1) + E_i^1 \tag{3}$$

Use of matrix notation enables us to write the relationship for the entire economy as equation (4):

$$x = \hat{U} (d + Ax) + e \tag{4}$$

x, d, and e are vectors whose ith elements are X_i, D_i, and E_i. $\hat{U}$ is a diagonal matrix of u ratios and A is the matrix of input-output coefficients.

We can express output in terms of other variables as equation (5) after some manipulations:

$$x = (I - \hat{U}A)^{-1} (\hat{U}d + e) \tag{5}$$

where I is an identity matrix.

Following the procedure and notations used above, we derive equation (6), which decomposes changes in output into changes in domestic final demand, exports, import substitution, and input-output coefficients:

$$\begin{aligned} \Delta x = {} & R^1 \hat{U}^1 (\Delta d) && \text{domestic demand expansion} \\ & + R^1 (\Delta e) && \text{export demand expansion} \\ & + R^1 (\Delta U)(w^2 + d^2) && \text{import substitution} \\ & + R^1 \hat{U}^1 (\Delta A) x^2 && \text{change in input-output coefficients} \end{aligned} \quad (6)$$

$R^1 = (I - U^1A^1)^{-1}$ is a Leontief inverse and w is a vector whose ith element is W_i. We will use equation (6) in section 4 to decompose output growth into four different components.

Three observations are in order on equation (6). First, unlike equation (2), each term in equation (6) takes account of both direct and indirect linkages between sectors, as it is multiplied by the Leontief inverse. In equation (2) the change in output in the ith sector is associated with the change in demand for the ith sector alone, but in equation (6) it is associated with the change in demand not only for its own sector but also for all the other sectors.

Secondly, the effect of changes in input-output coefficients is measured by total coefficients (A) and does not distinguish between changes in domestic and import coefficients. Therefore, given that the total coefficient remains the same, any changes in technology that substitute imported for domestic goods would be captured in the import substitution element.

Finally, there is an index number problem as shown by the time superscripts. More specifically, the decomposition can be defined either using the second period structural coefficients and the first period volume weights, or by using the first period coefficients and the second period weights. In the absence of an ideal weighting scheme between the two versions, we shall compute the values for both versions and present simple averages of the two.

3. THE SAMPLE ECONOMIES

Table 7.1 presents comparative economic indicators for the benchmark years of the fourteen sample economies. These economies are shown according to population size, from large to small. The sample economies vary widely in population and level of income, as well as in economic structure. China had the largest population, 992.8 million in 1981, while Israel had the smallest population, 3.1 million in 1972. Except for Mexico and Turkey, the growth rate of population was constant or declining during the sample periods. Per capita GNP in the sample also varies widely from a low of 65 dollars (China in 1957) to a high of 2769 dollars (Norway in 1969) in 1970 prices. Not only the level of income but also the rate of change in the level of income varied widely among the sample. Japan (1955–70), Korea (1963–73), Brazil (1970–75) and Norway (1953–61) achieved a rate of growth of over 7 per cent per annum, whereas the growth rate in China (1957–65), India (1960–73), Korea (1955–63) and Colombia (1953–66) was below 2 per cent. What is remarkable is that China, Korea, and Taiwan experienced substantial acceleration in the growth rate during the sample periods; the acceleration was particularly rapid in Korea, from 1.6 per cent in 1955–63 to 7.7 per cent in 1963–70.

Demand and production structures of the economies are presented in the last five columns in Table 7.1. As for the demand structure, the shares of imports, exports, and investment in GDP are presented, while the shares of primary and manufacturing production in GDP are presented to describe the production structure.

The importance of foreign trade in economic activities varies among the economies. Three factors account for the differences: the size of the economy, trade strategy, and natural resource endowments. In Table 7.1 the importance of foreign trade appears negatively correlated with the size of the economy measured by the population; the larger the economy, the less important the foreign trade. This finding, which is consistent with the earlier studies, including Chenery and Syrquin (1975), might be explained by a hypothesis that in large economies large domestic demand provides producers with an opportunity to attain optimum level of production in a number of commodities. This results in a low dependence on foreign economies for their supply.

This relationship is reinforced by the import substitution development strategy, often pursued in large economies. Such trade strategy has been adopted in a number of economies in the early stages of development, in order to save foreign exchange. Import substitution strategy is particularly attractive to large economies, because domestic demand is believed to be sufficiently large to attain the optimum level of production as discussed above. In addition, in a standard international trade theory with some strict assumptions, large economies can be shown to gain economic welfare from restricting foreign trade through terms of trade improvement. The reversal of these arguments may be presented to explain the large share of foreign trade in GDP in small economies that followed open economic trade policies.[4]

The type of trade strategies followed in the sample economies might be characterised as follows. China (before 1978), India, USSR, Indonesia, Mexico, Turkey (1953–68), and Colombia followed import substitution strategy. However, China (since 1978) and Turkey (1968–73) somewhat liberalised their trade regime subsequently. Brazil, Japan, Korea, Yugoslavia, Taiwan and Israel followed export promotion strategies coupled with varying degrees of control over imports. Generally the degree of import control became less restrictive over time. Norway followed a very liberal trade policy throughout the period.

Besides the size of the economy and trade strategy, the importance of trade is also influenced by natural resource endowments. The structure of the Indonesian economy fits in with this explanation. Despite its large size and its strategy of import substitution, the share of trade in GDP is high–around 20-30 per cent. The high trade share is obviously attributable to abundant oil resources, as about 75 per cent of its exports were crude oil and oil-related products in 1980.

The share of foreign trade in GDP increased over time in most economies; its share has increased very rapidly in Korea and Taiwan since the mid-1960s and also in China in 1975–81. The notable exception is India, whose shares of both imports and exports declined throughout the period, reflecting continous import substitution strategy. Several factors, including demand and supply factors, might be responsible for the increasing role of foreign trade in economic activities in the course of development. A supply-side factor might be that specialisation in manufacturing production increases with economic development, especially in the production of commodities whose production process is subject to scale economies such as petrochemicals, as a result of accumulation of human and physical capital. As for the demand-side factor, consumers tend to increase their demand for foreign varieties of goods, as their incomes rise. Coupled with these two factors, the volume of foreign trade grew rapidly during the period after the Second World War, especially during the pre-oil-crisis period, thanks to the reduction in trade barriers brought about by successful multilateral trade negotiations. In addition, export promotion strategies resulted in a remarkably fast

Table 7.1 Comparative economic indicators

Country	Year	Population: Number of people (million)	Population: Average annual growth rate[1] (per cent)	Per capita GNP: Level (1970 US$)	Per capita GNP: Average annual growth rate[1] (per cent)	Shares of GDP (%): Imports	Shares of GDP (%): Exports	Shares of GDP (%): Investment (per cent)	Shares of GDP (%): Primary Value added	Shares of GDP (%): Manufacturing Value added
China	1957	665.0		65		3.7	3.9	23.2	40.6	n.a.
	1965	746.8	1.5	75	1.8	3.1	3.4	25.0	41.1	30.9
	1975	930.5	2.2	118	4.6	5.2	5.0	31.6	34.8	37.9
	1981	992.8	1.1	151	4.2	8.6	9.2	27.8	36.5	37.5
India	1960	434.8		94		8.3	5.3	17.2	50.8	14.1
	1968	522.6	2.3	102	0.7	5.7	4.8	16.6	49.2	13.6
	1973	586.2	2.3	108	1.1	5.4	4.8	19.3	50.7	14.1
USSR	1959	210.5		n.a.		5.5	3.7	23.9	30.0	23.2
	1966	233.2	1.5	n.a.	n.a.	6.1	3.7	26.2	25.9	25.4
	1972	247.5	1.0	n.a.	n.a.	10.2	5.7	29.1	21.2	28.0
Indonesia	1971	118.6		157		17.0	14.4	15.8	52.8	8.4
	1975	130.2	2.4	190	4.9	22.2	22.9	20.3	51.3	8.9
	1980	146.3	2.4	243	5.0	22.2	30.5	20.9	50.5	11.7
Brazil	1970	95.8		580		7.4	7.0	25.5	13.1	28.2
	1975	108.0	2.4	839	7.7	11.4	7.4	32.1	12.7	29.5
Japan	1955	89.0		500		10.1	10.7	19.8	25.0	22.0
	1960	94.1	1.1	753	8.5	10.5	11.0	30.2	14.5	33.9
	1965	98.9	1.0	1 159	9.0	9.4	10.8	30.6	9.2	31.8
	1970	104.3	1.1	1 897	10.4	9.9	11.2	35.1	6.9	35.9
Mexico	1950	26.3		380		13.9	14.1	13.5	23.1	24.4
	1960	36.0	3.2	479	2.9	12.6	10.3	19.7	20.8	19.2
	1970	50.4	3.4	670	3.4	10.2	8.0	22.4	14.7	23.7

Turkey	1953	22.8		239		n.a.[2]	n.a.	12.8	49.6	10.9
	1963	29.7	2.7	319	2.9	9.9	5.5	14.5	40.8	14.1
	1968	33.5	2.4	377	3.3	7.3	5.3	17.2	32.7	17.5
	1973	38.3	2.7	461	4.1	10.0	7.3	17.3	28.9	17.0
Korea	1955	21.6		131		10.1	1.7	10.1	45.5	11.6
	1963	27.0	2.8	149	1.6	16.3	4.9	13.9	45.1	14.7
	1970	31.4	2.2	250	7.7	24.8	14.7	25.1	27.9	20.9
	1973	32.9	1.6	323	8.9	35.5	32.2	23.9	25.7	24.7
Yugoslavia	1962	18.8		469		17.2	16.1	30.9	30.7	25.1
	1966	19.6	1.0	581	5.5	20.5	19.5	25.5	30.6	21.7
	1972	20.8	1.0	781	5.1	23.1	21.0	26.3	24.3	26.1
Colombia	1953	12.5		274		14.6	15.6	16.8	39.3	14.6
	1966	18.4	3.0	330	1.4	15.3	12.3	17.0	31.6	18.0
	1970	20.6	2.9	369	2.8	16.3	14.6	20.9	29.7	17.3
Taiwan	1956	9.2		203		11.9	6.8	13.3	29.8	19.7
	1961	11.0	3.6	231	2.6	19.8	12.8	16.1	29.6	21.8
	1966	12.8	3.1	305	5.7	21.4	20.7	19.0	24.6	25.9
	1971	14.8	2.9	426	6.9	32.6	35.1	22.9	14.5	35.9
Norway	1953	3.4		1 171		42.5	38.1	30.2	15.4	27.5
	1961	3.6	0.9	2 028	7.1	43.1	40.2	29.9	9.7	22.2
	1969	3.9	0.8	2 769	4.0	38.9	41.6	24.7	6.6	21.7
Israel	1958	2.0		1 067		27.4	11.6	26.8	13.5[3]	22.3[4]
	1965	2.6	3.6	1 587	5.8	32.5	19.1	27.1	9.3	24.0
	1972	3.1	2.3	2 317	6.5	38.7	27.2	28.8	6.5	21.3

Note: 1. Growth rates refer to the period from the previous bench-mark years.
2. n.a. indicates no data available.
3. Only agriculture.
4. Manufacturing and mining.

Sources: Yearbook of National Accounts Statistics UN, various issues; Chenery, Robinson and Syrquin (1986); Central Intelligence Agency (1982); and Patterns of Development Research Project Data Bank, the World Bank.

increase in the export share in Korea and Taiwan after the mid-1960s, while rigorous open economic policies resulted in a greater role of foreign trade in China after 1978.

Turning to the share of investment in GDP, one finds that it generally rose over time, confirming the earlier findings.[5] Its increase was particularly fast in Japan, Korea and Taiwan. A positive correlation between investment share and level of income might be explained by various hypotheses. For example, according to the Keynsian hypothesis on the propensity to save, the rate of savings, which is the source of investment, increases with the increase in the level of income, as the marginal propensity to save out of increased income is greater than the average propensity to save.[6]

Production structure shifted from primary to manufacturing in all the economies except India and Norway.[7] A shift in the production structure from primary to manufacturing results from the interaction of demand and supply factors. Demand for manufactured goods increases more than that for primary goods as incomes rise, while the production of manufactured goods, which generally uses capital intensive production techniques compared to primary production, tends to expand faster than that of primary goods, as the rate of capital accumulation is likely to be faster than the rate of increase in labour inputs in the process of development. In the next section we will examine quantitatively the effect of demand factors on output growth and structural change in detail.

4. SOURCES OF ECONOMIC GROWTH AND STRUCTURAL CHANGE

Table 7.2 presents the results of computing the demand sources of output growth over the period covering two end years for the sample economies.[8] The last four columns in the table show the percentage contribution to changes in output of variations in domestic demand, exports, import substitution, and input-output coefficients; the figures across each row therefore sum to 100 per cent. Expressing the magnitude of each contribution in percentage shares enables us to compare the results across sectors and across economies.

Table 7.2 (column 1) also shows the growth rate of output for the primary, manufacturing, and services sectors as well as for the entire economy. The growth rates in the table indicate that all fourteen economies experienced significant structural changes in production

Table 7.2 Sources of growth for sample countries

Production category by economy	Average output growth rate	Sources of output growth: Domestic demand expansion	Export expansion	Import substitution	Changes in input-output coefficient
China (1957–81)					
Primary	3.9	86.8	9.9	-2.8	6.2
Manufacturing	8.2	79.8	10.9	-3.8	13.0
Services	5.6	104.0	7.9	-1.0	-10.9
Total	6.0	85.9	10.1	-3.0	7.0
India (1959–74)					
Primary	1.9	108.9	1.5	3.8	-14.3
Manufacturing	6.6	62.3	8.2	15.1	14.4
Services	5.1	95.2	4.9	-3.4	-3.4
Total	4.1	85.0	5.5	8.1	1.4
USSR (1959–72)					
Primary	2.8	141.4	11.3	-8.7	-43.9
Manufacturing	7.6	86.7	7.4	-5.8	11.7
Services	5.5	99.9	3.6	-2.0	-1.5
Total	6.0	95.1	7.1	-5.4	3.2
Indonesia (1971–80)					
Primary	5.0	67.6	65.2	-11.7	-21.1
Manufacturing	12.3	76.4	9.8	-2.5	16.4
Services	7.8	95.0	7.4	-1.1	-1.2
Total	7.8	82.4	21.4	-4.0	0.2

continued on page 154

Table 7.2 continued

Production category by economy	Average output growth rate	Sources of output growth: Domestic demand expansion	Export expansion	Import substitution	Changes in input-output coefficient
Brazil (1970–75)					
Primary	7.2	116.9	21.5	-2.9	-35.6
Manufacturing	13.0	84.9	8.8	-2.5	8.8
Services	9.6	99.5	3.2	-0.4	-2.3
Total	11.0	92.2	7.7	-1.8	1.9
Japan (1955–70)					
Primary	2.2	288.3	30.7	-93.4	-125.6
Manufacturing	13.3	76.3	17.7	-1.4	7.5
Services	11.4	91.2	7.5	-2.0	3.3
Total	11.5	85.4	13.9	-3.1	3.8
Mexico (1950–75					
Primary	4.8	100.9	5.5	-2.1	-4.3
Manufacturing	7.7	81.9	5.1	8.0	5.0
Services	6.4	96.5	1.6	1.0	1.0
Total	6.5	90.9	3.6	3.5	2.0
Turkey (1953–73)					
Primary	2.5	126.6	10.4	1.5	-38.5
Manufacturing	8.0	77.0	6.7	4.6	11.7
Services	6.7	87.1	6.4	0.4	6.1
Total	5.9	87.3	7.0	2.4	3.4
Korea (1955–73)					
Primary	5.7	111.2	27.5	-15.5	-23.2
Manufacturing	15.8	53.1	44.5	2.4	-0.04
Services	10.3	82.3	19.8	0.5	-2.6
Total	[illegible]	[illegible]	[illegible]	[illegible]	[illegible]

Yugoslavia (1962–72)					
Primary	2.6	162.0	63.7	-51.7	-74.0
Manufacturing	12.1	72.8	32.5	-15.4	10.1
Services	8.8	89.1	14.0	-3.6	0.5
Total	8.7	84.4	27.6	-13.2	1.3
Colombia (1953–70)					
Primary	4.5	57.0	40.6	2.2	0.3
Manufacturing	8.1	65.4	6.1	16.1	12.4
Services	5.5	91.2	6.4	0.8	1.6
Total	5.9	73.9	13.9	6.8	5.3
Taiwan (1956–71)					
Primary	7.1	85.6	51.7	-19.4	-18.0
Manufacturing	16.2	38.2	51.7	5.1	5.1
Services	9.7	79.3	23.7	0.5	-3.5
Total	12.0	55.3	43.4	1.2	0.2
Norway (1953–69)					
Primary	2.5	79.2	50.5	-35.8	6.1
Manufacturing	5.2	56.4	49.8	-18.0	11.8
Services	4.8	62.0	41.8	-5.2	1.3
Total	4.7	60.4	45.7	-12.3	6.1
Israel (1958–72)					
Primary	6.4	46.1	66.4	-5.5	-7.0
Manufacturing	12.5	69.3	41.9	-19.9	-8.8
Services	8.9	77.1	27.1	-3.1	-1.1
Total	9.9	72.0	35.7	-10.5	2.8

Source: World Bank Research Projects, Sources of Industrial Growth and Structural Change and Collaborative Research with China.

from primary to manufacturing. The shift was particularly rapid in Japan, Korea, Yugoslavia, and Taiwan, while it was very slow in Mexico, Colombia, and Norway. Other economies fall somewhere in between.[9]

The structural shift in production from primary to manufacturing is mainly caused by the corresponding shift in final and intermediate demand, as can be seen from the sign and magnitude of the contribution from domestic demand expansion and from changes in input-output coefficients for the primary and manufacturing sectors in Table 7.2. For the purpose of final consumption, demand for primary goods increases slower than that for manufactured goods, because the income elasticity of demand for primary goods is generally lower than that for manufactured goods. For the purpose of intermediate consumption, primary goods tend to be substituted by manufactured goods, as the relative price of manufactured goods to primary goods declines in the development process. This phenomenon may be seen by the negative contribution associated with changes in input-output coefficients for primary production, and by the positive contribution associated with the changes in input-output coefficients for manufacturing production.[10]

Domestic demand expansion contributed most to the increase in primary production in a large number of economies. A few exceptions are Indonesia, Colombia, and Israel, where the export demand expansion contributed substantially, as these economies were relatively well endowed with natural resources compared to their trading partners.

The sources of output growth in manufacturing reveal an interesting variation among the sample economies, particularly with respect to the size of the economy and with respect to the trade strategy followed in the economy. These points may be seen in Figures 7.1a to 7.1d, where percentage contributions of the four elements are plotted against per capita GNP. The contribution for sub-periods are shown for all the economies except for the USSR, whose per capita GNP in US dollars is not available.[11] The economies are numbered according to the population size, from large to small. In addition to these economies, the figures for two archetypes, which are obtained from the cross-country model developed by Chenery and Syrquin, are plotted.[12] Two archetypes shown here are large economies (L), defined as those with populations greater than 15 million in 1960 (or 20 million in 1970), and small, industry-oriented economies (SM), defined as those relatively specialised in the production and export of manufactured goods with populations less than 15 million in 1960 (or 20 million in 1970). In our sample, Taiwan, Norway, and Israel belong to the 'SM' group, while the rest belong to the 'L' group. In the figures unbroken lines indicate the relationship between the size of the contribution and the range of per capita GNP under study, whereas broken lines connect different sub-periods. For example, the contribution of domestic demand expansion in China between the income levels of 65 dollars and 75 dollars which, respectively, correspond to the years 1957 and 1965, was 88 per cent.

The magnitude of the contribution from domestic demand expansion appears to be positively correlated with the size, as the plotted lines for large economies are generally above those for small economies (Figure 7.1a). This finding is consistent with the patterns from the cross-country model, and also tends to support the hypothesis that, in large economies, large domestic demand provides producers with an opportunity to attain an optimum level of production. With the notable exceptions of Korea and Taiwan, the contributions of domestic demand expansion in the sample economies are greater than the values predicted from the cross-country model. The size of these deviations for individual economies is attributable mainly to the trade strategies followed in the economy, as will be discussed later. The contribution of domestic demand expansion appears to decline with an increase in per capita GNP.

One notices a remarkable increase in the contribution of domestic demand expansion from 1960–68 to 1968–73 in India. In contrast, the contribution of the changes in input-output coefficients in India declined significantly between the two periods (Figure 7.1d). The exceptionally large contribution from the changes in input-output coefficients during 1960–68, indicating a deepening of inter-industry linkages, seems to be attributable to a rigorous import substitution strategy, as it caused the producers to expand their production by increasing the use of intermediate inputs without paying attention to efficiency behind import protection.

Turning to the contribution of export expansion in Figure 7.1b, one finds that, in general, the magnitude of the contribution is negatively correlated with the size of the economy, reflecting that small economies rely more heavily on exports to make up for the limited size of the domestic market. A comparison with the patterns predicted for the archetypes reveals that the contributions in Korea, Yugoslavia, Taiwan, Norway, and Israel, all of which followed either export promotion strategies or liberal trade policies, are greater than

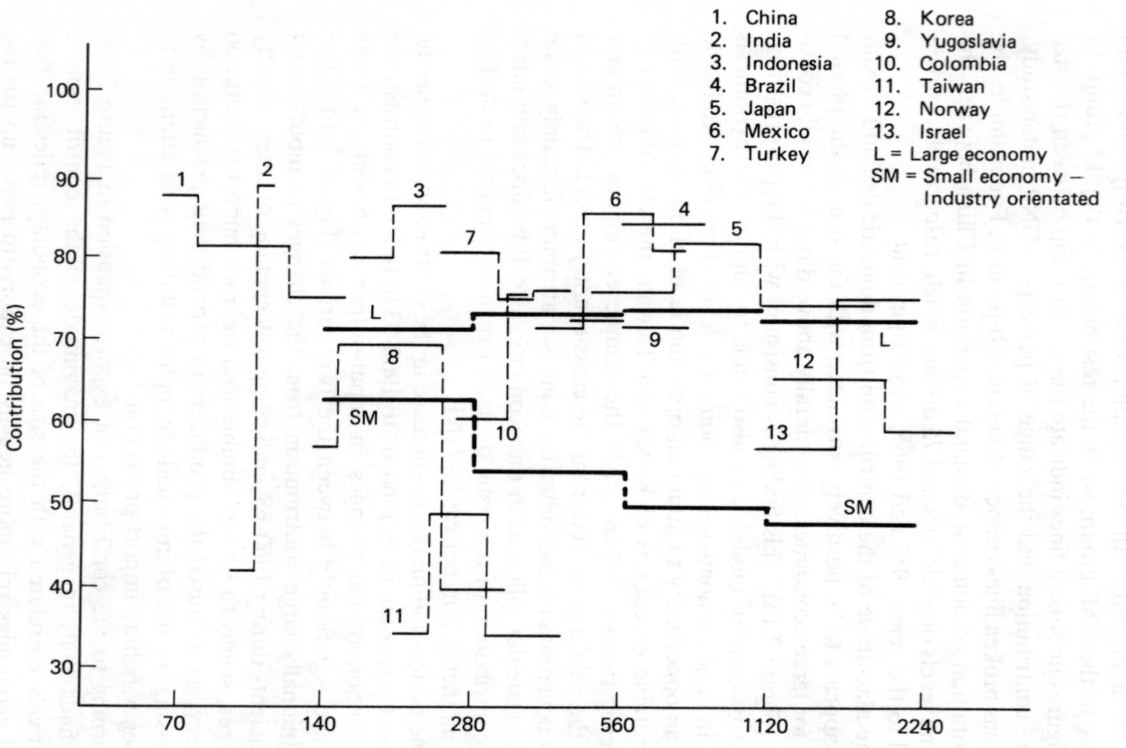

Figure 7.1a Domestic demand expansion

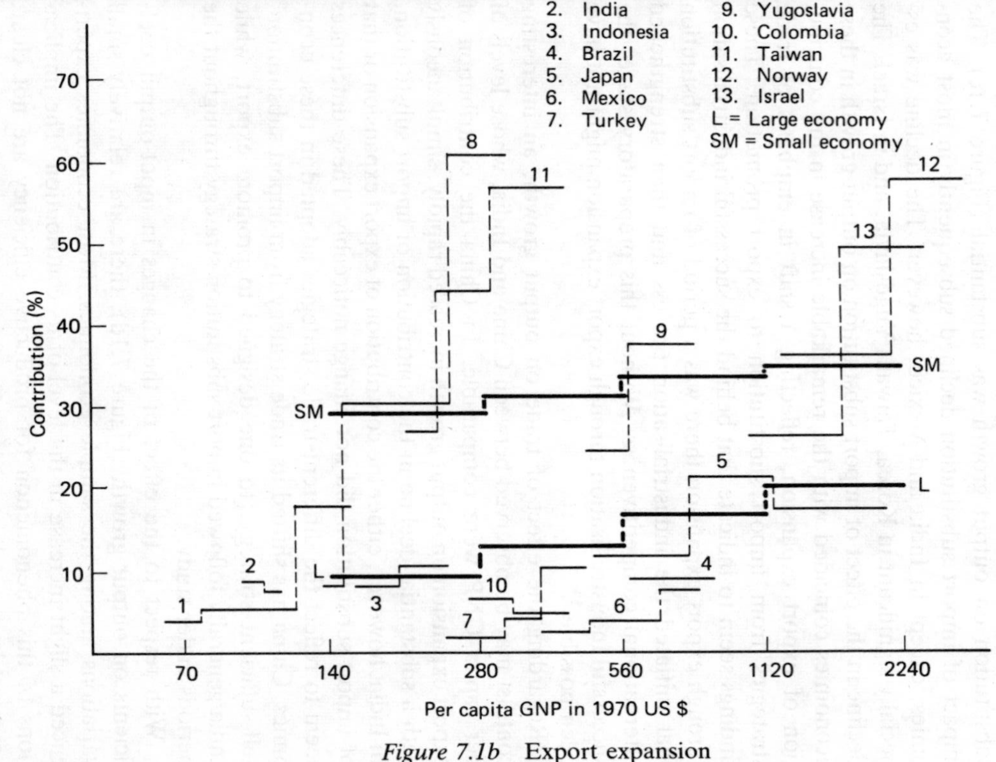

Figure 7.1b Export expansion

the predicted levels, while the contributions in Brazil, Mexico, Turkey, and Colombia, all of which followed import substitution strategies, are smaller than the predicted levels. Despite these differences in the size among the sample economies, the size of contribution of export expansion generally increased with per capita GNP.

In most economies there was a period when the effect of import substitution on output growth was substantial (Figure 7.1c). The impact of import substitution declined subsequently in most economies, except in India and Mexico, however. The decline was especially significant in Korea, Taiwan, Yugoslavia, and in Israel. The decline in the effect of import substitution on output growth in these economies coincided with the remarkable increase in the contribution of export expansion, reflecting a shift in emphasis of trade strategies from import substitution to export promotion. These findings seem to indicate that behind the successful industrialisation through export expansion there was a period of import substitution that initiated the industrialisation process and then strengthened international competitiveness. Without this preparatory stage, the successful industrialisation through export expansion might not have been possible.[13]

Regarding the effect of trade on output growth, an interesting contrast may be observed between China and India, whose levels of per capita GNP were comparable. In China the contribution of export expansion on output growth increased rapidly, simultaneously with a substantial decline in the contribution of import substitution. In India, however, either the contribution of export expansion or that of import substitution did not change noticeably. These differences seem to reflect the different trade strategies adopted in these economies. China has shifted its trade strategy from import substitution, self-sufficient strategy, to one designed to promote export, while India generally followed import substitution strategy throughout the periods under study.

With respect to the effect of the changes in input-output coefficients on output growth (Figure 7.1d), there are relatively small variations among the sample economies. Most economies experienced a slight increase in the level of its contribution. The implications of this phenomenon for productive efficiency are not clear, however; it may indicate rising efficiency due to extensive specialisation, or falling efficiency due to excessive use of intermediate inputs. In order to shed some light on efficiency issues, the sources of output

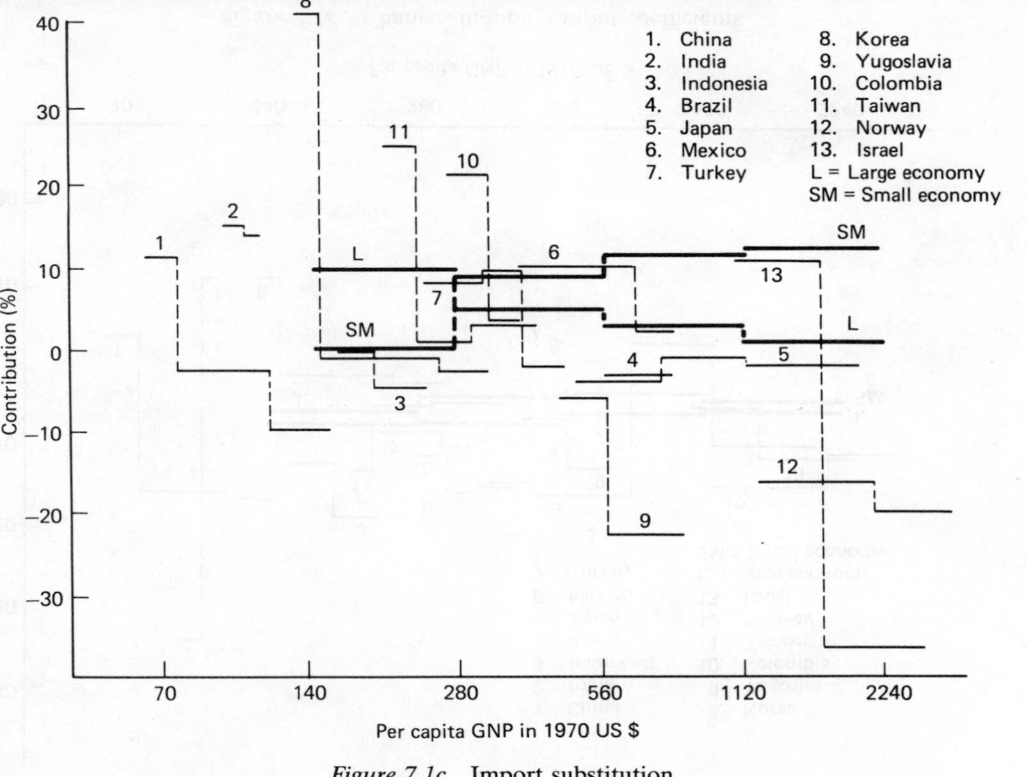

Figure 7.1c Import substitution

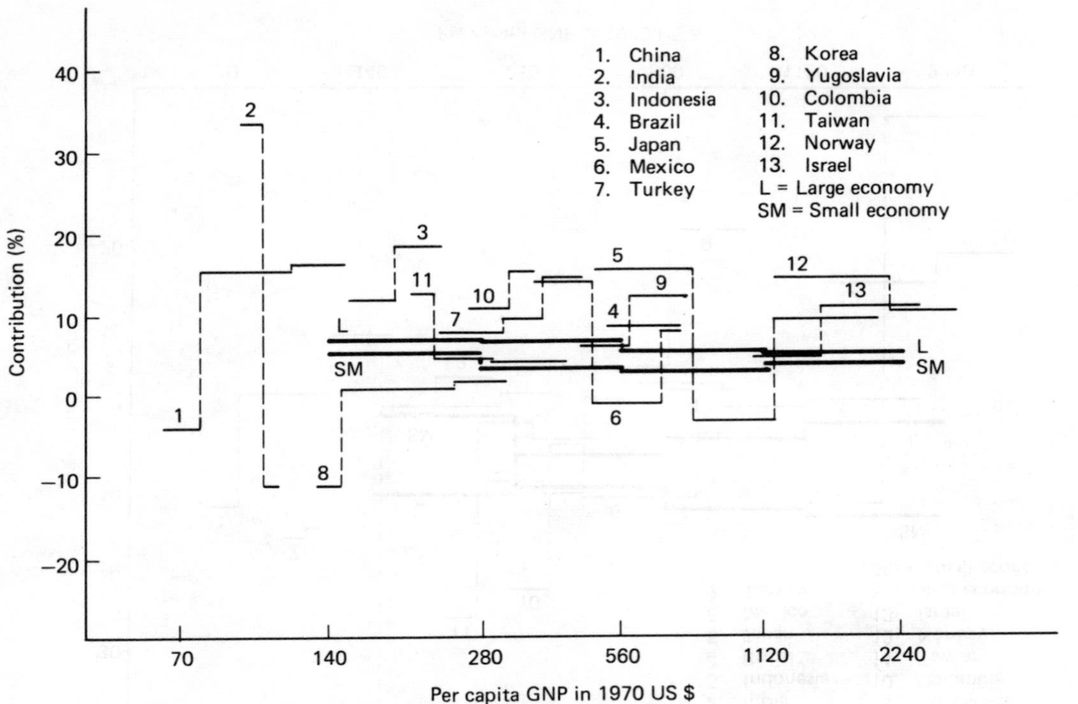

Figure 7.1d Changes in input-output coefficients

growth should be examined by considering supply factors such as labour, capital and intermediate inputs.

5. CONCLUSIONS

The analysis of demand side sources of growth used in this paper has proved to be useful in identifying various factors affecting growth and structural changes in an economy. It has also proved to be useful in relating those changes to the differences in the size of the economies and the trade strategies followed in different economies. Identification of special characteristics of individual economies regarding their size and their trade strategies was facilitated by incorporating 'average patterns' predicted from the cross-country model for comparison.

In this paper five large economies were added to the sample of nine economies, which had been analysed extensively under an earlier World Bank research project. We found here that the major impetus of output growth came from domestic demand expansion in large economies, which tended to follow import substitution strategies, as well as in those small economies that adopted import substitution strategies. In contrast, export expansion contributed significantly in small economies that adopted export promotion strategies. These findings are consistent with the earlier results.

The results obtained from the demand-side sources of growth analysis would be nicely complemented by a supply-side analysis to gain deeper understanding of the development process. To this end, a few studies have examined the effect of different trade strategies on total factor productivity growth.[14] According to these studies, the growth rate of total factor productivity is positively associated with export expansion, while it is negatively associated with import substitution. Various hypotheses, which might explain the factors responsible for these associations, have been presented. For example, expansion of exporting activities would enable producers to benefit from scale economies, or a lack of competition caused by import substitution strategies would lead to inefficient use of resources. These hypotheses have not yet been tested rigorously.

Moreover, several attempts have been made to examine the effect of both demand and supply factors on output growth and structural change simultaneously by using simulation models such as a dynamic input-output model or a computable general equilibrium model.[15] Although they have proved useful, the effectiveness of such applications in disentangling the development process seems to be limited by a lack of knowledge of the mechanisms through which such activities as production and accumulation are effected. Examples here would include those relationships presented above regarding export and scale economies, and competition and efficiency. Further studies on these and related subjects would increase our knowledge of the development process.

Notes

* This chapter reports some of the results from the World Bank Research Project on Collaborative Research with China (RPO 673–14). The author thanks participants at the Congress and Moshe Syrquin for helpful comments. The World Bank does not accept responsibility for the views expressed herein which are those of the author and should not be attributed to the World Bank or to its affiliated organisations. The findings, interpretations, and conclusions are the results of research project supported by the Bank; they do not necessarily represent official policy of the Bank. The designations employed, the presentation of material, and any maps used in this document, are solely for the convenience of the reader and do not imply the expression of any opinion whatsoever on the part of the World Bank or its affiliates concerning the legal status of any country, territory, city, area, or of its authorities, or concerning the delimitation of its boundaries, or national affiliation.

1. For a supply-side analysis, see, for example, Denison (1974).
2. For an analysis of the nine relatively small economies, Kubo and Robinson (1984), and Chenery, Robinson and Syrquin (1986) present and discuss the results from a World Bank project on sources of growth and structural change. For studies on large economies, see Ahluwalia (1985) on India, Urata (1988) on USSR, and Urata (1987) on China.
3. The original development of this analysis is due to Chenery (1960) and Chenery, Shishido and Watanabe (1962). The methodology applied in this paper is identical to the one applied in Dervis, de Melo, and Robinson (1982).
4. See Chenery (1979) for similar arguments.
5. See, for example, Chenery and Syrquin (1975), which found that the share of investment in GDP increased with the level of income from their study of about a hundred economies over the period of 1950–70.
6. For more discussions on this point, see Chenery and Syrquin (1975).
7. Production structure in India did not change much between 1960 and 1973, while in Norway both primary and manufacturing sectors lost their shares in GDP. A closer look at the developments in India reveals that the share of manufacturing in GDP increased sharply between 1960 and 1965, and then declined substantially afterwards before rising slowly in

1967. Several factors are responsible for the decline in the manufacturing share in the mid-1960s; a decline in public investment and a limited scope for further import substitution appear to have been important. Norway was already in an advanced stage in economic development where the share of the services sector started increasing at the cost of primary and manufacturing sectors.

8. The computation was conducted at about twenty disaggregated sectors for sub-periods, and then they were aggregated.
9. The degree of the rate of change in production structure differs depending on the indicator selected for the analysis. The difference may be clearly seen for India. As shown in Table 7.2, production in India shifted quite rapidly from primary to manufacturing in terms of output, while the production structure did not change much in terms of value added (Table 7.1). These differences are due to the increased importance of intermediate inputs in production as will be shown in the discussion of the results of the sources of growth and structural change below.
10. Deutsch and Syrquin (1986) also found a positive association between the level of income and the share of intermediate inputs in total inputs in manufacturing in their regression analysis of eighty-three input-output tables covering thirty economies over 1950–75.
11. The pattern of the changes in the sources of growth for the Soviet economy is similar to that observed for China, although the magnitude of the changes in most components are smaller in the Soviet economy.
12. See Chenery (1979), and Chenery and Syrquin (1980) for a detailed description of the model.
13. See Kubo and Robinson (1984) on these observations.
14. Nishimizu and Robinson (1984) examined the relationship between the growth rate of total factor productivity and the contribution of trade factors (i.e., export expansion and import substitution) on output growth for Japan, Korea, Turkey, and Yugoslavia, while Goldar (1986) attempted a similar analysis for India.
15. See Kubo, Robinson, and Urata (1986), and Chenery, Robinson, and Syrquin (1986) for application of a dynamic input-output model and a computable general equilibrium model, respectively.

References

Ahluwalia, I. J. (1985) 'Economic Growth and Structural Change in the Indian Economy 1959–60 to 1973–74', mimeo, the World Bank.

Central Intelligence Agency (1982) *USSR: Measures of Economic Growth and Development, 1950–80*, prepared for the use of the Joint Economic Committee, Congress of the United States, (Washington DC: US Government Printing Office).

Chenery, H.B. (1960) 'Patterns of Industrial Growth', *American Economic Review*, vol. 50, pp. 624–54.

Chenery, H.B. (1979) *Structural Change and Development Policy* (New York: Oxford University Press).

Chenery, H.B., Robinson, S. and Syrquin, M. (1986) *Industrialization and Growth* (New York: Oxford University Press).

Chenery, H.B., Shishido, S. and Watanabe, T. (1962) 'The Pattern of Japanese Growth, 1914–1954', *Econometrica*, vol. 30, pp. 98–139.

Chenery, H.B. and Syrquin, M. (1975) *Patterns of Development, 1950–1970*, (published for the World Bank by Oxford University Press).

Chenery, H.B. and Syrquin, M. (1980) 'A Comparative Analysis of Industrial Growth', in R.C.O. Matthews (ed.) *Economic Growth and Resources*, vol. 2, *Trends and Factors*., IEA (London: Macmillan; New York: St Martins Press).

Dervis, K., Melo, J. de and Robinson, S. (1982) *General Equilibrium Models for Development Policy* (Cambridge: Cambridge University Press).

Denison, E.F. (1974) *Accounting for United States Economic Growth 1929–1969* (Washington, DC: The Brooking Institution).

Deutsch, J. and Syrquin, M. (1986), 'Economic Development and the Structure of Production', mimeo, Bar-Ilan University, Israel.

Goldar, B. (1986) 'Import Substitution, Industrial Concentration and Productivity Growth in Indian Manufacturing', *Oxford Bulletin of Economics and Statistics*, vol. 48, no.2, pp. 143–64.

Kubo, Y. and Robinson, S. (1984) 'Sources of Industrial Growth and Structural Change: a Comparative Analysis of Eight Economies', in *Proceedings of the Seventh International Conference on Input-Output Techniques* (New York: United Nations).

Kubo, Y., Robinson, S. and Urata, S. (1986) 'The Impact of Alternative Development Strategies: Simulations with a Dynamic Input-Output Model', *Journal of Policy Modeling*, vol. 8, no. 4, pp. 503–29.

Nishimizu, M. and Robinson, S. (1984) 'Trade Policies and Productivity Change in Semi-Industrialized Countries', *Journal of Development Development Economics* vol. 16, pp. 177–206.

Urata, S. (1987) 'Sources of Economic Growth and Structural Change in China: 1956–81' *Journal of Comparative Economics*, vol. 11, no. 1, pp. 96–115.

Urata, S. (1988) 'Sources of Economic Growth and Structural Change in the Soviet Union' in Ciaschini, M. (ed) *Input-Output Analysis: Current Developments* (London: Chapman and Hall).

8 Korean Incentive Policies towards Industry and Agriculture*

Danny M. Leipziger
THE WORLD BANK

and

Peter A. Petri
BRANDEIS UNIVERSITY

In just three decades, South Korea has changed from an agrarian economy into a manufacturing powerhouse, capable of competing with Japan and the USA in high technology products and with Europe and Japan in the manufacture of ships and automobiles. It has done so in a way that has spread the benefits of growth widely, across both urban and rural households. From nearly 50 per cent of GDP in the early 1950s, agriculture's share fell to 14 per cent in 1984. Given the remarkable performance of Korean industry, it is easy to take for granted the rapidity and smoothness of Korea's sectoral transition. In many countries much smaller shifts in economic structure have created severe intersectoral frictions. This paper explores the circumstances and policies that made Korea's sectoral transition relatively painless, and concludes with an analysis of the current issues that confront Korea's intersectoral incentive policy.

1. KOREA'S ECONOMIC STRUCTURE

At the end of the Korean War in 1953, agriculture occupied two-thirds of Korea's workers and produced nearly half of its output. Its predominant role has several important roots. Most obviously, per capita incomes were low enough to ensure that much of domestic income was spent on staple foods. On top of an agriculturally-oriented domestic economy, a substantial agricultural export sector had been developed under the occupation in order to produce grain for Japan. In the later stages of the occupation, the Japanese also established a significant manufacturing base, but the disruption of capacity at the end of the war scaled back the manufacturing activity to a much greater extent than agriculture. Finally, since much of Korea's manufacturing capacity was located in the North, while rice farming was primarily practised in the South, the partition of Korea left the South with an even more agriculturally-oriented economy than was the case for the peninsula as a whole.

Yet agriculture is not favoured by Korea's economic endowments. Arable land per capita, at about 0.7 hectares per person, is just slightly above that of Japan, and is less than one-third that of China, India, and Thailand (Mason *et al.*, 1980). Indeed, Korea had managed to feed itself and to supply Japanese markets by practicing highly labour-intensive cultivation methods and reaching much higher yield levels than achieved by virtually all other producers except Japan. The prospects for bringing further land under cultivation were poor and, as food demand began to rise with the growth of the industrial economy in the 1950s and 1960s, Korea became an importer of about as much grain as had been exported, on average, under the Japanese.

In contrast to its poor land resources, Korea's literate and entrepreneurial labour force represented a very attractive endowment for industrial development. Rapid industrial growth began when policies designed to accelerate production and exports, to be described in further detail below, were put in place in the early 1960s. A comparison of Korea's sectoral development pattern with the international cross-sectional patterns calculated by Chenery and Syrquin (1975) confirms the unusually prominent role of agriculture in the post-war period. It also suggests that Korea's industrial sector experienced an unusually rapid and significant expansion in the mid-1960s. Specifically, the share of agriculture in GNP was approximately 45 per cent in the mid-1950s – half again as large as the 30 per cent share predicted by the cross-section model. The difference between actual and predicted agricultural shares narrowed to approximately 5 per cent in the early 1960s and remains at about 2 per cent today, still significantly larger than would be predicted on the basis of institutional norms. By contrast, the GDP share of manufacturing was around 15 per cent in the mid-1950s – about 5 per cent below its predicted rate. During the 1960s, the share of manufacturing crossed the cross-sectional prediction, and is now about 3 per cent above it (Urata, 1986; Chenery, Syrquin and Urata, 1986).

With continued rapid growth during the 1960s, the differences between industrial and agricultural performance widened and, as described in section 4, the government responded to these developments by abandoning its neutral approach to agriculture. However, in beginning to redress the gap between rural and urban incomes, it began a new trend of increasing effective protection of agriculture relative to other economic activities – a situation which persists today.

Despite the shift in intersectoral incentive policy in the early 1970s, the growth of industry continued to outpace agriculture by a very great margin. Korea's industrial growth rate was 6.3 times as high as its agricultural growth rate during the 1965–84 period – in contrast to the average ratio of 2.0 for other countries at a similar per capita income level (World Bank, 1986). The implications of this massive intersectoral shift can be judged also by its impact on migration. In 1950, only 18 per cent of the Korean population lived in urban areas – approximately the same as the developing country average of 17 per cent. By 1975, however, 51 per cent of the population was in cities, nearly twice the developing country average for that year, and not far below the average for developed countries.

The economy's transition from agriculture to industry proceeded particularly rapidly during the 1970s as Korea's dependence on agriculture declined both in terms of employment and value added (see Figure 8.1). By 1983 agriculture provided only 14 per cent of value added and 29 per cent of employment compared with 25 and 50 per cent, respectively, a decade before. Among the manufacturing subsectors, which in the aggregate now account for 30 per cent of value added and 22 per cent of employment, the largest gainers over the period were heavy industries, including metal products, machinery, primary metals, and chemicals. Light industries' performance has by comparison been sluggish. This trend reflects extensive intervention in industry in the 1970s, as well as the subsequent success of new exports like autos, ships, and electrical equipment.

The trend toward sophisticated manufactures is expected to continue: Korean estimates are that electronics and electrical equipment, now 11 per cent of manufacturing output, will grow to 18 per cent by 1996. The government is likely to support these trends, although its interventions will tend to focus on functional incentives to upgrade Korea's technological base. Research and development expenditures, which were just 0.9 per cent of GNP in 1980, are expected to reach 2 per cent by 1986, very close to the R&D

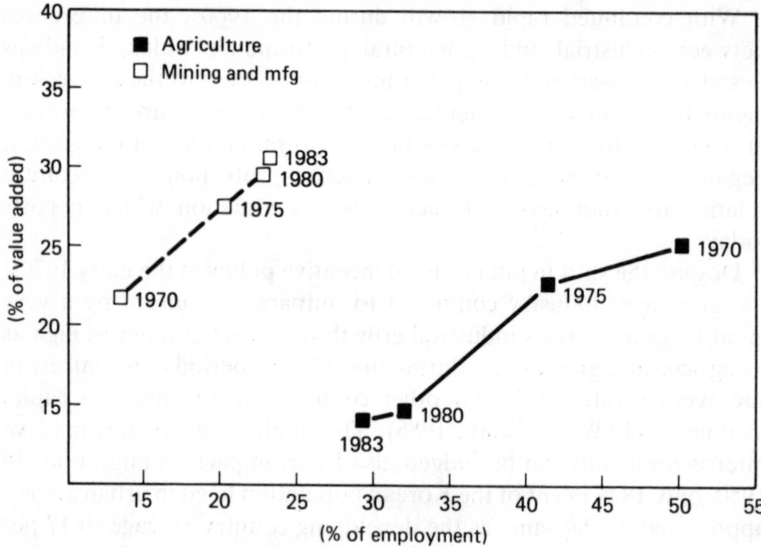

Sources: Based on data from Bank of Korea.

Figure 8.1 Korea value added and employment shares (1970, 1975, 1980, 1983)

proportions of the USA and Japan. At the same time, impressive gains have also been made in agriculture. The index of food production has risen by almost 140 per cent between 1960 and the mid-1980s, with the largest gains registered in rice, soybeans, corn, and livestock. Korean agriculture is now one of the most productive in the world, in terms of output per hectare cultivated, with rice yields second only to Japan. Fertiliser use is very high (second once again only to Japan) and there is considerable mechanisation in land preparation and threshing (but little in transplanting and harvesting). Rice still accounts for 44 per cent of area planted (compared to 39 per cent in 1961), while fruits, vegetables, and industrial crops (e.g., oilseeds, tobacco, and cotton) have combined to replace barley and wheat as the second most important set of crops. Its technical gains notwithstanding, Korean agriculture will continue to recede in importance as both a source of output and employment. Projections noted in Table 8.1 indicate that agriculture's contribution to GDP will be one-third lower by 1996, with its share taken over by manufacturing. In terms of employment, the projections foresee a drop from the current 25 to 14 per cent over the coming decade.

Table 8.1 Projected structure of production (percentages)

	Korea						Japan			
	GDP			Employment			GDP		Employment	
	1985	1991	1996	1985	1991	1996	1966	1982	1966	1982
Agriculture	14.0	11.5	9.3	27.1	18.0	13.6	8.8	3.5	24.2	8.9
Mining and manufacturing	30.4	34.1	37.2	24.2	25.0	26.2	33.5	30.3	25.2	24.7
Social overhead, construction, and services	55.6	54.4	53.5	48.7	57.0	60.2	57.7	66.2	50.6	66.4

Source: Korea Development Institute, Bank of Japan

Useful insights into the intersectoral transition can be gained by examining the relative movements of labour productivity in agriculture and manufacturing (see Table 8.2). Ideally, we could use differences between the marginal productivities of labour in industry and agriculture to quantify the extent of resource misallocation during the sectoral transition – i.e., the economy's deviation from the (efficient) state of equal marginal products in all sectors. Data on pure marginal products are not available, but value added per workers (average productivity) statistics, nevertheless, provide a hint of the evolution of productivity differences. Unlike the ratio of marginal productivities, the ratio of average productivities cannot be expected to remain at, or even approach, unity, since capital requirements in the two sectors may be quite different. But intersectoral differences in technology are relatively stable, and should result in relatively slow and smooth movements in the average productivity ratio. Transitory changes, on the other hand, are more likely to reflect differences in marginal productivities that arise with allocational imbalances between the two sectors.

The data reveal important movements in the productivity ratio during the most rapid stages of industrial transition, between 1960 and 1975. Specifically, the ratio of manufacturing to agricultural productivity describes an inverted 'U', indicating that the difference between industrial productivity and agricultural productivity may have peaked sometime during the mid-1970s. The improving relative position of industry reflects the more than doubling of value added

Table 8.2 Labour productivity comparisons (1980 constant prices)

	Agri. value added (A) (billion W)	Agri. workers (B) (thousand)	Value added per worker C = A/B (thousand W per worker)	Manuf. valued added (D) (billion W)	Manuf. workers (E) (thousands)	Value added per worker F = D/E (thousand W per worker)	Labour productivity ratio F/C
1970	4 966.5	4 916	1 010.3	2 822.7	1 395	2 023.4	2.00
1975	6 308.0	5 425	1 162.8	6 143.8	2 265	2 712.5	2.33
1979	6 862.1	4 887	1 404.2	11 393.7	3 237	3 519.8	2.51
1983	7 436.0	4 314	1 723.7	14 780.6	3 383	4 369.1	2.53
1984	7 453.2	3 909	1 906.7	16 930.4	3 492	4 848.3	2.54
1985	7 893.4	3 722	2 120.7	17 551.8	3 654	4 803.4	2.27

Sources: Bank of Korea, National Income Accounts; Bank of Korea, Economic Statistics Yearbook

per worker in manufacturing during the high growth years of 1970–80, compared to sluggish agricultural gains over the period.[1] Subsequently, very rapid declines in the agricultural work force served to substantially increase agricultural value added per man in the 1980–85 period.

This analysis supports the view that the impetus for the sectoral transition came from industry, where rapid gains in productivity created strong incentives for immigration. The productivity advantage of industry appears to have become quite large around 1970, leading not only to massive rural-urban migration, but also to government intervention in support of rural incomes. With time, the measured productivity advantage of industry diminished, mainly because of rapid outmigration, but also because the value of agricultural output was artificially raised by agricultural protection. Given the narrowing gap between the two sectors, protection is beginning to outlive its role as the redistributor of the benefits of industrial growth to agriculture. Instead, it may now be retarding the pace of the ongoing intersectoral transition, with negative implications for the overall efficiency of the economy.

2. EVOLUTION OF INDUSTRIAL POLICIES[2]

2.1 Industrial Takeoff (1961–73)

This period featured a unique, dual trade regime – aggressive promotion of an export-oriented sector and classic protection of major home markets. Korean policy-makers maintained close control over trade, exchange, and financial policy, as well as aspects of industrial decision-making. In contrast to other countries, however, they used these instruments in an integrated fashion to pursue the primary objective of export growth. The net effect of these somewhat conflicting policies was a trade regime that was slightly biased in favour of exports as a whole, but was almost neutral with respect to the composition of exports.

Industrial policy during this period involved neither functional interventions (addressing specific types of market failure) nor selective interventions (influencing the industry-specific compositon of economic activity), but rather a comprehensive incentive system designed to channel resources into export-oriented activities. Initially, the policy was implemented under direct bureaucratic control, but gradually it was reframed in terms of market-linked incentive schemes. The key precondition to outward oriented growth was the reform of the overvalued exchange rate regime. The elimination of this distortion in the early 1960s was critical to the success of the outward-oriented strategy.

The first instruments of export promotion were highly discretionary. Exporters were supported with multiple exhange rates, direct cash payments, permission to retain foreign exchange earnings for private use or resale, and the privilege to borrow in foreign currencies and to import restricted commodities under the so-called export-import link. This system granted exporters access, not only to foreign machinery and intermediate inputs for their own use, but also to scarcity rents in heavily protected domestic markets. Even as discretionary incentives were gradually replaced by more automatic instruments, exporters received significant concessions, such as wastage allowances permitting exporters to import, on preferential terms, greater amounts of intermediate inputs than required in production, concessional interest rates on export loans, preferential access to working capital, and tariff exemptions to direct and indirect exporters. In part, these interventions allowed Korean exporters to avoid some of the distortions involved in the protection of domestic markets. In part, they offset the advantages that protection afforded to domestically-oriented activities. Essentially, the policies amounted to off-budget subsidisation of exporting firms.

Support for exports was pervasively channelled through the state-controlled banking system. Government objectives were implemented through policy loans – bank loans explicitly earmarked for particular activities or industries, and lent passively by banks at interest rates below those charged for general lending purposes. Following explicit government directives, banks increasingly used export performance as the criterion of creditworthiness. Access to bank credit was extremely important, since the bank lending rate was substantially below the cost of borrowing in the alternative curb market.

As of the late 1960s, the principal features of the incentive regime were the following: (a) moderate overall protection of domestic markets, offset by special subsidies to exports; (b) approximately world market pricing of inputs and outputs across different export products; (c) high protection of the domestic market in industries with poor export prospects; (d) high protection of final consumer goods, relative to industrial raw materials and capital goods. Overall, protection did not significantly distort either the general level or the inter-industry pattern of export incentives. Consequently, the emerging export structure reflected comparative advantage more closely than is generally the case in protected economies.

2.2 The Heavy Chemical Industry Programme

Buoyed by the success of its policies during the take-off, the government next turned to accelerate structural change by directly promoting the development of the heavy chemical industry (HCI) sector. Support was mobilised using a broad range of policy instruments, including import protection and fiscal preferences, and most importantly, preferential allocation of credit. Government relied heavily on its control of the entire credit system to provide 'strategic' industries with access to substantially subsidised bank loans. The potential for subsidisation was great due to the complicated system of interest rate ceilings that governed the financial system throughout the 1970s. Real bank interest rates were negative throughout most of the 1970s, and the differential between bank rates and those charged in the informal credit markets represented a substantial discount for industries eligible for credit from government-controlled banks.

The National Investment Fund (NIF), established in 1974, lent as much as two-thirds of its portfolio to HCI projects. But the real impact of the NIF on credit allocations stemmed from its 'announcement effect' on bank lending practices. Using a conservative measure, it is estimated that roughly one-third of bank lending, the predominant source of formal domestic credit, went into'policy loans'. Strategic industries, such as chemicals, basic metals, and fabricated metal products and equipment, received favoured access to other bank lending as well, as compared to traditional industries producing either for the domestic market, such as food and beverages, or for export, such as textiles. Thus, the share of credit allocated to the three priority sectors virtually doubled from approximately one-third of total bank lending in 1973–74 to about 60 per cent in 1975–77. At the same time, light manufacturing industry's access plummeted.

Favoured sectors not only had better access to capital but also faced much lower average borrowing costs. Figure 8.2 shows the ratio of the average borrowing cost of heavy industry to that of light industry; a value of 1.0 would reflect neutrality in industrial finance. Beginning in 1974, the year of inception of the NIF and the year in which heavy industry began to claim a decidedly larger share of preferential loans, the gap in effective borrowing costs began to widen in favour of HCIs. Indeed, between 1975 and 1978, the height of the HCI push, the cost of borrowing averaged approximately 25 per cent lower for heavy than light industry. This disparity began to recede over the 1979–80 period of reform and is now approaching neutrality.

Microeconomic performance adds up to macroeconomic performance. Not only did the HCI drive result in substantial unusable capacity, but it also concentrated investment in the economy's most capital-intensive industries. This was in sharp contrast with the development strategy pursued in the 1960s. During the take-off, the rising share of exports in the economy led to a sharp initial decline in the capital-output ratio and then kept this ratio below its 1955–60 level of about 3.0 until 1970–75 (when it again reached that level). These changes underscore the significance of changes in the economy's output mix, since the overall capital stock substantially increased during the same period. The capital-output ratio took a sharp turn upward with the HCI drive, rising to almost 5.0 in 1975–80. A comparison of Korean capital-output ratios with similar measures for other newly-industrialising economics shows that Korean ICORs in the 1960s were extraordinarily low. The upturn of the 1970s moved the Korean ICOR closer to those of other economies; in the first half of the 1970s Korea was still near the bottom of the distribution, but by the latter half of the decide it had moved well above the average.[3]

The structural transformation brought about the HCI drive was, on the whole, consistent with emerging changes in Korean comparative advantage, but occurred too rapidly and at excessive cost. It is true that in some industries the strategy worked, and that in many others failures were due to external causes. But the great variance in HCI outcomes cannot be separated from the *ex ante* risk assumed by the HCI programme. Without the virtually unlimited government support that was offered to HCI investments, no private agent would have been willing to bear the obvious risks. These risks paid off in some cases, but on average produced low returns.

More fundamentally, the HCI programme again substituted bureaucratic judgement for market performance in determining the direction of economic growth. As we shall see, intervention in favour of agriculture also grew during this period. In contrast to previous years, access to credit was not sufficiently conditioned on the test of export performance. In shipbuilding and other industries the invest-

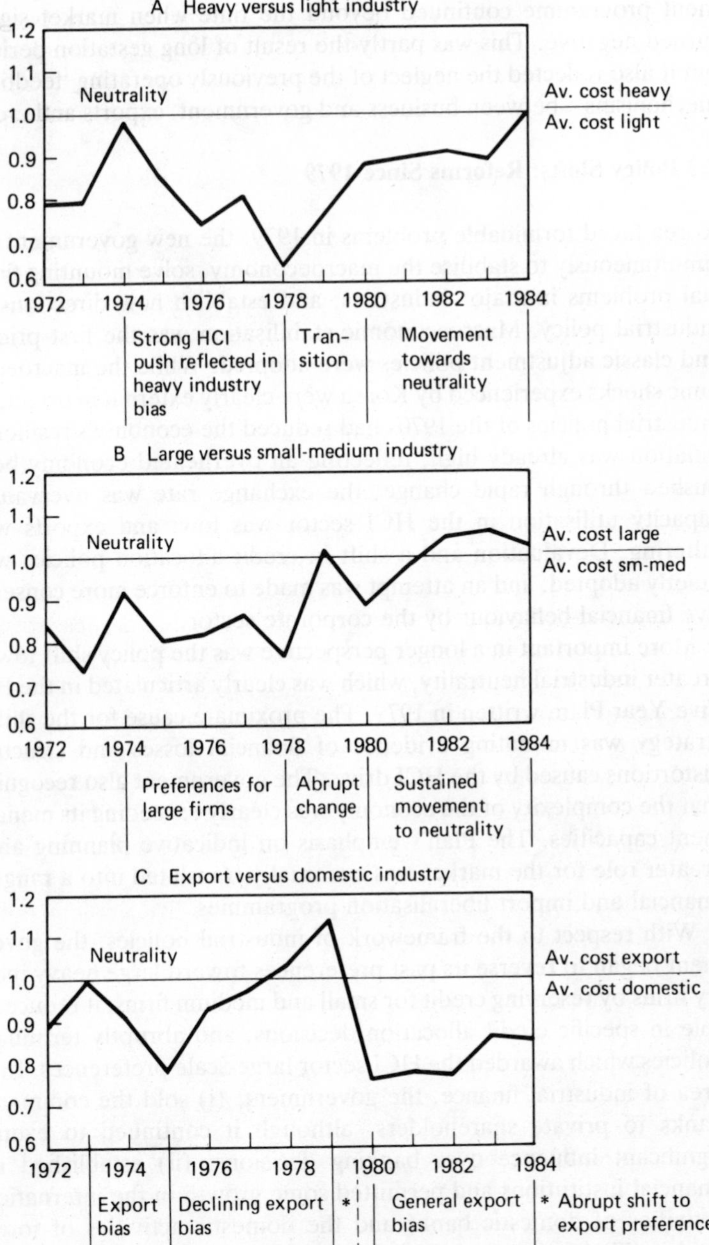

Source: World Bank (1987) *Korea: Managing the Industrial Transition* (Washington DC: World Bank).

Figure 8.2 Trends in industrial finance (average borrowing cost comparisons)

ment programme continued beyond the time when market signals turned negative. This was partly the result of long gestation periods, but it also reflected the neglect of the previously operating 'feedback' mechanisms – between business and government, exports and credit.

2.3 Policy Shifts: Reforms Since 1979

Korea faced formidable problems in 1979: the new government had simultaneously to stabilise the macroeconomy, solve mounting financial problems in major industries, and establish new directions for industrial policy. Macroeconomic stabilisation was the first priority and classic adjustment policies were adopted. While the macroeconomic shocks experienced by Korea were clearly external in origin, the industrial policies of the 1970s had reduced the economy's resiliency; inflation was already high, reflecting an overheated economy being pushed through rapid change; the exchange rate was overvalued; capacity utilisation in the HCI sector was low; and exports were faltering. Devaluation and a shift in credit allocation policies were quietly adopted, and an attempt was made to enforce more conservative financial behaviour by the corporate sector.

More important in a longer perspective was the policy shift toward greater industrial neutrality, which was clearly articulated in the Fifth Five Year Plan, written in 1979. The proximate cause for the shift in strategy was mounting evidence of financial losses and structural distortions caused by the HCI drive. The government also recognised that the complexity of the economy was clearly exceeding its management capacities. The Plan's emphasis on indicative planning and a greater role for the market was eventually translated into a range of financial and import liberalisation programmes.

With respect to the framework of industrial policies, the government began to reverse its past preferences toward large heavy industry firms by reserving credit for small and medium firms; it reduced its role in specific credit allocation decisions, and abruptly terminated policies which awarded the HCI sector large-scale preferences. In the area of industrial finance, the government: (i) sold the commercial banks to private shareholders, although it continued to exercise significant influence over banking decisions; (ii) established new financial institutions and permitted some growth in the international activities of domestic banks and the domestic activities of foreign banks; (iii) increased real interest rates, reducing the gap between the organised and unorganised sectors of the financial market; and

(iv) substantially reduced the scope of interest rate subsidies for particular borrowers.

With respect to protection, a bold plan was adopted to increase the liberalisation ratio from about 69 per cent in 1980 to 95 per cent in 1988 according to an 'advance notice' schedule (see Table 8.3). The government also revised tariffs in order to reduce product-to-product variations in protection; lowered the average legal tariff rate by approximately one-third, and committed itself to further reductions by 1988; and liberalised the system of controls over foreign direct investment. The trade liberalisation programme is proceeding smoothly with a steady addition to the list of automatically importable items which should be down to 367 by 1988 from about 1500 in 1983. There has been a minimum of offsetting interventions, such as tariff increases.

The general thrust toward neutrality notwithstanding, the government plays an active role in several areas of policy. Intervention since 1979 has focused on the restructuring of distressed industries, support for the development of technology, and the promotion of competition. An active role in these 'functional' areas can be regarded as consistent with the liberalisation effort, at least to the extent that it can be rigorously justified on the basis of market imperfections. Restructuring operations have become frequent in the wake of sharp reversals in world markets and the overambitious investment programmes of the 1970s. The firms and industries involved were large and highly leveraged, with their loans representing a significant share of commercial bank assets. These factors, coupled with the relative thinness of Korean capital markets, have drawn the government deploy into the restructuring process. The plans it has helped to orchestrate have involved mergers, capacity reduction programmes, as well as special financial reschedulings with commercial banks.

3. EVOLUTION OF AGRICULTURAL POLICIES

The trend of Korea's agricultural policies has been, in important ways, opposite to the trend of its industrial policies. The government has intervened in agriculture since the 1940s, but initially its role was limited and involved little developmental support. Today, the public sector's role is far more direct in agriculture than in industry, and goes beyond the provision of credit and investments to the operation of massive price support schemes for rice, barley, other grains and

Table 8.3 Import liberalisation programme automatic approval process

	Total items[a]	*Proportion of items subject to automatic approval*							
		1981	*1982*	*1983*	*1984*	*1985*	*1986*	*1987*	*1988*
Primary products	1 386	68.5	70.6	73.5	75.8	78.2	79.7	80.1	80.5
Chemical goods	2 182	93.4	94.0	94.6	95.0	95.6	97.7	99.1	99.6
Steel and metal products	802	88.9	89.7	90.9	92.8	95.6	99.4	100.0	100.0
Machinery	1 414	64.2	65.5	69.2	78.0	83.0	89.4	93.3	100.0
Electrical and machinery appliance, electronics	495	40.9	46.1	51.3	62.4	73.8	87.0	95.6	100.0
Textile (including leather garments)	1 089	65.4	68.4	79.9	90.3	93.1	96.1	96.9	98.8
Others	547	71.2	75.7	80.6	82.1	82.8	85.7	88.2	88.2
(Subset A)*	–	–	(44.5)	(48.8)	(72.4)[b]	(78.0)	(93.3)	(97.6)	(98.8)
Total (items)	*7 915*	*74.7*	*76.6*	*80.4*	*84.8*	*87.7*	*91.6*	*93.6*	*95.4*
Total (value)		*64.8*[c]			*72.2*[c]				

Notes:

* Subset A refers to those items on the monopoly list regulated by the Monopoly Commission, and defined as commodities whose sales exceed 30 billion won, and either a single producer whose market share exceeds 50% or the top three producers whose combined market shares exceed 70%.

[a] As of 1 July, 1984.

[b] Based on 254 items, rather than the 354 used as the basis prior to 1984. The ratio based on the larger set of market-dominating items was 52% in 1984.

[c] Estimate by Wontack Hong (1985).

Source: Ministry of Trade and Industry.

horticultural crops. In recent years, government purchases and sales of products have increased markedly, and now cover about 25 per cent of total rice production, 75 per cent of barley and 10 per cent of horticulture. The public sector's net loss on these acquired crops has risen *pari passu* with the size of its involvement. By 1985 the government was losing almost 20 000 won (some £23) per 80 kg bag of rice, for example, and had accumulated a stock of 3.7 million metric tons.

3.1 Degree of Government Intervention

The government's involvement in agriculture was first formalised with the Grains Management Law in February 1950. Initially, the aim of agricultural policy was to maintain low food prices for urban workers. This was accomplished with substantial PL 480 imports which permitted the government's own purchase prices to be below market prices. At times the government was faced with insufficient supply as farmers attempted to minimise sales, and it was obliged to requisition rice and barley administratively. But on the whole, government purchases and sales were small. Meanwhile, the government offered little developmental support – the share of bank loans going to agriculture fell, as a percentage of both total loans and agricultural value added, and even the extension system developed by the Japanese was allowed to decline, in part because of its unfavourable associations with the occupation. Indeed, until the early 1970s agriculture face 'benign neglect'.

By the late 1960s the combination of high growth in the urban manufacturing sector and sluggish agricultural development led to a major increase in income disparities between the urban and rural sectors (see Table 8.4) and a decision on income distribution (and political) grounds to bolster farm incomes and restore greater income equality.[4] Between 1967 and 1969, direct government expenditures on agriculture more than doubled, as did also the ratio of credit extended through the National Agricultural Co-operative Federation to agricultural GDP. Large irrigation projects were started, and a broad campaign to accelerate rural development was launched in 1971, under the name 'New Community Movement'.

In 1969 the government also took steps directly to improve farm revenues by raising the purchase prices for rice and barley. However, to avoid upward pressure on the general price level, the distribution prices for both commodities were held down. This was the beginning of the dual pricing system which remains in effect today. Since the

Table 8.4 Comparative farm and urban incomes

	Income per household			*Income per capita*		
	Farm (F)	*Urban(U)*	*F/U*	*Farm (F)*	*Urban(U)*	*F/U*
	(th. won per month)		*%*	*(th. won per month)*		*%*
First Five-year Plan (1962–66)	106	110	96	17	20	85
Second Five-year Plan (1967–71)	232	340	68	39	63	62
Third Five-year Plan (1972–76)	723	745	97	127	144	89
Fourth Five-year Plan (1977–81)	2 385	2 304	104	459[a]	563[a]	82
Fifth Five-year Plan						
1982	4 465	3 805	117	898	865	104
1983	5 128	4 368	117	1 028	1 016	101
1984	5 549	4 828	115	1 156	1 149	101

Note:
[a] The reported Gini coefficient (1981) measuring income inequality was 0.405 for urbanites and 0.356 for rural income earners. World Bank estimates (1984).
Source: Economic Planning Board.

inauguration of the dual price system, a substantial wedge has emerged between the release price and the acquisition plus handling price of foodgrains. Moreover, stockpile handling costs have risen to the point where they constitute a large portion of grain account deficits.

The results of the price support policy are visible today in the mounting deficit of the Grain Management Fund (GMF), which reached 1.55 trillion won in 1985 – excluding interest due on the bonds that had financed past deficits and the handling costs of the grain stockpile.[5] When adding the losses generated by the Fertiliser Fund (described below), the public policy expenditures associated with agricultural supports rise to 450 billion won in 1985 alone and some 2.25 trillion won cumulatively according to conservative estimates. These losses have been financed by borrowings from the Bank of Korea and through sales of grain bonds to the banking sector. In a new development, since 1984 some or all of the annual deficits incurred have been financed overtly through the budget, a procedure which may hasten their eventual elimination.

Government intervention is also evident in the protection of the domestic fertiliser industry. Fertiliser prices to farmers are 20–40 per cent above world prices, while producers receive prices as much as 60 per cent above world prices (Moon and Kang, 1986). This situation stems from earlier efforts to promote the fertiliser industry (initiated by the Fertiliser Control Law of 1962), which gave government the authority to set output as well as prices. At the same time, the government agreed to joint venture contracts with foreign parties guaranteeing them pre-specified profit rates.

The overall effect of price interventions in both input and output markets has been to raise domestic prices significantly above international prices. The premium of domestic (government purchase) prices over import prices, for example, has averaged 150 per cent since 1982, i.e., the domestic consumer is paying two-and-a-half times the world price for rice. A recent World Bank study has suggested that if the subsidy cum protection on rice were lifted in favour of free trade, national disposable income would rise by 2–3 per cent a year. This net gain involves large benefits to urban consumers as well as sharp reductions in farm incomes (Braverman *et al.*, 1983).

One way Korea might limit the decline of farm household incomes under agricultural liberalisation is to increase off-farm employment opportunities. Currently, about one-third of Korean farm income is derived from non-agricultural sources (of which two-thirds is from remittances and one-third is wage income), which is about 20 per cent higher than in the early 1970s, but still considerably below the comparable figure of 80 per cent for Japan in the early 1980s. It appears, therefore, that greater efficiency in agriculture (and correspondingly greater openness) could be achieved without excessive distributional consequences if the industrial sector were able to generate more jobs in secondary cities and rural areas.

3.2 Recent Initiatives

In light of the importance of non-agricultural income in rural areas, the government has repeatedly adopted policies to spur rural industrialisation. At present, 80 per cent of total manufacturing employment is in urban areas, mostly in Seoul and Pusan. Yet, despite the enactment of specific programmes, such as the Saemaul Factory Programme of 1973 and the Industrial Location Law of 1977,[6] the only major shift towards rural industrial location has occurred in conjunction with the development of industrial estates in the heavy and chemical industry push of the late 1970s. Even these shifts tended to result in the relocation of families, rather than in additional off-farm income for an existing population. Most recently the government has passed the Farm Household Income Source Law of 1983, which tries to develop rural industrial parks. This programme will provide infrastructure in rural employment zones and will offer preferential financial and tax treatment as well as technical and managerial assistance to firms agreeing to locate in the parts. Seven parks are currently under construction, but it remains to be seen whether the serious problems that have defeated earlier efforts can be overcome in this latest scheme.

Regardless of the pace of progress on rural industrial development, changing conditions economy-wide and in agriculture are putting current policies under increasing strain. Successes on the supply-side, spurred by irrigation, technical progress, and protection, have produced substantial agricultural surpluses. Given that average food consumption levels are already high and that population growth is slow, it is unlikely that demand for foodstuffs will grow at more than 1.5 per cent a year. Thus the growth of agricultural output (and incomes) will be limited to well below the expected 5–7 percent gains in manufacturing.[7] In the face of these disparate growth paths, the continuation of present policies would imply increasingly generous commodity price subsidies and/or substantially higher urban food prices. These policies would place Korea on the path of Japan and Western Europe – leading to high food costs, chronic food surpluses, and massive budgetary outlays.

The alternative is a smaller and more efficient agricultural sector, more open to imports and less important as a source of farm-household income. This strategy requires increased job creation in rural areas, as well as rapid absorption of new migrants in the urban industrial sector. Unfortunately, the reduction in agricultural employment opportunities coincides with other disturbing trends in labour markets, including: (i) a marked reduction in the employment elasticity of output, from an estimated .50 in the 1970s to roughly .25 in the 1980s and early 1990s (KDI, 1985) and (ii) a sharp shift in labour demand towards higher skilled workers, particularly in emerging industries in electronics and services.

4. CONCLUSIONS AND IMPLICATIONS FOR THE FUTURE

During the past three decades, South Korea experienced as rapid a transition from agriculture to industry, and from a rural to an urban economy, as any major country in history. This transition was remarkably smooth, as substantial income differentials between rural and urban households were largely avoided. Much of the transition occurred in a space of little more than a decade, between the early 1960s and mid 1970s. During this period the GDP share of Korea's manufacturing sector moved from a relatively low to a relatively high level (based on the experience of other economies at similar levels of development). Simultaneously, the share of agriculture fell sharply from an unusually high level to a level still somewhat above the relevant international experience (Chenery and Syrquin, 1975).

The evidence suggests that the sectoral transition was driven by the industrial sector, with little direct support from agriculture. Rapid industrial growth can be traced to a series of policy measures that encouraged the growth of exports and exposed a substantial part of the industrial sector to international competition. Capital and other resources for industrial growth were initially obtained from the urban sector and from foreign sources. Aside from relatively low food prices – due to the plentiful availability of food aid – there is no evidence of significant fiscal or financial transfers from agriculture to industry in the early stages of Korea's industrial take-off.

Once industrial growth was solidly established, increased demand for food also generated an acceleration of agricultural growth. However, due to the poor land resources of the Korean economy, agriculture's ability to respond to rapidly growing food demand was severely limited, and much of the new demand spilled into agricultural imports. As a result, agricultural output and income growth fell far short of the levels achieved in the industrial sector. Although agricultural growth was solid by international standards, the difference between agriculture and industrial growth rates was much higher than is usually observed. The increasing income differential arose between the two sectors despite massive population movements from rural to urban areas.

The government responded to widening rural-urban differentials and rapidly growing agricultural imports by adopting policies to subsidise and protect agriculture. At first these policies tended merely to offset advantages previously offered to industrial producers and exporters. For example, the improvement of agricultural financing in the early 1970s merely returned the ratio of agricultural loans to agricultural value added to its late 1950s level. Later, agricultural policy began to provide unusually high protection to agriculture as compared to other sectors of the economy. The government's purchase price of grains was well below its sales price in the 1950s and

1960s, moved above the sales price in 1970, and subsequently delivered an increasing subsidy to farmers. In recent years, the policy has led to a substantial build-up of grain reserves and budgetary deficits.

The increase in agricultural protection is now taking place in parallel with a decline in industrial protection and in direct government support for industry. Due to the failures of massive, selective intervention in the 1970s, the government is gradually reducing its role in industrial decision-making in favour of markets and private financial institutions. Serious liberalisation programmes are under way in both trade policy and industrial finance. These liberalisations will tend to erode the advantages previously afforded to industrial firms, including those oriented to the export market. The net effect of these policy changes is to shift the balance of protection in favour of agriculture.[8] In this context, agricultural protection is no longer simply distributing the rewards of rapid development, but is becoming a drag on efficiency and continued growth. The costliness of continued protection of agriculture to the Korean economy is becoming increasingly apparent, not only in the direct sense of suboptimal resource allocation, but also in the context of diplomatic trade pressures.[9]

The mounting problems of current agricultural policy, coupled with a general tendency to withdraw from direct intervention in the economy, suggest that future policies will need to favour the reduction, and perhaps eventual elimination, of agricultural subsidies.[10] Political pressure from the USA, Korea's most important trade partner, is pointing the same way. Yet the elimination of agricultural protection will undoubtedly create employment problems; repeated efforts at accelerating rural non-agricultural development have met with little success, and the increasingly sophisticated structure of Korean industry may make it more difficult to absorb workers released by agriculture than was the case while labour-intensive industries dominated the industrial sector (Castaneda and Park, 1986). Inasmuch as decisions on these and other economic issues are likely to be subject to much greater political scrutiny in the future than in the past, the government may opt for more direct interventions to achieve welfare objectives in rural areas.

In sum, Korea managed its transition smoothly for two major reasons. First, the remarkably strong performance of industry directly helped to accelerate agricultural growth, initially through demand effects, and later by supplying capital and other resources to increase production capacity. Industry also provided employment for a massive inflow of workers into urban areas, and thus helped to raise agricultural incomes by relieving the sector of its marginally-employed workers. Secondly, when especially rapid industrial growth did lead to increasing income differentials between rural and urban households, the government quickly adopted policies that helped to increase agricultural incomes and distribute the benefits of industrial growth. These policies were adequate in reducing intersectoral income differentials, but were not strong enough – at least in their initial years of operation – to eliminate the signals directing the movement in resources from agriculture to industry.

Now that the bulk of the intersectoral transition of its economy is complete, Korea faces new challenges in removing the protectionist policies that served a useful welfare function during earlier periods. With continued rapid growth, this last phase of the transition is likely to be also completed effectively, judging by the example of recent policy changes in the industrial sector. The main lesson of the Korean experience, with regard to the ease with which the agricultural-industrial transition occurs, is that the problems that arise in the transition have solutions – as long as the leading sector of the economy provides as substantial a margin of success as did Korean industry during the last twenty-five years.

Notes

* All views expressed are the personal opinions of the authors and do not necessarily reflect those of the World Bank.

1. The disparity is somewhat overstated in 1980, which was a year of exceptionally poor harvests, so that 1979 data are used in Table 8.2.
2. Parts of this section reflect work by the authors in the context of analysing changes in Korean industrial policy (see Petri (1986); Leipziger (1986).
3. A sample of ten newly-industrialising countries shows an average ICOR of 3.4 over the decade of the 1960s and 4.4 in the 1970s (World Bank estimates).
4. See Mason *et al.* (1980) and Moon and Kang (1986) for a fuller description of the chronology of events in Korea, and Ahluwalia (1976) for a cross-country perspective on income distribution trends during the development process.
5. Estimates of the total GMF deficit incurred over the 1980–85 period have reached 2.5 trillion won or close to $3 billion.
6. Neither the 1973 Saemaul Factory Programme (part of a larger community development programme), nor the Folkcraft Development Pro-

gramme, nor the Farm Household Side Business Programme has been able to achieve the efficiencies of scale, requisite quality, and access to markets needed for success. Skills have been lacking in rural areas and local infrastructure has been inadequate (see Song and Ryu, 1986, for details).

7. At present only 4.3 per cent of agriculture's output is exported, accounting for a mere 1.7 per cent of Korea's 1984 export bundle. Except for speciality items, Korea is unlikely to substantially improve on this performance in the future.
8. The extent of this distortion can be seen in the wide discrepancies between domestic and import prices for many agricultural commodities, including rice and beef (three times world prices), barley and soybeans four times world prices), and peanuts, red pepper, and garlic (five to ten times world prices) (USDA sources).
9. Korea recently agreed, for example, to begin to open very slowly its cigarette market, a domestic government monopoly, to US tobacco-makers, and will, in the light of persistent trade imbalances between the USA and Japan, be subject to further pressures to liberalise agriculture.
10. It is noteworthy in this context that 270 of the 367 items which will not be liberalised by 1988, at the end of the current reforms, will be primary commodities.

References

Ahluwalia, M. (1976) 'Inequality, Poverty, and Development', *Journal of Development Economics*, vol. 3, pp. 307–42.

Braverman, A., Ahn, C. Y. and Hammer, J. S. (1983) 'Alternative Agricultural Pricing Policies in the Republic of Korea'. World Bank Staff Working Paper no. 621, Washington, DC.

Castaneda, T. and Park, F. K. (1986) 'The Role of Labor Markets', paper presented at KDI-World Bank Conference on Structural Adjustment in a Newly Industrialized Country: Lessons from Korea, Washington, DC (June).

Chenery, H. B. and Syrquin, M. (1975) *Patterns of Development: 1950–1970* (Oxford University Press)

Chenery, H. B. Syrquin, M. and Urata, S. (1986) *Patterns of Development: 1950–83* (Oxford University Press).

Hong, W. (1985) 'Import Restriction and Import Liberalization in Export-Oriented Developing Economy: In Light of Korean Experience', Mimeo (Seoul University).

Korea Development Institute (KDI) (1985) 'Long Range Forecasts for Manpower Supply and Demand'. mimeo (Seoul).

Leipziger, D. M (1986) 'A Re-examination of Korean Trade Policy', mimeo (Washington, DC World Bank).

Mason, E. S. *et al.* (1980) *The Economic and Social Modernization of the Republic of Korea* (Cambridge, Mass.: Harvard University Press).

Moon, P. Y. and Kang, B. S. (1986) 'A Comparative Study of the Political

Economy of Agricultural Pricing Policies: The Case of South Korea', mimeo.

Petri, P. (1986) 'The Legacy of Korean Industrial Policy', mimeo Washington, DC: World Bank

Song, D. H. and Ryu, B. S. (1986) 'Agricultural Policies', paper presented at KDI-World Bank Conference on Structural Adjustment in a Newly Industrialized Country: Lessons from Korea, Washington, DC (June).

Urata, S. (1986) 'Korea's Industrial Structure', mimeo, World Bank.

World Bank (1984) *Republic of Korea: Agricultural Sector Survey*, (Washington, DC: World Bank).

World Bank (1986) *World Development Report* (New York: Oxford University Press).

World Bank (1987) *Korea: Managing the Industrial Transition*. (Washington, DC: World Bank).

9 Sector Proportions and Growth in the Development of the Nigerian Economy

T. Ademola Oyejide
UNIVERSITY OF IBADAN

1. INTRODUCTION

The Nigerian economy has undergone tremendous changes over the last three decades. Not only has its growth experience been uneven but the rate of growth has also fluctuated widely. Thus, the growth process has been accompanied by sharp sectoral changes, some of which were partly policy-induced while others have resulted from fortuitous and external factors.

According to the country's Development Plan documents and other official statements, the overall objective of policy over this period has been the acceleration of aggregate economic growth based on a balanced expansion of the key sectors of the economy. But, at different times during this period, emphasis has been placed by government, explicitly or implicitly on the rapid expansion of particular sectors at the expense of others. In the 1950s, as the country was being prepared for political independence from Britain, the primary focus of government policy and expenditure commitments was the development of infrastructures and public sector services. In the 1960s an import-substitution-industrialisation strategy was adopted, and policy was directed towards the growth of the manufacturing sector via tariff protection and an industrial incentive system. The 1970s witnessed perhaps the most significant structural changes in the economy. Increased oil production and rapid increases in oil prices radically transformed the economy; the fairly broad-based agricultural economy became considerably less diversified, as the oil sector dominated production and trade structures. The consequences and legacies of Nigeria's oil boom continued to influence the country's economy in the 1980s. Because of these legacies, and the prolonged deterioration of the oil sector from 1981, the 1980s have been characterised by difficult external and internal imbalances.

This paper presents an analysis of the economy's growth and sectoral structure over the period 1950–84. The analysis is designed to determine the 'sources of growth', identify the sectoral shifts which have accompanied and influenced the growth process, and determine the impact of the oil boom on the sectoral structure. In addition, the paper offers an examination of the key policy interventions which may have influenced the pattern and speed of economic growth and determined the economy's sectoral structure.

The rest of the paper is organised as follows: Section 2 discusses some general patterns of sectoral shifts which may be expected to occur in the process of development and income growth; and relates these to Nigeria's choice of development strategy and corresponding broad policy objectives. Section 3 presents an analysis of Nigeria's growth performance in both aggregate and sectoral terms while Section 4 offers a source of growth decomposition. Section 5 identifies significant sectoral shifts and uses the constant market-shares model to examine the observed structural changes. Finally, Section 6 shows how the policy environment responded to, influenced, or accommodated economic growth and sectoral shifts.

2. GROWTH AND STRUCTURAL CHANGE

The Chenery-Syrquin model (Chenery, 1979; Chenery and Syrquin, 1975; Wood, 1986) offers a framework for analysing growth and structural changes in an economy and provides several general empirical results on this subject. These results demonstrate a systematic pattern of structural change for a wide variety of countries spanning different historical periods. For the large middle-income countries – a category in which Nigeria has been classified from the mid-1970s – the most significant structural changes include a decline in the share of agriculture in total exports, and in total output as income increases. Corresponding to this is a marked increase in the share of manufactured products in both exports and gross output. Similarly, the Chenery-Syrquin model predicts that the share of total nontradable sector in GDP would rise substantially with income.

Changes also occur in the sectoral distribution of employment. Income growth is influenced by labour productivity and high per capita income growth rates result both from the shifting of labour from low to high productivity sectors, and by increasing labour productivity within each sector of the economy. Since the model predicts that labour productivity increases much faster in manufacturing than in the agricultural and service sectors, the shift in employment tends to be away from agriculture, although because of its below-average productivity growth, agriculture's share of employment falls less rapidly than its share of output.

The sectoral shifts predicted by the Chenery-Syrquin model should serve as a good starting point for an empirical analysis of economic growth and structural change in Nigeria, particularly during the 1950–70 period. But the emergence of oil as the predominant sector during the 1970–84 period could have introduced additional changes in both the growth and structure of the economy. Hence, a second analytical framework for analysing growth and structural change may be more appropriate for this latter period.

The 'Dutch disease' model (Corden and Neary, 1982; Gelb, 1981; Edwards and Aoki, 1983) analyses the overall growth and structural-change implications of asymmetic sectoral growth corresponding to a discovery of oil or a sharp increase in its price. The model provides several basic generalisations[1] relating to the effects of a resource boom, particularly on the relative size of sectors and the real exchange rate. The model argues that a booming sector such as oil, generates an increase in income by bringing additional foreign exchange into the economy, and thus raises domestic demand which, in turn, places pressure on domestic prices. The excess demand for nontradables forces up their prices compared with other products, whereas the increased demand for tradables is met by increased imports. The consequent fall in the price of tradables relative to the price of nontradables – or an appreciation of the real exchange rate – provides the signal for resources to be drawn away from the non-oil tradable sectors into non-tradables. In other words, the model predicts that an oil boom would cause a real exchange rate appreciation accompanied by a relative contraction of the non-oil tradable sectors. Since the increased oil revenue largely accrues to government, these results are likely to be intensified where the pattern of government expenditures is biased towards infrastructure, services and other non-tradables.

A combination of the generalisations derived from the Chenery-Syrquin and the 'Dutch disease' models implies that, while historical patterns of development indicate that agriculture's share of total production and employment would decline as income grows, this decline could occur more prematurely, and be more abrupt, when an oil boom is part of the growth process. The result with respect to the manufacturing sector is not so straightforward because two tendencies appear to work in opposite directions; historical development patterns suggest an increase in the manufacturing sector's share of overall output whereas the existence of an oil boom would lead to a decline in the share of the non-oil tradable sectors, i.e., both agriculture and manufacturing. Both factors are expected to cause an increase in the share of non-tradables in output as income grows. Thus, the predicted decline in the share of agriculture and increase in the share of nontradables in total output are common to both models, while they differ with respect to changes in the share of manufacturing.

A related issue in this connection is Nigeria's choice of development strategy, especially as it relates to the question of intersectoral balance as an essential part of sustained development. This strategy has been seen in Nigeria primarily as a function of the nature of the desirable linkages which need to be developed between agriculture and manufacturing industry. At independence, features of the Nigerian economy reflected the result of the processes set in motion during the colonial period. Hence, the role of colonial policies in shaping the economic structure influenced the development strategy pursued in the post-colonial period. More specifically, colonial policy focused largely on the promotion of export crops among Nigerian peasants so that the choice of post-colonial long-term development strategy was viewed primarily in terms of whether to opt for a deepening of the prevailing pattern of specialisation by concentrating on agricultural exports or to seek to restructure the economy so as to develop the manufacturing sector and the possibilities for non-agricultural exports.[2] The ultimate choice was, in some sense, predetermined. The desire to reduce dependence on agricultural exports was strong, and the sharp decline in international terms of trade for agricultural exports around the period of independence only intensified this desire.[3] The choice was made therefore to reduce the country's dependence on a few agricultural export commodities and to diversify the economy in general by promoting rapid industrialisation. The process of structural economic change implied by this choice is consistent with the historical pattern of development, i.e., relative expansion of the share of manufacturing, and decline of the share of agriculture, in the GDP. Although this development strategy was not abandoned explicitly, during the 1970s, with the emergence of the oil boom, it appears to have been modified. The new direction was expressed in the desire to use the new oil wealth to modernise the economy by rapidly improving the state of the country's social and physical infrastructure and by establishing heavy manufacturing industry while simultaneously achieving satisfactory growth performance of agriculture and light manufacturing so as to ensure a broadly-based economic progress.[4] It can be implied therefore that the development strategy sought to strengthen the general directions of the historical pattern of development by enhancing rapid expansion of manufacturing and, at the same time, counteract the presumed negative impact of the oil boom on non-oil tradables, i.e., the agricultural and manufacturing sectors.

There are some clear differences between the development strategies adopted in the 1960s compared to those of the 1970s. In the earlier period, the primary objective was to achieve an expansion of the manufacturing sector, largely at the expense of agriculture. As an import-replacing strategy, the main focus was on light manufacturing to produce consumer goods. In the 1970s, however, the strategy had the objective of enhancing the growth of both agriculture and manufacturing industry to reduce reliance on the oil sector. Within agriculture, the emphasis was to encourage increased production of food rather than export crops; in manufacturing, the focus was on heavy industry.

3. ECONOMIC GROWTH PERFORMANCE

Nigeria's GDP increased by about 4 per cent per annum in real terms between 1950 and 1960. In the first half of the decade, growth was spurred largely by exceptionally high export prices in spite of the relatively low rate of gross domestic investment (which was about 6 per cent of GDP in 1950 rising to 10 per cent in 1955). In the second half, however, growth was due more to rising investment. By 1960, gross investment was about 15 per cent of GDP while public investment had risen from 30 to 44 percent of total investment during the decade. But the structure of public investment was skewed in favour of infrastructure and social services which accounted for between 70 and 80 per cent of the total.

Nigeria's first decade as an independent country was not an entirely happy one. The fragile political system was sorely tested in the second half of the decade; the internal conflicts which started early in 1966 culminated in a brutal civil war between mid-1967 and January 1970. GDP in real terms is estimated to have grown at an average annual rate of 5.5 per cent during the first half of the 1960-1970 decade, but declined sharply between 1966 and 1969 before picking up again 1970 with the restoration of peace. This growth performance reflects the fact that the war caused major damage to production assets, with industrial and power installations, roads and transport facilities suffering the most due to serious disruptions of the transport system, over-use of existing facilities and lack of maintenance.

Between 1950 and the mid-1960s, practically all the funds required for investment were raised within Nigeria. High agricultural export prices particularly between 1947 and 1954, enabled the country to maintain substantial export surpluses and the accumulation of large foreign exchange reserves. In addition, the marketing board system was used to transfer a large part of the 'windfall' arising from high agricultural export prices to the government. As a result, more than 85 per cent of the approximately ₦800 million of public investment between 1955 and 1965 was financed without recourse to foreign assistance. Rather the required funds were derived from current budget surpluses (34 per cent), surplus of marketing boards (22 per cent), earnings of statutory corporations (11 per cent), surpluses accumulated before 1955 (10 per cent), and domestic borrowing mainly from banks (9 per cent).

During the second half of the 1960–70 decade, finance for public investment became a problem. Export prices were falling, previously accumulated surpluses had been used up, and rapidly increasing government recurrent expenditure was making it increasingly difficult to achieve current budget surpluses. Trade deficits had increased from 1955 onwards, and capital inflows were not sufficient to finance them. The exigencies of the civil war increased the government's current expenditures in 1967 by nearly 50 per cent over 1966, and in spite of a reduction in investment expenditure the overall deficit increased sharply (by 47 per cent between 1967 and 1968). Disbursements made under existing loan agreements and grants financed only about 25 per cent of the deficit, with the balance being financed by government borrowing from the banking system. It is clear, therefore, that the government funded the war effort largely through deficit financing, virtually all of which was covered by domestic borrowing; the government's overall deficit increased from ₦96 million in 1966 to ₦400 million in 1969 while its indebtedness to the banking system over the same period increased by nearly ₦520 million. Thus, internal debt nearly trebled over the war period, while external public debt increased by less than 10 per cent.

Despite the upsurge in deficit-financing, particularly in the last three years of the period, it is clear that Nigeria's economic growth between 1950 and 1970 was fuelled mainly by agricultural exports which generated income domestically, provided the major part of government revenues, and supplied the foreign exchange for importing intermediate and capital goods needed to launch industrialisation.

The most striking feature of the Nigerian economy between 1970 and 1973 is its remarkably rapid recovery from the dislocation caused by the civil war. The trends in these three years indicate that the war did not permanently damage the economy, but simply delayed the full impact development forces which were already emerging in the 1960s. This recovery is explained partly by the developments in the oil sector (sharp increases in both quantity and price of crude oil exports) and partly by an element of 'catching up' as the economy responded to the war period. Thus, in the two years immediately following the war, GDP increased by about 50 per cent in real terms and even during 1969–70, continued investment activities and substantial revival of production outside the war-affected areas had already resulted in a level of GDP higher than the maximum pre-war level of the whole economy. The powerful resurgence of economic activity is underlined by the experience in 1971–72 when investment expenditures increased by about 33 per cent, exports by 20 per cent, imports by 22 per cent and consumption expenditures by 11 per cent. Unlike the 1950s and 1960s, when growth was widely based and included substantial growth of agriculture and manufacturing industry, economic growth during this period was founded on the oil and government sectors; this growth resulted mainly from the phenomenal growth of the oil sector. The oil boom of 1973–74 enabled Nigeria to embark upon a massive development process, which placed a strong emphasis on infrastructure, social services (particularly education) and heavy industry.

During the main oil-boom period between 1974 and 1978, GDP grew at an average annual rate of 6 per cent in real terms. But, as in the early 1970s, the growth continued to be concentrated in the service and infrastructure (especially construction) sectors. The commodity producing sectors failed to grow commensurately with GDP and domestic demand, agriculture in particular, stagnated. Hence, the rapidly growing income was spent more and more on imports. The fragility of the oil-dominated economy has been vividly demonstrated since 1978. A softening of the world oil market in 1977–78 sent GDP down to an average annual growth rate of -1.3 per cent between 1978 and 1981, and as the oil market worsened further in mid-1981, the growth performance of GDP declined further to an average annual rate of -4.7 per cent between 1981 and 1984.

The financial side of general economic performance during 1970–84 is also mixed. Shortly after the massive oil price increases in 1973–74, all aspects of the financial situation improved dramatically; the savings rate (i.e., percentage of GDP) rose from an average of 15 per cent in 1965–70 through 23.3 per cent in 1970–73 to over 30 per cent in 1973–78. Correspondingly the investment rate (percentage of GDP) was also close to 30 per cent between 1973 and 1981. This high rate of investment, however, was not accompanied by a high rate of economic growth, partly because the investment mix favoured large infrastructural and heavy industrial projects with large gestation periods, and partly because the investment was apparently not highly productive.[5] The implicit economy-wide incremental capital output ratio (ICOR) is estimated to have risen from less than 3 in the 1960s to about 6 in the 1970s as the high level of capital formation appears to have strained the capacity of the construction sector and increased construction costs significantly (at about 20 per cent per annum) during the 1970s.

As the increased oil revenues accrued largely to the government, its financial position improved dramatically in the first half of the 1970s. The budget deficits of the 1965–70 period were replaced by substantial surpluses. But the rapid expansion of public sector expenditures soon changed the pictures; between 1970 and 1976, federal capital expenditures increased 40-fold, state capital expenditures also grew by a factor of 16, while total public sector expenditures accounted for about 50 per cent of GDP in 1976 compared to about 20 per cent in 1970. The accumulated internal and external surpluses made possible by the oil boom did not therefore, last long; sizable budget deficits emerged as early as 1976. Similarly, balance of payments deficits re-emerged in 1976; while the value of imports increased by almost 300 per cent between 1974 and 1976, the value of exports hardly grew at all. Hence, external reserves began a downward trend. The financial situation became critical in 1978; the deficit on current accounts reached ₦ 2.4 billion which had to be financed by borrowing from the Eurodollar market for the first time on a large scale ($1.75 billion), as well as by further drawing down foreign exchange reserves. Increased oil exports (both in quantity and price terms) in 1979–80 turned the situation around, but only temporarily because export revenues declined rapidly in 1981. Nigeria's apparent inability to reduce imports and government expenditures sufficiently to match declines in oil revenues during the 1980s has resulted in a massive build-up of internal and external debt. Total internal debt grew from just over ₦1 billion in 1970 to about ₦8 billion in 1980 and to over ₦25 billion in 1984; approximately 75 per cent of the internal debt was owed to the banking system, while the debt service on internal debt was equivalent to 20 per cent of government revenue in 1984. Similarly, public and publicly guaranteed external debt (outstanding and disbursed) rose from less than $0.5 billion in 1970 to $4.4 billion in 1980 and almost $12 billion in 1984, the debt service ratio (percentage of export earnings) rose from 4.2 per cent in 1970 to 25.6 per cent in 1984, while external debt as a proportion of GDP increased from 0.7 per cent to 5 per cent between 1970 and 1984.

4. SOURCES OF GROWTH

Implicit in the analysis of Nigeria's economic growth performance presented in section 3 is the suggestion that agricultural exports played an important role in the growth process during the 1950–70 period, while oil exports played a similar role in the subsequent (1970–84). In this section the suggestion is examined more closely within the framework of 'sources of growth' generated through a production function approach (Robinson, 1971; Hwa, 1983). Rates of growth of GDP have been shown in general to be positively correlated with export growth rates in various studies (Michalopoulos and Jay, 1973; Krueger, 1977; and Michaely, 1977). In addition, several studies which have examined the contributions made by capital and labour to GDP growth rates in combination with exports have also found a positive relationship between exports and economic growth (Balassa, 1978; Feder, 1982).

The model behind this exercise assumes a Cobb-Douglas production function:

$$Y = AK^{\alpha}L^{\beta}e^{\log R} \quad (1)$$

in which Y is GDP, A is a scale parameter, K is capital stock, L is labour force and $\ln e^{\log R}$ represents the rate of productivity change over time. Taking time derivatives of equations (1) yields:

$$\dot{Y} = \alpha\dot{K} + \beta\dot{L} + \dot{R} \quad (2)$$

Assuming that the rate of productivity change is positively related to the rate of export growth in the form

$$\dot{R} = C + \sigma\dot{E} + \varepsilon \quad (3)$$

Where C is a constant and ε is assumed to be a randomly distributed residual. Then, equations (2) and (3) combine into:

$$\dot{Y} = C + \alpha K + \dot{\beta} L + \dot{\sigma} E + \varepsilon \quad (4)$$

where:

- $\dot{Y}$ = average annual GDP growth rate;
- $\dot{K}$ = average annual rate of growth of capital (proxied by the average gross domestic investment/GDP %;
- $\dot{L}$ = average annual rate of growth of the labour force (proxied by population growth rate);
- $\dot{X}$ = average annual rate of growth of exports.

Exports are separated into two major components in the estimated regression equations; agricultural exports growth rate ($\dot{X}_A$) here is used for the 1950–70 period while oil exports growth rate ($\dot{X}_o$) is used in the 1970–80 period. The regression results[6] are:

(a) *1950–70 period*

$$\dot{Y} = \underset{(4.7193)}{0.7638} + \underset{(2.0013)}{0.1476\dot{K}} + \underset{(2.1044)}{0.7050\dot{L}} + \underset{(2.7003)}{0.0744\dot{X}}$$

$$R^{-2} = 0.5214; \text{ D.W.} = 1.1971 \quad (5)$$

(b) *1970–84 period*

$$\dot{Y} = \underset{(-3.9746)}{-0.1596} + \underset{(1.9986)}{0.0711\dot{K}} + \underset{(2.3560)}{0.5513\dot{L}} + \underset{(2.4881)}{0.0342\dot{X}}$$

$$R^{-2} = 0.5311; \text{ D.W.} = 1.9173 \quad (6)$$

Table 9.1 Sources of economic growth

	1950–1970			1970–1984		
	Average growth rate (%)	*Share (%)*	*Mean value*	*Average growth rate (%)*	*Share (%)*	*Mean value*
Accounting for GDP growth						
Factors of production	3.94	74.34	–	3.18	83.68	–
Capital	2.32	43.77	15.7	1.64	43.16	23.1
Labour	1.62	30.57	2.3	1.54	40.52	2.8
'Residual'	1.36	25.66	–	0.62	16.32	–
GDP	5.30	100.00	–	3.80	100.00	–
Accounting for residual						
Agricultural exports	0.96	18.11	12.9	–	–	–
Oil exports	–	–	–	0.78	20.53	22.7
Others	0.40	75.55	–	–0.62	–4.21	–
'Residual'	1.36	25.66	–	0.62	16.32	–

Source: Derived from regression results on p.199, following Robinson (1971).

Both regression results show that exports, along with labour and capital, contribute positively and significantly (statistical level of significance: 5 per cent) to the growth of GDP, although the three variables included in the regression equations only account for just 50 per cent of the variations in GDP growth rates.

'Sources of growth' are derived on the basis of these regression results[7] and presented in Table 9.1.

A comparison between the two periods reveals some major differences. The average GDP growth performance was, for instance, substantially better in 1950–70 (5.3 per cent), than in 1970–84, (3.8 per cent) in spite of the much larger resources available during the latter period. Although the combined relative contribution of capital and labour was higher in 1970–84 (84 per cent) than in 1950–70 (74 per cent), the actual contribution to GDP growth rate was greater in the first period (3.94 per cent) compared to the second (3.18 per cent). Although the investment rate was substantially higher in the second than the first period, its contribution to growth in the second period was less than half of its contribution in the first period. Similarly, while agricultural exports provided 0.96 per cent of the average annual growth rate of GDP in 1950–70, oil exports made a smaller (0,78 per cent) contribution to GDP growth rate in 1970–84. The rather wide fluctuations in oil exports, the large role of the public sector made possible by increased oil revenues, and the negative impact of the oil boom on non-oil tradables together created a pattern of investment and consumption spending which was apparently not conducive to substained long-term growth.

5. SECTORAL SHIFTS—EMPIRICAL ANALYSIS

The growth processes whose manifestations are analysed in sections 3 and 4 above were accompanied by significant sectoral shifts. The discussion in section 2 provides some generalisations about the pattern of structural changes which characterise growth processes. The primary objective of this section is to identify and analyse the major sectoral shifts which occurred in the Nigerian economy between 1950 and 1984 in relation to the general patterns shown in section 2. The analysis focuses specifically on sectoral shifts in the structure of production, employment and trade.

Sectoral shifts relating to the structure of production can be analysed in terms of both gross output and value added. Table 9.2 presents a decomposition of GDP (current prices) in terms of five key sectors: agriculture, oil and mining, manufacturing, infrastructure and services. It is clear from this table that agriculture's share of total GDP declined consistently between 1950 and 1980. During the first decade of this period, agriculture's share fell from 69 to 61 per cent, it declined further to 36 per cent in 1970 and 19 per cent in 1980, before rising again to 26 per cent in 1984. Except for the relative upsurge between 1980 and 1984, the decline in agriculture's share appears to have followed the historical pattern of development enunciated in the Chenery-Syrquin model. The rate of decline was particularly sharp in the 1970–80 period (41 per cent), compared to the earlier decade 1960–70 (12 per cent). The oil boom was probably responsible for the accelerated rate of decline, just as the oil slump in the 1980s enabled the agricultural sector to regain some of its lost relative share. The oil sector exploded between 1970 and 1980 as its share of GDP rose from 8 to 33 per cent in 1980 before dropping back to 21 per cent in 1984 as the slump in the world oil market took hold.

Manufacturing remained a disappointment over the 1950–84 period; although its share of GDP rose from 2 to 7 per cent between 1950 and

Table 9.2 Structure of GDP (current prices)

	1950		*1955*		*1960*		*1965*		*1970*		*1975*		*1980*		*1984*	
	₦ (m)	*(%)*	*₦ (m)*	*(%)*	*₦ (m)*	*(%)*	*₦ (m)*	*(%)*	*₦ (m)*	*(%)*	*₦ (m)*	*(%)*	*₦ (m)*	*(%)*	*₦ (m)*	*(%)*
Agriculture	815.2	68.8	1 013.5	62.8	1 316.0	61.3	1 574.7	52.6	2 554.4	36.4	5 354.7	25.6	8 832.0	19.2	12 165.7	26.0
Oil and mining	15.4	1.3	19.4	1.2	26.0	1.2	143.0	4.8	568.0	8.1	4 668.4	22.3	15 072.4	32.8	9 923.0	21.2
Manufacturing	22.5	1.9	38.7	2.4	107.6	5.0	214.6	7.2	313.0	4.5	1 170.4	5.6	2 354.4	5.1	2 372.5	5.1
Infrastructure	93.6	7.9	192.0	11.9	226.6	10.5	348.6	11.6	746.0	10.6	2 551.6	12.2	5 638.1	12.3	5 916.8	12.7
Services	238.3	20.1	350.2	21.7	472.5	22.0	712.2	23.8	2 832.0	40.4	7 211.9	34.5	14 053.7	30.6	16 394.6	35.1
Total GDP	1 185.0	100.0	1 613.8	100.0	2 148.7	100.0	2 993.1	100.0	7 013.4	100.0	20 957.0	100.0	45 950.6	100.0	46 772.6	100.0

Note: Infrastructure = construction + transport + communication + electricity + gas + water.
Services = public administration + defence + other services.
Percentage columns do not always sum to 100 due to rounding.
Source: Central Bank of Nigeria, *Annual Reports*.

1965, it fell back to an average of 5 per cent through the rest of the period. Thus, it did not quite follow the historical pattern of development by increasing its share of GDP rapidly even before the 'Dutch disease' effect of the oil boom apparently choked it off in the 1970s. The nontradable sectors, i.e., infrastructure and services, taken together, accounted for 28 per cent of GDP in 1950; this share increased to 33 per cent by the end of the decade. Expansion of the share of nontradables in GDP was more rapid in the next decade and by 1970 the share stood at 50 per cent. It appears that the significant growth in the relative share of nontradables originated primarily in the sharp increase in the size of the public sector which accompanied the civil war, although the subsequent oil boom and the fact that increased oil revenues accrued largely to the government helped not only to sustain but also enhance this relative share.

The structure of value added reveals similar significant changes. Data on value added is more aggregative and less complete; only three sectors can be separated and data are available only from 1970. The sharp decline in agriculture's share noted previously with respect to GDP is also revealed for value added in Table 9.3. In the same way, the share of industry (i.e., oil, mining and manufacturing) roughly doubled between 1970 and 1975 before falling to less than 40 per cent in 1982. The service (i.e., infrastructure and other services) sector's share shows a different pattern; it actually fell slightly between 1970 and 1975 before rising to 39 per cent in 1982.

The structure of GDP and value added reflects only one aspect of the sectoral shift in the economy. Sectoral employment shares illustrate another important aspect of these changes. As Table 9.4 shows, however, the dramatic structural changes implied by the structure of GDP are less pronounced in the case of the sectoral distribution of the labour force. Thus, although agriculture's share declined over the 1960–85 period, it dropped much less than its share of GDP or value added. In accordance with the findings of the Chenery-Syrquin model, however, the predominant shift in the pattern of employment has been from agriculture ato infrastructure and the services sector.

Significant differences in the sectoral shares of output and employment imply substantial differences in sectoral labour productivity. Per capita income grows partly as workers are shifted from low to high productivity sectors, and partly by raising labour productivity within each sector. Table 9.5 provides data on sectoral and labour productivity changes. Productivity increased in all sectors, particularly between 1970 and 1975. Industry's productivity increased by

Table 9.3 Structure of value added

	1970 $ *(million)*	*(%)*	*1975* $ *(million)*	*(%)*	*1982* $ *(million)*	*(%)*
Agriculture	3 284	50.0	9 897	30.4	15 464	21.8
Industry	1 062	24.4	15 032	46.2	26 709	39.0
Services	1 683	25.6	7 636	23.4	27 849	39.2
Total	6 569	100.0	32 565	100.0	71 022	100.00

Source: Federal Office of Statistics, *Annual Abstract of Statistics*, several years.

463.5 per cent, services by 276.1 per cent and agriculture by 177.4 per cent over this period. Thus, industry's labour productivity grew almost twice as fast as services and about twice as fast as agriculture. Increases in labour productivity between 1975 and 1982 were lower for all sectors; and within this period, productivity in the service sector grew by 160.5 per cent compared to the industry's 49.4 per cent and agriculture's 62.3 per cent. Expressed in another way, productivity in agriculture declined from approximately 40 per cent to 10 per cent of the industrial sector's labour productivity between 1970 and 1975. Labour productivity in the services' sector started at about 80 per cent of industry's in 1970, declined to 53 per cent in 1975 before rising to 92 per cent in 1982. This pattern of sectoral labour productivity differences appears to be broadly consistent with sectoral changes in the structure of employment discussed earlier. It is also roughly consistent with general historical pattern of development in which the gap (in labour productivity) between industry and agriculture widens substantially as income increases.

An analysis of the structure of exports provides a sharper picture of another aspect of the structural changes. Looking first at the ratio of output of each sector which is exported, Table 9.6 demonstrates the performance of agriculture and manufacturing sectors relative to the oil and mining sector. The impact of the oil-boom is clearly demonstrated by the rapid increase in the share of exports out of the output of all tradables up to 1981; this share rose from an average of 32 per cent in 1960–65 to an average of 73 per cent in 1978–81, before falling back to 53 per cent during 1981–84. In sharp contrast, the share of agricultural output which is exported fell steadily from 28 per cent in

Table 9.4 Structure of employment

	1960 Labour force (million)	(%)	*1970* Labour force (million)	(%)	*1975* Labour force (million)	(%)	*1982* Labour force (million)	(%)
Agriculture	14.8	71.0	16.8	70.0	18.3	61.2	19.6	59.8
Industry	2.1	10.0	3.1	12.9	5.2	17.4	6.2	18.9
Services	3.9	19.0	4.1	17.1	6.4	21.4	7.0	21.3
Total	20.8	100.0	24.0	100.0	29.9	100.0	32.8	100.0

Source: Federal Office of Statistics, *Annual Abstract of Statistics*, several years.

Table 9.5 Labour productivity (value added per worker)

	1970		1975		1982	
	$	% of Average	$	% of Average	$	% of Average
Agriculture	195	71.4	541	49.7	878	40.6
Industry	513	189.7	2 891	265.5	4 320	199.5
Services	406	148.7	1 527	140.2	3 978	183.7
Economy-wide Average	273	–	1 089	–	2 165	–

Source: Federal Office of Statistics, *Annual Abstract of Statistics*, several years.

the early 1960s to roughly 3 per cent in the 1980s. Similarly, manufactured exports as a proportion of the output of manufactures declined uniformly over the same period, falling from 13 per cent in 1960–65 to 1 per cent in 1981–84.

A more direct evidence of the changing structure of merchandise exports is provided in Table 9.7. The value of total merchandise exports rose sharply, particularly during the 1970–80 period; but this increase was essentially due to the oil boom. In comparison, the value of agricultural exports rose by about $100 million between the early and mid-1960s, then it more or less stagnated at approximately $450 million annually between 1965 and 1975, rose above $600 million briefly and fell back to its 1962 value of about $370 million by 1984. In terms of shares, the decline of agricultural exports is even more dramatic; while it accounted for over 80 per cent of merchandise exports in 1962, its share fell to 37 per cent in 1970 and even more sharply to 3 in 1984. Export of manufactures has never been significant of course; hence, as the oil and mining sector's contribution to exports expanded, it quickly replaced agriculture as the dominant export sector; while it accounted for only 11 per cent of exports in 1962, its share went up to 32 per cent in 1965 and by 1970 it contributed over 60 per cent, and has accounted for over 90 per cent annually between 1975 and 1984.

The discussion so far has established that substantial sectoral shifts accompanied the growth of Nigeria's economy between 1950 and 1984. An application of the constant-market-shares model to analyse the structural shifts would assist in providing further insights. This model examines structural changes within a specified period. Sup-

Table 9.6 Share in exports in tradable output

	1960–65 (%)	*1965–70* (%)	*1970–73* (%)	*1973–78* (%)	*1978–81* (%)	*1981–84* (%)
1. Share of exports in tradables output	31.6	38.0	45.7	68.6	73.4	52.6
2. Share of exports in agricultural output	27.6	26.9	14.1	10.7	6.5	3.1
3. Share of exports in manufacturing output	12.6	8.3	6.6	5.9	4.8	1.0

Note: Tradables = agriculture + manufacturing + oil + mining
Source: Central Bank of Nigeria, *Annual Reports*, various years.

Table 9.7 Structure of merchandise exports

		1962	*1965*	*1970*	*1975*	*1980*	*1984*
Agriculture:	Value ($ mil)	372.7	481.5	447.6	459.2	622.3	377.9
	Share (%)	81.1	65.3	36.5	5.8	2.3	3.1
Oil & mining:	Value ($ mil)	52.3	238.9	765.6	7 485.7	24 744.8	11 568.0
	Share (%)	11.4	32.4	62.4	93.8	97.3	96.8
Manufacture:	Value ($ mil)	34.3	16.9	14.7	38.5	71.4	2.4
	Share (%)	7.5	2.3	1.1	0.4	0.3	—
Total merchandise exports ($ mil)		459.3	737.3	1 227.9	7 983.4	25 438.5	11 948.3

Source: Central Bank of Nigeria, *Annual Reports*, various years.

pose that values of GDP at the beginning and end of the period are represented by GDP_0 and GDP_1 respectively. Suppose also that at the beginning of this period, the value of agricultural output is related to GDP during the same time interval so that agriculture (AS_0) contributes b_0 per cent of GDP_0. Then, at the end of the period, the value of agricultural output ought, hypothetically, to be $AS^*_1 = b_0$ per cent of GDP_2, if no structural change has occurred during the period. But if any structural change has taken place, the actual value of agricultural output at the end of the period (AS_1) will be different from its hypothetical value (AS^*_1). In fact, the total (actual) change (TC) in the value of agricultural output can be decomposed into two parts:

$$TC = AS_1 - AS_0 \quad (7)$$
$$= (AS_1 - AS^*_1) + (AS^*_1 - AS_0) \quad (8)$$

such that the structural shift effect (SSE) is captured by the first component ($AS_1 - AS^*_1$), and the overall economic growth effect (EGE) equals the second component ($AS^*_1 - AS_0$). In essence, SSE can be identified as the change in a particular sector's share of total GDP that results from shifts in the distribution of that GDP among its component sectors. EGE, in comparison, reflects the impact on a particular sector of changes in the overall size of the GDP. Thus, given a set of constant relative sectoral shares, particular sectoral values may increase or decrease as a result of general expansion or contraction of the GDP. Total change for a given sector is made up of these two parts. The decomposition achieved by this simple analytical technique is obviously not perfect; it implicitly relies on a 'normal' growth pattern that assumes proportional growth rates in each sector and compares this with situations where sectoral growth rates may, in fact, be uneven. But it does produce a general indication of the direction and effects of sectoral shifts even where growth is non-proportional.

The effects of structural change on sectoral output in both value and percentage terms are presented in Table 9.8. The results reveal a clear pattern with respect to agriculture. This sector suffered substantial output losses between 1950 and 1980, and gained between 1980 and 1984. Thus, agriculture's output in 1955 was ₦97 million (or 10 per cent) below its normal growth projection from 1950. This loss of output increased in subsequent years and was particularly high between 1965 and 1980. The reversal between 1980 and 1984 is a

Table 9.8 Output effects of structural change by sector in value and per cent

	1950–55	*1955–60*	*1960–65*	*1965–70*	*1970–75*	*1975–80*	*1980–84*
Agriculture							
Value (₦ million)	–96.8	–33.6	–260.1	–1 120.6	–2 273.6	–2 931.4	3 185.4
Per cent	–9.6	–2.6	–16.5	–43.9	42.5	–33.2	26.2
Oil & mining							
Value (₦ million)	–1.6	0.2	107.1	213.4	2 970.9	4 825.4	5 418.4
Per cent	8.2	0.8	25.1	40.7	63.6	32.0	–54.6
Manufacturing							
Value (₦ million)	8.0	56.0	64.9	–192.0	227.3	–218.8	–12.9
Per cent	20.7	52.0	69.8	–61.3	19.4	–9.3	–0.5
Infrastructures							
Value (₦ million)	64.5	–29.1	34.3	–67.6	330.0	32.1	163.8
Per cent	33.6	–12.8	30.7	–9.1	12.9	0.6	2.8
Services							
Value (₦ million)	25.8	6.2	53.7	1 162.8	–1 254.7	–1 799.1	2 082.2
Per cent	7.4	1.3	7.5	41.1	–17.4	–12.8	12.7

Source: Central Bank of Nigeria, *Annual Reports*.

reflection of the oil slump, which had the effect of stemming, if not reversing, the flow of resources from agriculture. The oil and mining sector gained significantly from structural changes, particularly between 1965 and 1980. In the 1965–70 period, for instance, oil and mining sector's gross output was ₦213 million or 41 per cent above its normal growth projection; its relative gain from structural change was 64 per cent in 1970–75 and 32 per cent in 1975–80. This trend was reversed in 1980–84 during which period the sector suffered a 55 per cent loss of output.

The generalisations of both the Chenery-Syrquin and 'Dutch disease' models appear to have been confirmed with respect to agriculture's structural-change induced relative decline in the 1950–80 period; the negative impact of the 'oil syndrome' apparently intensified the established pattern of decline. The manufacturing sector's output gains from structural change from 1950 to 1965 also seem to be consistent with the historical development pattern. But the pattern of gains and losses exhibited by this sector between 1970 and 1980 is not entirely in agreement with the simple generalisations of the 'Dutch disease' model to the effect that non-oil tradables would decline as a result of the oil boom. In fact, the cumulative effects of structural change on the manufacturing sector turn out to be essentially neutral during the 1970–80 period; while the sector suffered a small (0.5 per cent) loss during the oil slump (1980–84) period. Taken together, the output gains and losses of the infrastructure and service sectors during 1970–80 do not appear to be consistent with the expected 'Dutch disease' results. Losses suffered by the service sector over this period completely swamp the gains recorded by the infrastructure sector; and both sectors gained substantially from structural change in 1980–84 after the oil boom had subsided.

Turning to the employment effects of structural change, it is clear, from the results provided in Table 9.9, that agriculture was the net loser through 1960–82. Agricultural sector's level of employment was 1.2, 14.2, and 2.6 per cent below its 'expected' level in 1970, 1975 and 1982 respectively. The clear gainer from structural changes was industry (i.e., oil, mining and manufacturing) whose employment level was 22.6, 25.0 and 8.1 per cent above the 'expected' in 1970, 1975 and 1982. Services gained (20 per cent) in employment in 1970–75, but had lost (12.2 per cent) earlier (1960–70), while structural change had no employment effect on the sector in 1975–82.

The analysis presented in this section shows that at the level of a comparison of sector shares of output, employment and trade over

Table 9.9 Employment effects of structural change

	1960–70 Labour force (million)	(%)	1970–75 Labour force (million)	(%)	1975–82 Labour force (million)	(%)
Agriculture	–0.2	– 1.2	–2.6	–14.2	–0.5	–2.6
Industry	0.7	22.6	1.3	25.0	0.5	8.1
Services	0.5	– 12.2	1.3	20.0	0.0	0.0

time, the Nigerian experience broadly fits the main generalisations of the Chenery-Syrquin and 'Dutch disease' models, particularly with respect to agriculture. The share of this sector fell, as the Chenery-Syrquin model predicts, with increases in income. The rate of decline also increased during the period of the Nigerian oil boom, as expected. An attempt to pin down the effects of structural change on different sectors with respect to output and employment, using a version of the constant-market-share model, broadly confirms this general result with respect to the agricultural sector; but does not provide robust support for a structural-change induced relative expansion of nontradables implicit in the 'Dutch disease' model.

6. POLICY INTERVENTION

Changes in the structure of an economy are desirable to the extent that they contribute to economic growth, perhaps by increasing the rate of capital accumulation or by shifting resources towards sectors in which productivity is higher than average or rising faster. In such cases, deliberate policy measures may be designed to encourage and enhance the changes. There may be other cases, however, in which sectoral shifts, induced perhaps by a short-lived oil boom, are not consistent with a country's sustainable long-term development so that policy measures are aimed at excluding or slowing down the contraction of non-oil tradable sectors or reducing the expansion of non-tradables. Nigeria's choice of development strategy in the 1960s appears to have been influenced by the assumption that certain sectoral shifts were desirable and that they could be induced or strengthened by policy. The advent of the oil boom in the 1970s presented more difficult problems of economic management. Most of the sector-specific policies are apparently directed towards protecting non-oil tradables to ensure that the effects of sectoral shifts did not result in their contraction either relatively or absolutely. But many of the economy-wide policies were, in reality if not by intention, pointed in the opposite direction. Hence, an analysis of policy intervention with respect to economic growth and sectoral shifts in Nigeria during 1950–84 needs to focus on the effects of the interactions between sector-specific and macroeconomic policies.

In the early 1960s, the clear policy objective in Nigeria was to promote industrialisation, using resources drawn from agriculture to finance the development effort. It would not be surprising therefore to find that some sector-specific policies favoured the manufacturing sector while others discriminated against agriculture. Policies penalising agriculture did not originate in the 1960s, of course. Helleiner (1964) has shown quite clearly that agriculture was heavily taxed through the 1940s and 1950s – between 1941 and 1961 total withdrawals by taxation as a proportion of potential producer income for the major agricultural crops were: cocoa (32 per cent), groundnuts (25 per cent), palm kernels (28 per cent). What is more significant in the 1960s is that this policy was continued even though export prices were falling. During the 1960–70 period, cocoa producers received only 65 per cent of cocoa export earnings; similar proportions for other major cash crops over the same period are: groundnuts (58 per cent), cotton (20 per cent), palm oil (59 per cent), and palm kernel (51 per cent). It is clear therefore that agricultural marketing and pricing policy through 1950–70 was not aimed towards creating or improving producer incentives; rather it was decidedly pointed in the direction of withdrawing resources from the agricultural sector for the purposes of financing government spending and the development of other sectors, particularly manufacturing, infrastructures and social services. It is important to note, however, that this policy applied only to the export subsector of agriculture.[8] Domestic agricultural food production was not subject to government control through a marketing board system and hence escaped a similar tax burden.

Sector specific policy towards agriculture changed radically in the 1970s. The advent of the oil boom and increasing government concerns about poor agricultural performance led to reforms of the marketing board system in 1973 and 1977. All crop taxes were abolished in 1973, the boards were further reorganised in 1977, and the crop producer price-fixing authority was centralised, with the objective that crop producer prices would be fixed without providing for trading surpluses. At the same time, food crop producers were to be assisted through a system of guaranteed minimum prices, while very generous input (especially fertiliser) subsidies were established.

From the late 1950s, Nigeria gave a prominent place in its development goals and plan to policies directed towards stimulating the rapid growth of manufacturing industries. This, in practice, meant the pursuit of an industrialisation strategy based essentially on import substitution which has relied heavily on tariff and non-tariff protection. Following the classic import substitution tariff profile (Oyejide, 1975, Bertrand and Robertson, 1981), the tariff structure escalated from capital goods to consumer goods; nominal tariffs on capital goods averaged 5 per cent, intermediate goods averaged 20 per cent, while the tariff rates on consumer goods varied from 30 to 150 per cent. In addition, the system on industrial incentives included tax holiday and tax relief, accelerated depreciation allowance and import-duty refund on imported raw materials. In the 1965–70 period, balance-of-payments problems associated with the civil war led to a widening of the use of tariff and nontariff measures. Even though the increased use of import compression measures during this period was not necessarily directed towards encouraging industrial expansion via tariff protection, in practice it had the same effect. Imports were substantially liberalised between 1970 and 1973 as many of the tariffs and restrictions imposed during the civil war were rolled back. Increased oil revenues made liberalisation easy, and the desire to reduce inflation worked in the same direction. The protective nature of the tariff structure remained basically in place through the 1970s, however, and the increased balance-of-payments pressures since 1981 have helped to sustain high levels of protection for manufacturing industries.

The intention of policy intervention (at the sector-specific policy level) with respect to the manufacturing sector since the late 1950s has been to provide it with special incentives aimed at its rapid expansion. This policy intention has been pursued consistently before and after the oil boom. In the pre-oil period the apparent objective was to induce sectoral shifts in favour of the manufacturing sector while, after the advent of oil, the implicit objective has been to protect an important traded goods sector. The intention of sector specific policy intervention with respect to agriculture has changed over time. Up to 1970, agriculture was viewed essentially as a reservoir from which resources should be withdrawn for the development of other sectors; this changed, however, as the oil boom took hold and agriculture was beginning to be treated as a critical tradable sector that needed to be protected against sectoral shifts emanating from the 'Dutch disease' phenomenon.

The agricultural sector appears, at first sight, to have benefited from the more favourable policies directed towards it. As a result of the reforms carried out between 1973 and 1977, the nominal producer prices for most agricultural export crops were more or less the same as, or substantially above, their average export values (at the official exchange rate). During the 1977–84 period withdrawal of resources through export and sales taxes as well as marketing board 'surplus' ceased, and crops such as groundnut, cotton and palm oil were subsidised, at least in nominal terms. In addition, the trends in nominal producer prices were reversed. In general, average producer prices had declined between 1955–59 and 1960–69, the rates of decline ranging from 8 per cent in the case of palm kernel to 28 per cent for cocoa. But, between 1960–69 and 1970–76, average producer prices for each crop doubled; they increased further two to three-fold between 1970–76 and 1977–84. Furthermore, substantial tariff protection was extended to many agricultural (particularly food) crops with effect from the mid-1970s.

The signal produced by relative price trends appears to have been the opposite of that implied by a consideration of nominal price trends alone. First, the guaranteed minimum prices for food crops were not effective since they were never more than about 50 per cent of corresponding retail prices. Secondly, real crop producer prices (obtained by using the consumer price index to deflate nominal producer prices) actually fell or remained constant for most crops over the 1970–84 period. Thus, while nominal crop producer prices were raised substantially during the 1970s and the 1980s, their rates of increase were insufficient to compensate for the general effect of inflation, so that nominal price increases resulted in little or no production incentive. Similarly, the agricultural sector received a lower level of protection (35 per cent) on the average than manufacture.

In addition to sector-specific policies, general macroeconomic policies and developments have influenced growth and sectoral shifts in the economy. Up to the mid-1960s, Nigeria had pursued generally sound fiscal and monetary policies. Pressures to resort to inflationary deficit financing mounted at this time. Previously accumulated reserves were exhausted, export crop prices were falling, and the civil war sharply increased government's current expenditures. A deficit

Table 9.10 Some indicators of macroeconomic policy

	Budget deficit (−) or surplus (+) (₦ mill.)	*Deficit or surplus as % of GDP*	*Current account balance (₦ mill.)*	*Rate of growth of M1 (%)*	*Inflation rate (%)*	*Nominal exchange rate (index 1970=100)*	*Real exchange rate (index 1970=100)*
1970	−188.8	−1.6	−262.9	43.8	13.6	100.00	100.00
1971	+36.2	+0.4	−289.6	4.2	16.1	99.79	92.52
1972	−82.7	−0.9	−225.0	11.5	2.8	92.11	93.40
1973	+188.9	+1.7	−5.5	24.0	5.5	92.11	99.58
1974	+1 247.6	+6.7	3 086.6	89.7	12.5	88.03	95.06
1975	−1 435.8	−6.8	26.2	48.3	33.8	86.16	78.38
1976	−1 870.1	−6.9	−223.5	48.3	24.2	87.72	64.43
1977	−2 134.2	−6.7	−656.4	43.8	19.5	90.24	66.03
1978	−1 185.0	−3.5	−2 404.6	−8.2	18.6	88.92	61.43
1979	−757.0	−1.9	−1 003.0	20.5	11.1	84.38	60.89
1980	−143.0	−0.3	2 789.5	50.1	11.4	76.52	59.20
1981	−4 734.0	−8.9	−3 593.2	5.6	20.9	85.93	52.90
1982	−4 524.0	−8.0	−4 851.9	3.1	7.7	94.25	52.88
1983	−6 650.0	−11.0	−3 030.5	12.4	18.9	101.28	44.16
1984	−3 984.0	−5.2	405.0	8.2	39.6	106.99	31.30

Sources: Federal Office of Statistics, *Annual Abstract of Statistics*, several years; Central Bank of Nigeria, *Annual Reports and Statements of Accounts,* several years; IMF, *International Financial Statistics*.

on the government's current account appeared for the first time in 1967, and in spite of reductions in investment expenditures, the overall deficit increased by 42 per cent from 1967 to 1968. Between 1967 and 1969, the government's internal debt (largely borrowing from the banking system) increased sixfold. This heavy deficit financing was accompanied by an increase in money supply, and eventually, beginning from 1969, prices began to rise significantly.

Table 9.10 provides data on some major indicators of the general direction of macroeconomic policy during 1970–84. It is clear from the trend in budget deficits that government spending ran out of control, the deficit financing practices of the civil war years were continued, and sharp increases in oil revenues in 1973–74 and 1979 had little impact on the trends of budget deficits and current account balances. The deficits were financed by borrowing internally and from abroad (including short-term trade credits); money supply increased sharply as did the rate of inflation. In addition, since Nigeria maintained a more or less fixed nominal exchange rate for a non-convertible currency with extensive capital controls, a significantly appreciated real exchange rate was inevitable. The effect of the high nominal currency overvaluation has been to reduce substantially the competitiveness of the non-oil tradables (i.e., agriculture and manufacturing) sectors as the negative impact of macroeconomic policies on producer incentives, particularly in agriculture, overwhelmed the effects of an otherwise favourable set of sector-specific policies. Thus the intention of sector-specific policies to protect both agriculture and manufacturing industries from the sectoral shifts expected to accompany the oil boom was negated by inappropriate macroeconomic policies.

Notes

1. These refer to the broad results of the basic model, some of which are qualified further in the rapidly growing literature.
2. This theme is elaborated further in Oyejide (1975).
3. Prices of major export crops fell by 20 to 30 per cent between 1954 and 1962.
4. This view emerges quite clearly from Nigeria's Third National Development Plan 1975–80.
5. Part of the problem was hastily and badly designed projects poorly executed, as well as official corruption.
6. Numbers in parentheses are *t*-values. The regression equations have been estimated using the ordinary least squares method.

7. Following Robinson (1971).
8. It should be noted also that the export subsector constituted only 20 per cent of total agriculture.

References

Balassa, B. (1978) 'Exports and Economic Growth: Further Evidence', *Journal of Development Economics*, vol. 5, no. 2 (June) pp. 181–9.

Bertrand, T. and Robertson, J. (1981), *The Structure of Industrial Incentives in Nigeria, 1979–80*, World Bank.

Chenery, H.B. (1979) *Structural Changes and Development Policy* (New York: Oxford University Press).

Chenery, H.B. and Syrquin, M. (1975) *Patterns of Development, 1950–1970* (London: Oxford University Press).

Corden, W.M. and Neary, J.P. (1982) 'Booming Sector and Deindustrialization in a Small Open Economy', *Economic Journal*, vol. 92 (December) pp. 825–484.

Edwards, S. and Aoki, M. 1983 'Oil Export Boom and Dutch Disease, A Dynamic Analysis', *Resource and Energy*, vol. 5, pp. 1–24.

Feder, G. (1982) 'On Exports and Economic Growth' *Journal of Development Economics*, vol 11, pp. 59–73.

Gelb, A. (1981) *Capital Importing Oil Exporters: Adjustment Issues and Policy Choices*, World Bank Staff Working Paper no. 475 (August).

Helleiner, G.K. (1964) 'The Fiscal Role of the Marketing Boards in Nigerian Economic Development, 1947–61', *Economic Journal*, pp. 582–610.

Hwa, E.-C. (1983) *The Contribution of Agriculture to Economic Growth: Some Empirical Evidence*, World Bank Staff Working Paper no. 619 (November).

Krueger, A.O. (1975) *Foreign Trade Regimes and Economic Development: Liberalization Attempts and Consequences* (Cambridge, Mass.: Ballinger).

Michaely, M. (1977) 'Exports and Growth: A Empirical Investigation', *Journal of Development Economics*, vol. 4, no. 1 (March) pp. 49–53.

Michalopoulos, C. and Jay, K. (1973) *Growth of Exports and Income in the Developing World: A Neoclassical View*, USAID Discussion Paper no. 28, (November).

Oyejide, T.A. (1975) *Tariff Policy and Industrialization in Nigeria* (Ibadan: Ibadan University Press).

Robinson, S. (1971) 'Sources of Growth in Less Developed Countries: A Cross-Section Study', *Quarterly Journal of Economics*, vol. 95, no. 3 (August) pp. 391–408.

Wood, A. (1986) *Growth and Structural Change in Large Low-Income Countries*, World Bank Staff Working Paper no. 763 (February).

10 A Study of Linkages between Agriculture and Industry: the Indian Experience*

Isher Judge Ahluwalia
CENTRE FOR POLICY RESEARCH, NEW DELHI

and

C. Rangarajan
RESERVE BANK OF INDIA, BOMBAY

There has been extensive literature exploring the interrelationships between agriculture and industry in the Indian economy, including some attempts at quantification of some of the linkages.[1] While the full complexities of these interrelationships cannot be modelled within the confines of a model, this study attempts to analyse certain crucial interlinkages between the two sectors within the framework of a macroeconomic model.

The two central issues addressed by the model are (i) the role of the variations in the terms of trade between agriculture (or parts of it) and non-agriculture (or parts of it) in influencing the demand for industrial consumption goods, and (ii) the possibility of a differential impact of the movements in the terms of trade on investment by different sectors. It is worth noting that while the demand side impact of an increase in investment on the output of the heavy industries is explicitly incorporated, the model does not trace the impact of increases in investment, particularly public investment, on creating infrastructural capacities and releasing supply bottlenecks in both agricultural and industry.

A similar model was earlier specified and estimated for the period 1961 to 1972 by Rangarajan (1982). The model presented in this study retains the principal specifications in the Rangarajan model but makes agricultural outputs endogenous along the lines specified in the macroeconomic model by Ahluwalia (1979). The model for the present study is estimated over a twenty-year period from 1960–61 to 1979–80. Another feature worth noting is that the model uses data on value added at constant prices for the use-based sectors of industry rather than the industrial production indices. The latter are known to suffer from serious limitations.[2]

Section 1 presents an overview of the trends in the growth of agriculture and industry covering the period form 1960–61 to date. Section 2 sets out the principal linkages between the two sectors. Section 3 presents the model – its specifications as well as the OLS estimates derived from using time series data for the period from 1960–61 to 1979–80. Section 4 analyses three simulations of the model with a view to tracing the effect of specified changes in the agricultural sector on the endogenous variables of the model. More specifically, the model is used alternatively to trace the effect of 'no drought' in 1974–75 and an increase in the trend growth rate of the output of foodgrains and non-foodgrains. The impact of an increase in the terms of trade is also analysed.

1. OVERVIEW OF TRENDS

An analysis of the trends in the growth of agriculture and industry in the Indian economy since the mid-1950s clearly reveals a continuing dominant role played by agriculture in the economy. The share of agriculture in the Net Domestic Product at 1970–71 prices was as high as 58 per cent in 1956–57 and it declined to 49 per cent in 1970–71 and 39 per cent in 1980–81 (Table 10.1). However, even in 1980–81 agriculture was by far the largest contributor to the Net Domestic Product.

A bird's eye-view of the relative rates of growth in the major sectors of the economy from 1956–57 to 1983–84 is provided in Table 10.2. Whether we take the period from the beginning of the Second Five Year Plan to 1983–84 or the twenty-year period ending in 1980–81, the growth of value added in agriculture was around 2 per cent per annum and that in industry was between 4.5 and 5 per cent per annum. The long-term growth of industry exceeded that of agriculture by 2.7–2.8 per cent per annum. The long term rate of growth of the Net Domestic Product at 1970–71 prices was of the order of 3.5 per cent per annum. As Table 10.2 shows, the first three years of the 1980s recorded a higher growth rate than the long-term average both for agriculture and for industry and this was reflected in

Table 10.1 Share of major sectors in net value added (per cent)

	1956–57	*1960–61*	*1970–71*	*1980–81*
Agriculture	58.4	51.2	49.2	39.2
Industry	12.4	15.4	15.2	19.0
Manufacturing	11.2	13.9	13.4	16.1
Registered	(6.3)	(8.0)	(8.3)	(10.1)
Unregistered	(4.9)	(5.9)	(5.1)	(6.0)
Electricity	0.3	0.5	0.9	1.5
Mining	0.8	1.0	0.9	1.4
Construction	4.4	4.7	5.4	5.1
Railways	1.3	1.9	1.5	0.9
Other services	23.5	26.8	28.7	35.8
Total	100.0	100.0	100.0	100.0

Source: National Accounts Statistics.

---

the Net Domestic Product increasing at over 5 per cent per annum during this period.

Over the two decades from 1959–60 to 1979–80, the output of foodgrains increased at a compound annual rate of 2.5 per cent, and when adjustments are made for the low base of 1979–80, the increase in the first four years of the 1980s has been of a slightly higher order. The evidence on non-foodgrains production suggests some slowdown in the growth after the mid-1960s.

The evidence on industrial growth is also characterised by an improvement since the mid-1970s. The recovery from 1976–79 was set back by the poor performance at the turn of the decade and an improvement after that. The growth in value added in industry at constant prices during the decade ending with 1965–66 was of the order of 6.5 per cent per annum and this declined sharply to 3.5 per cent per annum in the subsequent decade. The second half of the 1970s was characterised by a pick up so as to record a growth rate of 4.6 per cent per annum and the period from 1980–81 to 1983–84 witnessed a further increase in growth to 6 per cent per annum.

The annual growth rates for value added in agriculture and industry as well as the growth rates for the major subsectors of agriculture and industry are presented in Table 10.3. For the agricultural sector, these data highlight the cyclical variation in growth. Some important

Table 10.2 Trends in net domestic product at 1970–71 prices
(compound growth rates)
(per cent per annum)

	1956–57 to 1983–84	*1960–61 to 1980–81*	*1980–81 to 1983–84*
Agriculture	2.2	1.9	3.5
Industry	5.0	4.6	6.2
Manufacturing	4.7	4.4	5.8
Registered	5.3	4.9	7.3
Unregistered	3.8	3.8	3.3
Electricity	9.7	9.3	7.2
Mining	5.2	4.0	9.9
Construction	3.7	3.7	1.8
Railways	3.4	3.5	0.1
Other services	5.4	5.1	7.7
Total	3.6	3.4	5.3

Source: National Accounts Statistics.

---

observations can also be made for the industrial sector. The first half of the 1960s was the only the period when the growth rate was consistently high. The performance in the subsequent decade was disappointing, while the period since then has shown a pick up in growth. Figure 10.1 presents the annual growth rates in value added in industry against two-year moving averages of the growth rates in value added in agriculture. It shows the transmission of the cyclical movements of agricultural growth to industry, albeit with lower amplitude of fluctuations.

The variations in the terms of trade between agriculture and non-agriculture are shown in Figure 10.2. A number of methodologies have been evolved in the Indian literature to develop time series on the terms of trade between agriculture and non-agriculture. The graphs show the variations in four such series. First, the relative price of foodgrains and manufactures (referred to in the model as the foodgrains terms of trade) is defined as the ratio of the wholesale price indices of foodgrains and manufactures. An alternative series has been prepared by Thamarajakshi (1977) using as elements wholesale price indices and weights derived from intersectoral trade flows between agriculture and industry. Kahlon and Tyagi (1983) present a

Table 10.3 Growth of net value added in agriculture and industry (per cent per annum)

Year	*Agriculture*			*Industry*				
	Foodgrains	*Non-foodgrains*	*Value added*	*Basic goods*	*Capital goods*	*Interm. goods*	*Consumer goods*	*Total*
1960–61	7.1	10.7	6.3	10.9	17.0	–3.6	8.4	8.6
1961–62	0.8	–1.3	0.7	16.7	11.8	13.0	10.8	9.0
1962–63	–3.5	2.0	–2.7	30.3	18.2	14.3	0.7	8.0
1963–64	1.8	3.1	2.6	19.9	13.9	8.2	1.0	8.7
1964–65	10.5	12.3	9.3	11.8	21.7	–1.0	9.1	7.2
1965–66	–19.6	–10.7	–14.9	–0.3	11.9	–1.2	6.9	1.9
1966–67	1.7	–3.4	–1.6	–3.4	6.3	–1.5	3.3	–1.1
1967–68	28.0	12.4	16.2	–7.3	3.7	5.7	–4.8	1.8
1968–69	–1.4	–2.0	0.6	13.8	–4.5	6.5	6.1	4.5
1969–70	6.9	6.3	6.5	9.8	17.8	4.4	22.1	9.4
1970–71	8.6	4.9	8.8	19.9	3.7	9.6	–2.0	0.7
1971–72	–1.3	2.1	–0.8	–9.0	20.7	8.6	–1.1	2.9
1972–73	–8.2	–7.6	6.9	9.3	7.2	7.4	9.4	4.4
1973–74	7.8	14.3	8.0	–1.6	2.4	2.1	1.2	5.7
1974–75	–5.4	1.3	–2.7	2.4	1.6	–8.0	2.2	4.4

continued on page 224

Table 10.3 continued

Year	Agriculture			Industry				
	Foodgrains	Non-foodgrains	Value added	Basic goods	Capital goods	Interm. goods	Consumer goods	Total
1975–76	22.0	1.6	11.3	2.0	–0.6	3.0	1.5	2.9
1976–77	–9.1	–2.5	–6.3	12.8	13.7	5.6	10.9	8.7
1977–78	15.5	11.8	15.1	2.1	4.9	8.7	8.7	5.9
1978–79	4.3	2.9	2.7	16.6	10.6	12.4	11.4	10.1
1979–80	–17.6	–10.0	–13.7	–5.8	–0.6	4.4	0.6	–1.3
1980–81	19.8	7.1	13.5	—	—	—	—	0.2
1981–82	2.2	13.4	4.1	—	—	—	—	6.2
1982–83	–3.3	–4.7	–3.6	—	—	—	—	7.3
1983–84	18.6	3.8	10.6	—	—	—	—	5.1
1984–85	–3.7	5.7	—	—	—	—	—	—

Note: Value added in agriculture and industry at 1970–71 prices are taken from the National Accounts, while value added for the use-based sectors of industry are derived from the Annual Survey of Industries data using wholesale price indices at the disaggregated level as deflators. The production indices for foodgrains and non-foodgrains are obtained from Economic Survey, Ministry of Finance.

Source: National Accounts, Annual Survey of Industries, Economic Survey.

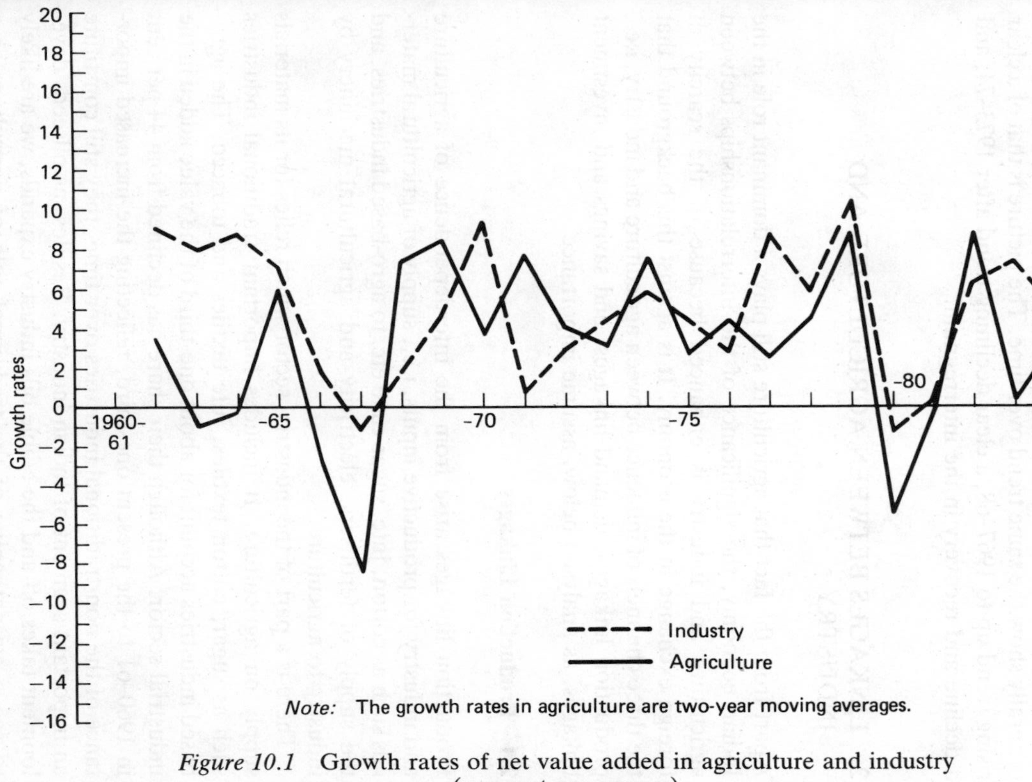

Note: The growth rates in agriculture are two-year moving averages.

Figure 10.1 Growth rates of net value added in agriculture and industry (per cent per annum)

third series with farmgate prices of agricultural products combining these prices using a set of weights obtained from the 26th round of the NSSO Consumer Expenditure for Cultivator Households – Rural (July 1971-June 1972). Finally, the overall terms of trade between agriculture and non-agriculture can be defined as the ratio of the implicit price deflators from national accounts. However, these series generally show the same trend over time. The picture is that of a clear rising trend up to 1967–68, a clear declining trend after 1973–74, and a decline and recovery in the interregnum.

2. LINKAGES BETWEEN AGRICULTURE AND INDUSTRY

Apart from the fact that agriculture still plays a dominant role in the Indian economy, the significance of the interrelationships between agriculture and industry is enhanced because of the scarcity of foreign exchange in the economy. It is against this background that the three channels of influence between agriculture and industry, i.e., production linkages, demand linkages, and savings and investment linkages, as analysed below, assume importance.

2.1 Production Linkages

Production linkages arise from the interdependence of agriculture and industry for productive inputs, i.e., supply of agricultural materials such as cotton, jute, sugar-cane etc. to agro-based industries, and the supply of fertilisers, electricity and agricultural machinery by industry to agriculture.

There is a part of the industrial sector which relies for its materials supply on agriculture. It includes important traditional industries such as sugar, cotton textiles, jute textiles and tobacco. The agro-based industries account for about one-third of the value added in the industrial sector. Although their share has declined from 44 per cent in 1960–61 to the present one-third, reflecting the increased importance of the non-traditional industries over time, they still constitute an important segment of Indian industry. Besides, as oilseeds development takes off and the edible oils industry expands, we are likely to see a strengthening of the traditional role of agriculture as a supplier of inputs to industry.

The production linkages between agriculture and industry can best

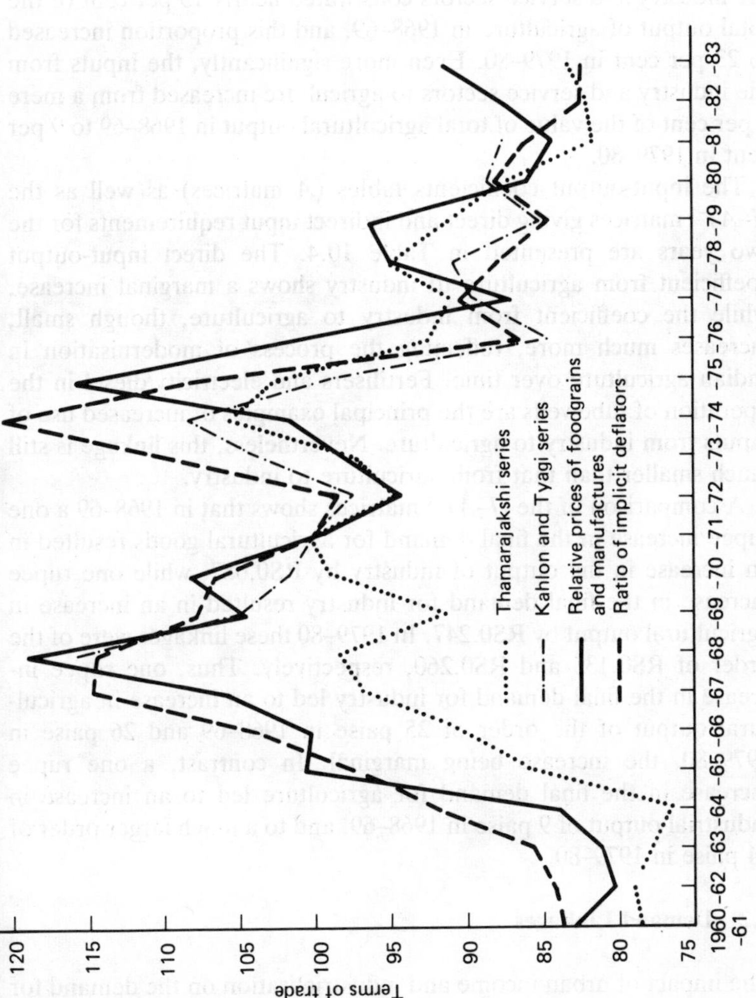

Figure 10.2 Terms of trade between agriculture and non-agriculture

be illustrated through the input-output tables for the economy. Comparing the inter-industry transactions tables at a three sector level of disaggregation for 1968–69 and 1979–80 at 1968–69 prices, it appears that the flow of output of agriculture and allied activities to the industry and service sectors constituted nearly 13 per cent of the total output of agriculture in 1968–69, and this proportion increased to 23 per cent in 1979–80. Even more significantly, the inputs from the industry and service sectors to agriculture increased from a mere 6 per cent of the value of total agricultural output in 1968–69 to 9 per cent in 1979–80.

The input-output coefficients tables (*A* matrices) as well as the $(I\text{–}A)^{-1}$ matrices giving direct and indirect input requirements for the two years are presented in Table 10.4. The direct input-output coefficient from agriculture to industry shows a marginal increase, while the coefficient from industry to agriculture, though small, increases much more, reflecting the process of modernisation in Indian agriculture over time. Fertilisers and electricity/diesel in the operation of tubewells are the principal examples of increased use of inputs from industry to agriculture. Nevertheless, this linkage is still much smaller than that from agriculture to industry.

A comparison of the $(I\text{–}A)^{-1}$ matrices shows that in 1968–69 a one rupee increase in the final demand for agricultural goods resulted in an increase in the output of industry by RS0.087, while one rupee increase in the final demand for industry resulted in an increase in agricultural output by RS0.247. In 1979–80 these linkages were of the order of RS0.135 and RS0.260, respectively. Thus, one rupee increase in the final demand for industry led to an increase in agricultural output of the order of 25 paise in 1968–69 and 26 paise in 1979–80, the increase being marginal. In contrast, a one rupee increase in the final demand for agriculture led to an increase in industrial output of 9 paise in 1968–69, and to a much larger order of 14 paise in 1979–80.

2.2 Demand Linkages

The impact of urban income and industrialisation on the demand for food and agricultural raw materials is generally recognised. The impact of rural income on industrial consumption goods, on the other hand, has not received adequate attention. There are certain industrial consumption goods, i.e., clothing, footwear, sugar and edible oils (accounting for about a quarter of the value added in consumer

Table 10.4 Production linkages
(per cent per annum)

	I. *Matrix of input-output coefficients*					
	1968–69			*1979–80*[1]		
	Agriculture	Industry	Services	Agriculture	Industry	Services
Agriculture	0.1813	0.1266	0.0171	0.1604	0.1300	0.0390
Industry	0.0434	0.3328	0.1325	0.0682	0.3465	0.1049
Services	0.0162	0.1347	0.0954	0.0197	0.1489	0.0956

	II. $(I - A)^{-1}$					
	1968–69			1979–80[1]		
	Agriculture	Industry	Services	Agriculture	Industry	Services
Agriculture	1.236	0.247	0.059	0.214	0.260	0.083
Industry	0.087	1.562	0.230	0.135	1.601	0.191
Services	0.035	0.237	1.141	0.049	0.269	0.139

Note:
1. At 1968–69 prices.
Source: Planning Commission and CSO.

goods) for which rural consumption is over three times the urban consumption.[3]

The terms of trade between agricultural commodities and industrial manufactures plays an important role in influencing agricultural incomes and therefore agricultural demand for industrial consumption goods. If the terms of trade move in favour of agriculture, those who buy agricultural commodities are adversely affected but those who sell agricultural commodities benefit. An increase in the terms of trade in favour of agriculture (particularly food prices) adversely affects the demand for non-food items in the urban areas. The cross-elasticity of demand is negative among lower income groups in urban areas where food consumption is a sizeable part of the total budget. In rural areas, the effects on lower income groups will be the same as in urban areas. However, in the case of the rural upper income groups, the negative effect on demand arising from an increase in the terms of trade in favour of food can be offset by the increase in income resulting from the improvement in agricultural prices. Thus, the overall effect of the changes in the terms of trade will be a combination of the effects for all population groups. In the case of some commodities for which separate data were available on rural consumption and urban consumption, Rangarajan (1982) had shown that the terms of trade had a positive and significant effect on rural non-food expenditures and a negative, but statistically insignificant, effect on the urban non-food expenditures.

2.3 Savings and Investment Linkages

The effect of the agricultural sector on savings in the economy is not limited to private savings which is influenced by the levels as well as the distribution of incomes in the agricultural sector. Government savings are also affected, not only by agricultural incomes, but also by the foodgrains terms of trade. This is because government revenues tend to be indexed to the price of manufactures while government expenditures are related more to foodgrains prices because of the dearness allowances payable to government employees. As foodgrains terms of trade increase, both factors tend to depress public savings. Thus, while agricultural and industrial value added influence government savings positively, the foodgrains terms of trade tend to have a negative effect.

Investment by the private corporate sector is influenced by both agricultural and industrial output. An increase in the terms of trade between agriculture and non-agriculture may, however, have a negative effect on private corporate investment for two reasons. First, an increase in the foodgrains terms of trade may push up the product wage and hence squeeze profitability. Secondly, an increase in the non-foodgrains terms of trade may have a cost-push effect through the materials price rise for the agro-based industries. The effect of the terms of trade on household savings is analogous to their effect on the demand for industrial consumption goods. It is worth noting that the household sector is defined to include individuals and non-corporate business – both urban and rural. While rural savings may be influenced positively, urban savings may be affected adversely.

3. THE MODEL

In designing a macroeconometric model to study the interactions between agriculture and industry, the important linkages have been explicitly built in. As agricultural income increases, it generates demand for industrial consumer goods, e.g., sugar, textiles, edible oils, etc. As terms of trade between agriculture and industry move in favour of agriculture, this affects the demand for industrial consumer goods. It also has a dual effect on investment, i.e., through affecting the generation of surpluses in agriculture and through affecting profitability in industry. Thus, investment by the 'household' sector (which includes agriculture) may increase, while that by the private corporate sector may decline. Terms of trade between agriculture and industry also influence public investment through their influence on public saving. The dependence of industry on agriculture for materials is another important interaction which has been modelled. As for the dependence of agriculture on industry for the growing use of fertilisers and electricity, it was not possible to model this explicitly. The effect of the new technology in agriculture, however, is captured by the use of an irrigated area as a proxy for these inputs in the production function for agriculture.

The model focuses upon four principal sectors of the economy, i.e., foodgrains and commercial crops subsectors of agriculture, and consumer goods and heavy industries (basic and capital goods) subsectors of industry. Because there is substantial scope for changing cropping patterns, reflecting changing expectations on the relative price of foodgrains and commercial crops, the allocation of gross sown area between the two subsectors of agriculture is explicitly modelled. Output of foodgrains and commercial crops is then determined by the gross sown areas, the area under irrigation, and the rainfall index for the respective crops. The output of consumption goods is directly determined by certain income factors, the terms of trade between agriculture and industry, and infrastructural constraints. The output of basic and capital goods is determined by gross investment in the economy, import of capital goods, and infrastructural constraints. The determination of the three major components of gross investment and the differential impact of the terms in these components is explicitly modelled. Once output in the four principal sectors is determined, value added in agriculture as well as value added in industry are related to the value added in the respective subsectors. The value added in agriculture and industry together constitute the material product which in turn determines national income.

The equation system used in the model is presented in Table 10.5 (the symbols used related to the sources of data are shown in the Appendix). Equations 1 to 8 relate to the agricultural sector. Area under foodgrains is specified in equation 1 as a function of the expected relative price of foodgrains and commercial crops (defined as the one period lagged relative price) and total gross sown area, both variables exercising a positive influence. Area under commercial crops is derived residually through the identity presented in equation 2. Equations 3 and 4 specify the production functions in the two subsectors of agriculture. Output of each subsector is a function of the gross sown area, irrigated area and a rainfall index for that subsector. Consumption of fertilisers had to be dropped as an explanatory variable because of problems of multicolinearity. However, the inclusion of gross irrigated area as an explanatory variable captures the effect of the use of fertilisers and other such inputs since farmers tend to cultivate their land relatively intensively when irrigation is available. It is worth noting that the use of a rainfall index in the production function postulates a simple relationship between rainfall and output but does not incorporate the effect of the distribution of the rainfall over the year. Once the outputs of foodgrains and commercial crops are determined, value added in agriculture is derived through a simple statistical relationship as presented in equation 5.

The relative price of foodgrains and manufactures (referred to in the model as the foodgrains terms of trade) is specified in equation 6 to be a function of supply factors represented by the one period

Table 10.5 Estimated equations (1960–61 to 1979–80)

Eq. no.	*Dependent variable*	*Independent variables*									R^2	D-W
1.	*AF*	6.7137	+	0.6803**A*	+	3.3294 (*TTF/TTNF*)−1					0.94	1.23
		(1.0)		(16.9)		(1.7)						
2.	*AC*	*A* − *AF* − *AO*										
3.	*QF*	−167.6905*	+	1.4252 *AF*	+	0.6810* *IRF*	+	0.3368* *RIFG*			0.94	2.32
		(2.3)		(1.7)		(3.3)		(2.2)				
4.	*QC*	−16.7594	+	1.2217 *AC*	+	0.5147* *IRNF*	+	0.3633* *RINF*			0.88	1.19
		(0.4)		(0.7)		(9.5)		(4.0)				
5.	*YA*	7.2851*	+	0.5357* *QF*	+	0.2897* *QC*					0.99	1.38
		(2.5)		(12.1)		(4.6)						
6.	*TTF*	162.9835*	+	0.0213 (*Y-YA*)/*POPNA*	−	0.4778* (*QF/POP*) −1					0.31	1.28
		(6.2)		(1.4)		(3.2)						
7.	*TTNF*	122.9231*	+	0.3314* *YIC*	−	0.5883 *QC*−1	−	0.0633* *MRMI*			0.10	1.68
		(5.7)		(2.2)		(1.9)		(1.0)				
8.	*TT*	0.4720* *TTF*	+	0.5312* *TTNF*							0.99	1.12
		(10.4)		(10.6)								
9.	*YIC*	17.1149	+	0.6304**YI* −1	+	0.1960 *YA* −1	−	0.1894**TTF*−1	+	0.2118*EG*	0.98	2.00
		(0.7)		(2.3)		(1.0)		(2.0)		(1.3)		
10.	*SPU*	−376.3574	+	27.2436**Y*−1	−	8.5096 *TTF*−1					0.84	0.80
		(0.8)		(9.6)		(1.7)						

continued on page 234

Table 10.5 continued

Eq. no.	Dependent variable	Independent variables							R^2	D-W
11.	*IPU*	1560.3960* (8.3)	+	0.5118* *SPU* (2.1)	+	2.3771* *BMI* (3.4)	+	15.5147 *T* (0.7)	0.90	1.36
12.	*IPC*	−1823.5234* (4.5)	+	2.0000*Y*−1 (1.1)	−	13.6840**TT*−1 (2.8)			0.25	1.85
13.	*IHH*	−4047.0781* (3.0)	+	21.2970**TTF*−1 (2.7)	+	12.8868 *TTNF*−1 (1.1)	+	39.8549**Y*−1 (8.9)	0.87	1.56
14.	*I*	*IPC* + *IHH* + IPU								
15.	*YIBK*	8.7782 (0.8)	+	0.0114**I* (2.8)	−	0.1079**MKI* (2.1)	+	0.2920 *EG* (1.9)	0.98	1.17
16.	*YI*	22.4612* (6.3)	+	0.3647**YIC* (2.7)	+	0.4325*(*YIBK* (4.0)			0.98	0.87
17.	*Y*	0.4316**YA* (22.3)	+	0.5652**YI* (32.8)					0.99	1.67

Notes: *t*-statistics are given in parentheses below the coefficients.
* indicates that the coefficient is significantly different from zero at the 0.05 level of significance.

lagged per capita output of foodgrains and demand represented by the one period lagged per capita income originating in the non-agricultural sector. While the supply variable is expected to have a negative effect on the terms of trade, the demand variable will exert a positive influence. Equation 7 represents the relative price of commercial crops and manufactures. The supply factors are represented by one-year lagged output of commercial crops (since what is normally available in the market is the output of the previous year) and the import of materials. The demand factor is represented by the gross value added in consumer goods' industries. The latter is expected to affect the numerator, i.e., the price of manufactured consumption goods. The sign of the coefficient of this variable depends on the relative strength of the effect of value added in consumption goods on the price of materials and the price of manufactured goods. Equation 8 is simply an identity that treats the terms of trade between agriculture and industry as a weighted average of the ratio of food prices to the price of manufactured goods and the ratio of commercial crop prices to the price of manufactured goods. The weights have been determined on the basis of historical experience.

Gross value added in consumer goods industries is specified in equation 9 as a function of demand represented by income as well as the terms of trade and supply constraint represented by a variable representing electricity generation. The income variables representing demand are value added in agriculture and value added in industry of the previous period. While the income factors exert a positive influence, the foodgrains terms of trade has a negative effect. As food price rises in relation to manufactures, the demand for consumption goods goes down because of the pre-emption of expenditure on food out of the budget. The supply constraint exerted by the availability of electricity generation is incorporated by the inclusion of this variable in equation 9.

In order to trace the differential impact of the terms of trade on investment by institutional sectors, the model specifies three investment functions separately for the public sector, the private corporate sector, and the household sector, respectively. The effect on public investment is through public savings. The latter is specified in equation 10 to be a function of national income and the terms of trade for foodgrains both lagged one year. While the increase in income is expected to increase government revenues and therefore public savings, the foodgrains terms of trade are expected to affect public savings negatively as discussed in the preceding section. Investment by public sector as specified in equation 11 is positively related to public savings and market borrowing while a time trend is added to take account of the government's tendency to increase public investment over time. Investment by the private corporate sector and investment by the household sector are both specified in equations 12 and 13 respectively, as functions of lagged national income and lagged terms of trade variable. While the effect of national income is positive in both cases, that of terms of trade is asymmetric, as is to be expected, and as was discussed in the preceding section.

Value added in heavy (basic and capital goods) industries is specified in equation 15 to be a function of gross investment, import of capital goods and electricity generation. The demand variable is represented by gross investment, while the electricity variable represents the infrastructural constraint on the output. The negative effect of the imports of capital goods on the value added in the heavy industries is also part of the specification.

Equation 16 is a statistical relationship relating total value added in industry to the value added in consumer goods and heavy industries. It is treated as a stochastic equation because value added in the intermediate goods industries is assumed to be related to the value added in consumer goods and in heavy industries. Equation 17 is a stochastic equation relating value added in the economy to the value added in agriculture and industry, thereby assuming that value added in services moves in consonance with value added in material product. An alternative specification for equation 17 could be that total income in the economy is a function of total investment. This allows for a direct effect of the level of investment in the economy on the overall level of economic activity. Money is not introduced in the model because behaviour is linked to relative price ratios.

The seventeen equations presented in Table 10.5 include 11 behavioural or reduced form equations, 4 statistical association type relationships and 2 identities. The equations have been estimated with Ordinary Least Squares using annual observations for the period 1960–61 to 1979–80. The signs of the coefficients in the various equations are in agreement with *a priori* expectations, and the coefficients are generally statistically significantly different from zero. The explanatory power of most of the equations is good, although the foodgrains terms of trade equations has a relatively low R^{-2}.

4. SIMULATIONS

The performance of the model over the sample period was tested by a static as well as a dynamic simulation of the model. In the dynamic simulation, the lagged endogenous variables are given only for the initial period, and the solutions for the subsequent periods use the generated values for the lagged endogenous variables. In the static simulation the actual lagged values of the relevant endogenous variables are used throughout. Both static and dynamic solutions are useful in analysing the properties of a model. The static solution can be used to examine how well a model replicates the behaviour of an economy over the sample period. By using actual values rather than the generated values for lagged endogenous variables, the static solution prevents the errors from cumulating over time. On the other hand, the dynamic solution yields a time path which explicitly incorporates the inter-period linkages by using values for the endogenous variables as generated by the model. We have also simulated the model to study the impact of the changes in certain exogenous variables on the other variables in the system. These simulations are designed to study the impact of an improved agricultural performance on the other variables in the model. The impact of an improvement in the terms of trade for agriculture is also analysed.

The broad picture of the 'goodness of fit' of the model can be obtained by analysing the average percentage error (percentage deviation of the actual from the solution value) over the sample period for both the static and the dynamic simulations. Table 10.6 presents the mean absolute percentage error as well as the root mean square error for all the endogenous variables. While the former statistic assigns equal weights to all the errors, the latter measure assigns a proportionately higher weight to a larger error than to a smaller one. The dynamic simulation shows higher errors as is to be expected, but the pattern generally remains the same under both the simulations. The model performs well, judged by the fact that the mean absolute percentage error exceeds 10 per cent only in the case of two variables under the static simulation and three variables under the dynamic simulation. Value added in industry which is a crucial endogenous variable in the model has mean absolute percentage error of less than 3 per cent for the static simulation and less than 3.5 per cent for the dynamic simulation. Investment in the private corporate sector has large errors. This is not surprising in view of the poor fit of this equation. Figures 10.3 to 10.6 present the results of the

Table 10.6 Errors in endogenous variables

Error/variable	Static		Dynamic	
	Mean absolute percentage error	Root mean square percentage error	Mean absolute percentage error	Root mean square percentage
AF	0.68	0.80	0.76	0.89
AC	3.08	3.61	3.46	4.08
QF	3.18	3.96	3.10	3.91
QC	2.94	3.55	2.95	3.97
YA	2.46	3.00	2.40	2.96
TTF	6.96	8.18	7.21	8.90
TTNF	6.43	7.81	5.99	8.28
TT	5.06	6.15	5.32	6.21
YIC	2.47	3.32	3.89	4.51
SPU	11.21	14.28	14.65	17.66
IPU	7.35	8.88	7.34	9.32
IPC	12.28	17.37	16.45	21.52
IHH	7.99	10.40	11.99	15.18
I	4.45	5.75	5.15	6.22
YIBK	4.85	7.81	4.84	7.76
YI	2.90	3.29	3.40	3.81
Y	1.57	2.00	1.71	2.14

dynamic simulations of the basic model for the major income variables and the terms of trade.

The first re-simulation that was attempted was to study the impact on the economy of raising the rainfall index for the 1974–75 which was a drought year. In effect, the attempt was to see how the economy would have behaved if the rainfall in that year had been normal. The rainfall indices for foodgrains and commercial crops were raised from their actual levels of 81.6 and 82.5, respectively, to their average levels for the period, i.e., 96.6 and 95.1, respectively.

The effects of the improved rainfall scenario on the endogenous variables of the model were studied by obtaining percentage differences between the value of the endogenous variables in the 'improved rainfall' simulation, and the values of the endogenous variables in the base dynamic simulation. These results are presented in Table 10.7. With better rainfall, as the outputs of foodgrains and commercial crops increase, the value added in agriculture increases by a little over 4 per cent in 1974–75. In that year, the influence on national income merely reflects the better performance of agriculture. With better availability of foodgrains, the foodgrains terms of trade declined by 3.5 per cent and the non-foodgrains terms of trade by 2.5 per cent in the following year.

The increase in investment in 1975–76 in both the private corporate sector and the household sector reflects the lagged effect of the increase in income of the preceding year on account of better rainfall. Since the public saving is higher for the same reason and affects public investment in the same direction, total investment in the economy increases and leads to an increase in the value added in heavy industries of the order of nearly 1 per cent. In 1976–77, on the other hand, the asymmetric effect on the investment of the private corporate and the household sectors is due to the dominant asymmetric effect of the decline in the terms of trade in the preceding year. Because of the relatively high elasticity of private corporate investment with respect to the terms of trade, this investment is significantly higher. Household investment in 1976–77, on the other hand, is lower, reflecting the lagged effect of the decline in the terms of trade in 1975–76. Given the large weight of the household investment in the total, the latter declines and the value added in heavy industries is also lower, albeit marginally.

Value added in industrial consumption goods is higher by 0.7 per cent and 0.9 per cent in 1975–76 and 1976–77, respectively, and the increases are smaller in the succeeding years. It is worth noting that

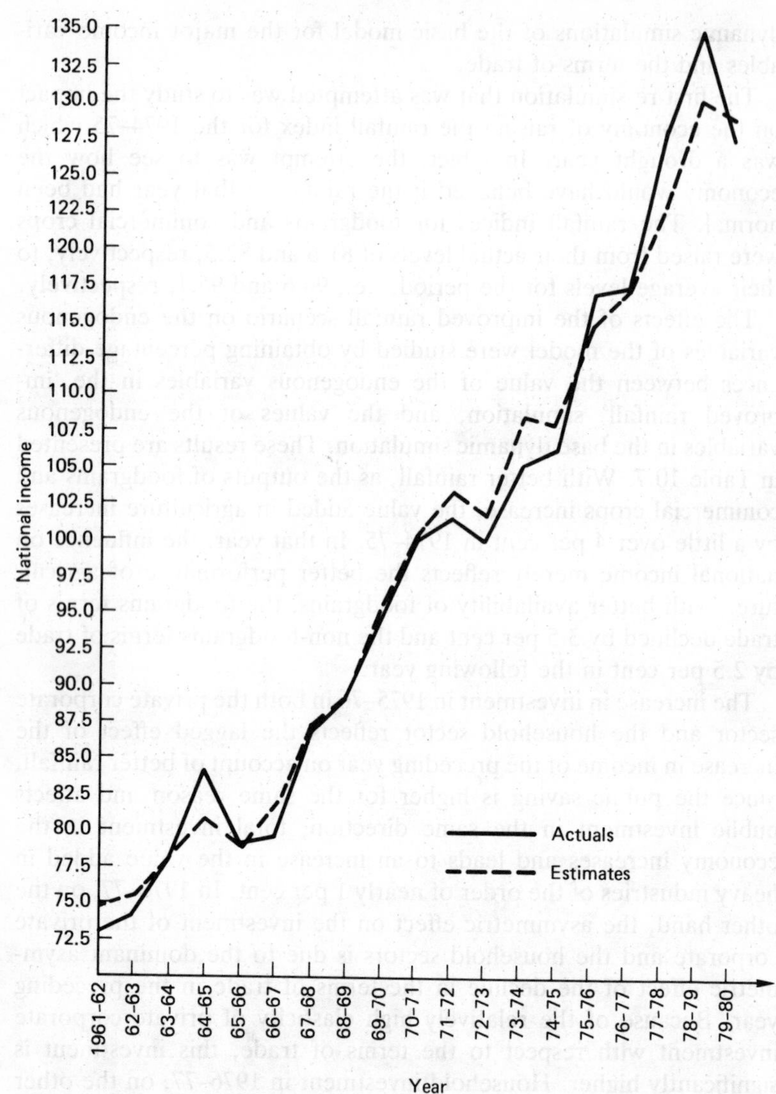

Figure 10.3 National income – actuals and estimates from dynamic simulation of the model (Base: 1970–71=100)

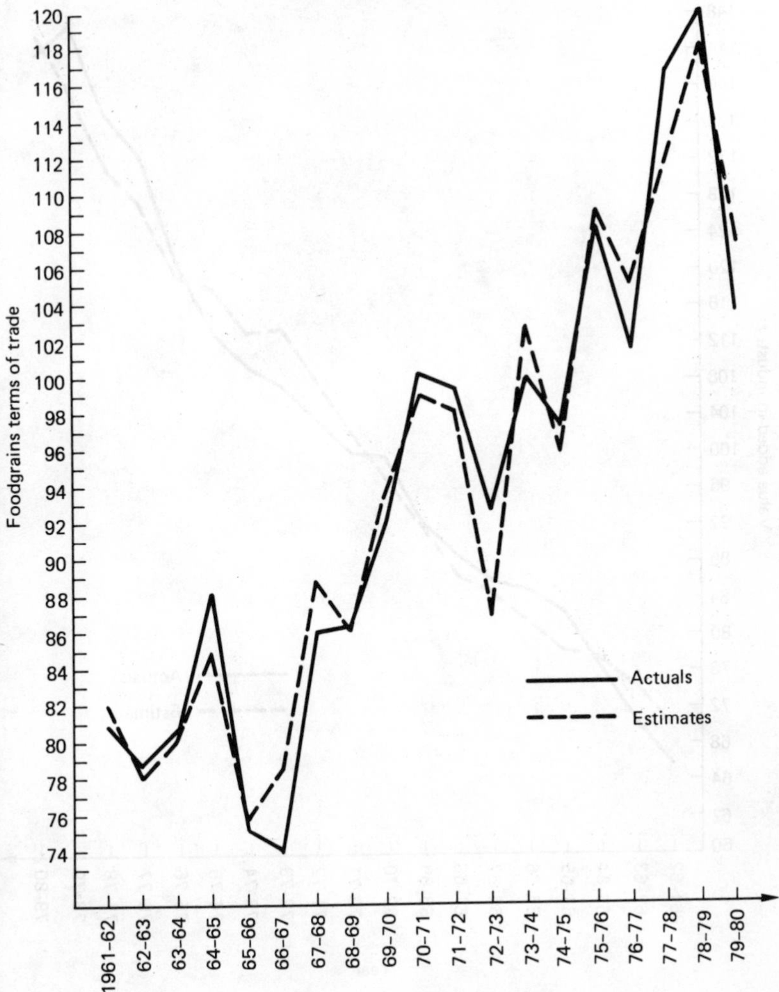

Figure 10.4 Value added in agriculture – actuals and estimates from dynamic simulation of the model (Base: 1970–71=100)

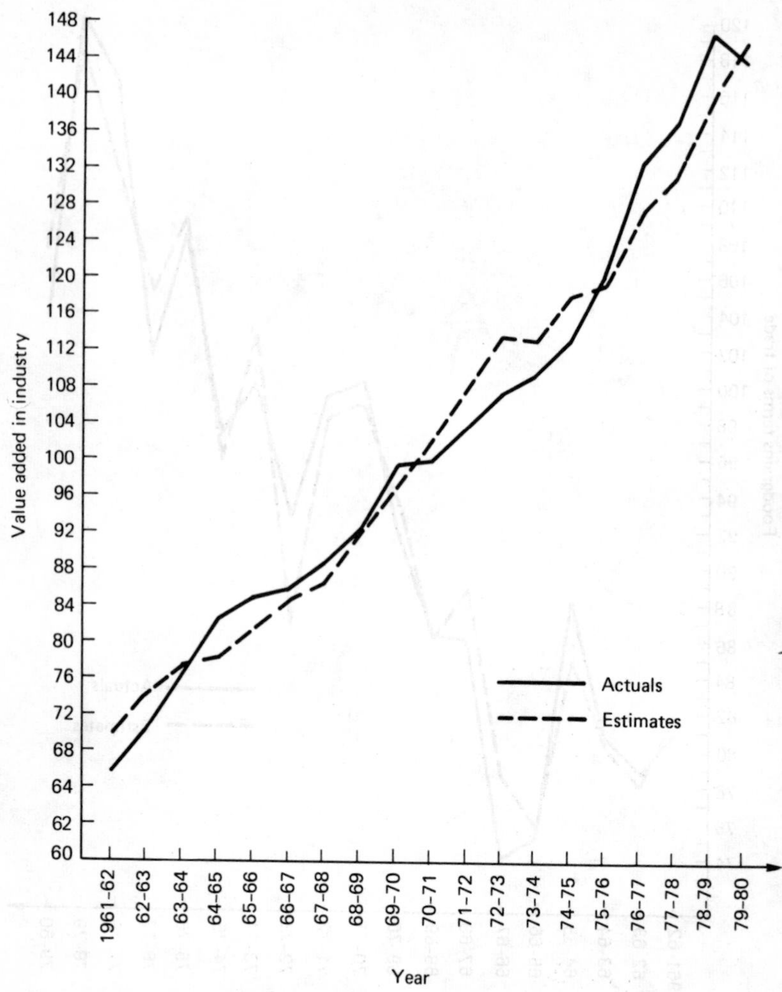

Figure 10.5 Value added in industry – actuals and estimates from dynamic simulation model (Base: 1970–71=100)

the effect on the industrial consumer goods is spread out, while that on heavy industries is more concentrated in the first year. The total effect of a 4.2 per cent increase in agricultural income on industrial value added through generating demand for industrial consumption goods is of the order of 1.8 per cent. The effect of the improved

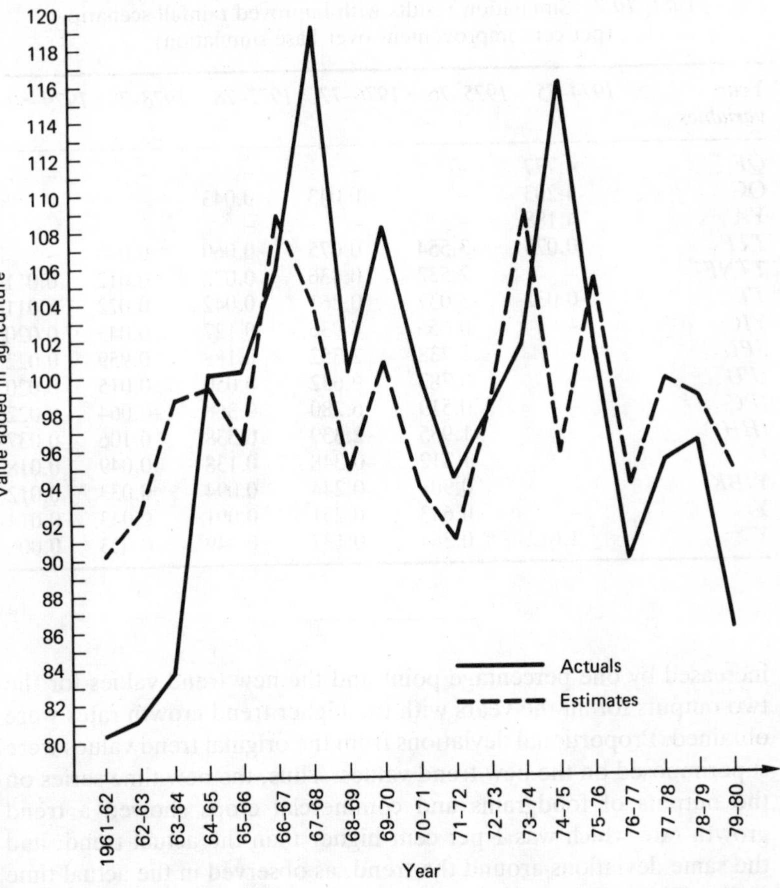

Figure 10.6 Foodgrains terms of trade – actuals and estimates from dynamic simulation of the model (Base: 1970–71=100)

rainfall scenario on national income is to raise it by 1.6 per cent in 1974–75 and by 0.38, 0.14, 0.05, 0.02, and 0.01 per cent, respectively, in the subsequent five years.

A second simulation is attempted to analyse the impact of an increase in the trend rate of growth of agriculture on the economy. To study this impact the following procedure was adopted. The trend rate of growth in the output of foodgrains and commercial crops was

Table 10.7 Simulation results with improved rainfall scenario (per cent improvement over base simulation)

Year/ variables	*1974–75*	*1975–76*	*1976–77*	*1977–78*	*1978–79*	*1979–80*
QF	4.777	–	–	–	–	–
QC	4.203	–	0.043	0.043	–	–
YA	4.188	–	–	–	–	–
TTF	–0.073	–3.554	0.075	0.060	0.030	–
TTNF	–	–2.532	0.436	0.022	0.012	0.011
TT	–0.032	–3.037	0.261	0.042	0.022	0.011
YIC	–	0.656	0.936	0.127	0.043	0.020
SPU	–	2.738	2.352	0.188	0.959	0.022
IPU	–	0.787	9,632	0.050	0.015	0.020
IPC	–	0.519	6.280	–0.368	–0.064	–0.022
IHH	–	1.955	–2.339	0.338	0.106	0.037
I	–	1.312	–0.348	0.138	0.049	0.018
YIBK	–	0.901	–0.244	0.094	0.033	0.012
YI	–	0.643	0.251	0.091	0.043	0.014
Y	1.612	0.384	0.137	0.049	0.023	0.008

increased by one percentage point and the new trend values for the two outputs for all the years with the higher trend growth rates were obtained. Proportional deviations from the original trend values were superimposed on the new trend values. Thus, the new time series on the outputs of foodgrains and commercial crops showed a trend growth rate which was 1 per cent higher than the actual trend, and the same deviations around the trend, as observed in the actual time series. Equations 1 to 4 of the model relating to the area and the output of foodgrains and commercial crops were suppressed. Instead, the new data series on the outputs of foodgrains and commercial crops were used, and the model was simulated to determine the impact of a higher trend growth in agricultural output on the endogenous variables.

Table 10.8 presents the trend growth rates of some selected variables in the new simulation and the base simulation. When value added in agriculture increases at a rate of growth which is higher by 1 per cent over its trend growth, the effect on the value added on consumer goods is to raise their growth rate by 0.5 per cent per annum. While the effect of the agriculture-industry linkages is widely recognised, another effect through the savings-investment linkages

Table 10.8 Trend rate of growth of selected variables (per cent per annum)

Variable	*Base simulation*	*Simulation with higher agricultural trend growth*
Y	3.3	3.9
YA	2.2	3.1
YI	4.1	4.3
YIC	4.7	5.2
YIBK	5.9	6.7
SPU	6.1	7.2
I	3.8	4.2

on the heavy industries has received little notice. The simulations reported in Table 10.8 show that this effect is relatively stronger. The growth rate of heavy industries is higher by 0.8 per cent points. It is worth noting that in the Rangarajan (1982) model which covered the period from 1961 to 1972, the effect of an increase in the trend growth rate of agriculture was greater on industrial consumption goods than on basic and capital goods. In the present study which covers a longer period, and the linkages are specified in terms of value added rather than gross output, the effect on heavy industries is relatively stronger.

The overall effect on the growth of value added in the industrial sector, however, is only 0.2 per cent per annum which seems to be on the low side. In the Rangarajan (1982) model covering the period from 1961 to 1972 where the linkages were specified in terms of agricultural production and industrial production, a one per cent increase in the growth of agriculture led to a 0.5 per cent increase in the growth of industrial production. Admittedly, the period of this study is extended to cover the decade of the 1970s, and the relationships are modelled in terms of the growth in value added (value added in Indian industry has grown at a slower rate than the value of output), but we expected the growth in value added in industry to increase somewhat more than by the 0.2 per cent simulated by the model.

As far as the overall effect on an increase in the trend rate of growth on agriculture on national income, Rangarajan (1982) had found the rate of growth of national income to increase from 3.1 per cent per annum to 3.8 per cent per annum due to the 1 per cent increase in the trend growth rate of agriculture. In the present study the growth rate of national income increases from 3.3 per cent per annum to 3.9 per cent per annum due to the same factor. The effect on national income is only a little smaller than that simulated by the Rangarajan (1982) model. This must imply a relatively larger effect of an increase in the agricultural trend growth rate on the services sector which has not been explicitly specified in either of the two models.

A third simulation of the model was attempted in which the foodgrains terms of trade was raised in 1970–71.[4] The results (not reported here) showed that the negative effect on the value added in consumer goods industries offset the positive effect on the value added in the heavy industries. The net effect on the economy was negligible. The model shows that an attempt to generate industrial demand from agriculture by raising the foodgrains terms of trade is self-defeating.

In the end, it is useful to draw some broad conclusions from the simulations of the model which was designed to explore certain important linkages between the agricultural and the industrial sectors of the Indian economy. A one-time increase in agricultural value added (due to better weather) leads not only to an increase in the growth of value added in consumer goods industries but also to that in the basic and capital goods industries. An increase in the trend rate of growth of agriculture (due to technological change, for example) also leads to a positive effect on the growth of consumer goods industries as well as heavy industries, although the total effect on the industrial growth rate is a little on the low side. Though not studied explicitly, the result also indicates a positive effect on the tertiary sector. An increase in the foodgrains terms of trade, on the other hand, has a negligible effect on the growth of the industrial sector or that on the economy at large.

Appendix Index to the symbols used in the model

Symbol	Description	Unit	Source
Endogenous variables			
AF	Gross cropped area under foodgrains	Hectares (m)	1
AC	Gross cropped area under commercial crops	Hectares (m)	1
QF	Production of foodgrains	Index, base triennium ending 1969–70 = 100	1
QC	Production of commercial crops	Index, base triennium ending 1969–70 = 100	1
YA	Value added in agriculture at 1970–71	Index, base 1970–71 = 100	2
TTF	Terms of trade: foodgrains versus manufactures	Index, base 1970–71 = 100	3
TTNF	Terms of trade: non-foodgrains versus manufactures	Index, base triennium ending 1970–71 = 100	3
TT	Terms of trade: agriculture v. non-agriculture	Index, base 1970–71 = 100	3
YIC	Value added in consumer goods industries at 1970–71 prices	Index, base 1970–71 = 100	3,4
SPU	Gross savings of the public sector deflated by implicit deflator for gross investment	Rupees (crores)	2
IPU	Gross domestic fixed capital formation by public sector at 1970–71 prices	Rupees (crores)	2

IPC	Gross domestic fixed capital formation by private corporate sector at 1970–71 prices	Rupees (crores)	2
IHH	Gross domestic fixed capital formation by household sector at 1970–71 prices	Rupees (crores)	2
I	Gross domestic fixed capital formation at 1970–71 prices	Rupees (crores)	2
YIBK	Value added in basic and capital goods industries at 1970–71 prices	Index, base 1970–71 = 100	3,4
YI	Value added in industry at 1970–71 prices	Index, base 1970–71 = 100	2
Y	Value added by all sectors at 1970–71 prices	Index, base 1970–71 = 100	2
Exogenous variables			
A	Gross cropped area	Hectares (m)	1
AO	Gross cropped area under other crops	Hectares (m)	1
IRF	Gross irrigated area under foodgrains	Index, base 1970–71 = 100	1
IRNF	Gross irrigated area under non-foodgrains	Index, base 1970–71 = 100	1
RIF	Rainfall index for foodgrains	Index, base 1970–71 = 100	5
RINF	Rainfall index for non-foodgrains	Index, base 1970–71 = 100	5
MRMI	Imports of inedible crude materials except fuel at 1968–69 prices	Index, base 1970–71 = 100	7

MKI	Import of capital goods at constant prices	Index, base 1970–71 = 100	7
EG	Electricity generated	Index, base 1970–71 = 100	8
BMI	Market borrowings by government deflated by investment deflator	Index, base 1970–71 = 100	7
POP	Population	Rupees (crores)	2
POPNA	Population non-agricultural	Rupees (crores)	6
T	Time trend	–	

Sources:

1. Estimates of area and production of principal crops in India, Directorate of Economics and Statistics, Ministry of Agriculture and Irrigation, Government of India.
2. National Accounts Statistics.
3. Index Number of Wholesale Prices in India, Office of the Economic Adviser, Ministry of Industrial Development, Government of India.
4. Annual Survey of Industries.
5. National Council of Applied Economic Research
6. Ahluwalia (1979).
7. Report on Currency and Finance, Reserve Bank of India.
8. Economic Survey, Ministry of Finance, Government of India.

Notes

* The authors wish to thank V.W.Varghese for his assistance in the preparation of this paper.

1. Rudra (1967), Chakravarty (1974), Mellor (1976), Raj (1976), Mitra (1977), Thamarajakshi (1977), Ahluwalia (1979), 1985), Chakravarty (1979), Krishna (1982), and Rangarajan (1982).
2. The shortcomings of the industrial production indices as well as the method of compiling the data on value added for the use-based sectors are presented in Ahluwalia (1985).
3. NSSO Tables on Consumer Expenditure, 28th Round, October 1973 to June 1984.
4. This was done by raising the constant term in the foodgrains terms of trade equation by one in 1970–71.

References

Ahluwalia, I.J. (1979) *Behaviour of Prices and Outputs in India: A Macroeconomic Approach*, (Macmillan Company of India Ltd).

Ahluwalia, I.J. (1985) *Industrial Growth in India – Stagnation since the Mid-Sixties* (Oxford University Press).

Chakravarty, S. (1974) 'Reflections on the Growth Process in the Indian Economy', Hyderabad Administrative Staff College of India.

Chakravarty, S. (1979) 'On the Question of Home Market and Prospects for Indian Growth', *Economic and Political Weekly*, Special number (August).

Kahlon, A.S. and Tyagi, D.S. (1983) *Agricultural Price Policy in India*, (Allied Publishers).

Krishna, R. (1982) 'Some Aspects of Agricultural Growth Price Policy and Equity in Developing Countries', *Food Research Institute Studies*, vol. 18, no. 3.

Mellor, J.W. (1976) *The New Economics of Growth : A Strategy for India and the Developing World* (Ithaca, New York: Cornell University Press).

Mitra, A. (1977) *Terms of Trade and Class Relations* (London: Frank Cass).

Raj, K.N. (1976) 'Growth and Stagnation in Indian Industrial Development', *Economic and Political Weekly*, (26 November).

Rangarajan, C. (1982) 'Agricultural Growth and Industrial Performance in India', IFPRI Research Report no. 33 (Washington, DC: International Food Policy Research Institute)(October).

Rudra, A. (1967) *Relative Rates of Growth – Agriculture and Industry* (Department of Economics, University of Bombay)

Thamarajakshi, R. (1977) 'Role of Price Incentives in Stimulating Agricultural Production in a Developing Economy', in Ensminger Douglas, (ed.) *Food Enough or Starvation for Millions* (Rome:FAO).

Part III

Industry versus Agriculture in East European Economies

11 The Development of the Agro-industrial Complex in the USSR

L.V. Nikiforov
INSTITUTE OF ECONOMICS OF THE ACADEMY OF SCIENCES OF THE USSR, MOSCOW

1. STAGES OF ECONOMIC DEVELOPMENT

Several stages may be identified in the economic development of the USSR which differ significantly from each other in terms of the nature of the interrelationship between the rates of industrial and agricultural growth, the impact on the general rates of economic development, and the utilisation of the productive and social potentials. Four stages may be distinguished:

1. Before the socialist revolution. Leaving aside periods of conflict and war and their aftermath, the country's economy was predominantly dependent on agriculture, which was at the stage of simple co-operation of labour, with some manufacture. It did not provide adequate momentum for development and general economic growth. Agricultural development was insulated from industrial development, partly because of the high share of subsistence agriculture.

2. The years 1925–39. This was the stage characterised by an increasingly larger impact of industry upon general economic development, the conversion of the economy to the basis of socialist principles, and industrialisation, all of which affected the national growth rates. The rates of agricultural development were overtaken by those for industry due to the creation of the sectors for production of industrial inputs and for further indirect and complex production processes. At the same time, the dependence of agriculture on industrial development increased because of its transition to the use of machines and the establishment of a system of large-scale socialist agricultural production.

3. 1950–65. The industrial and agricultural development rates of growth became increasingly interdependent, which was a most important factor in the dynamic growth of the national economy. Mechanisation and the use of chemicals increased, as did the share of agricultural output processed by industry. Thus the agricultural and industrial growth rates became closely tied.

4. 1965 to the present. This is the period of the industrialisation of agriculture by mechanisation and industrial technologies, and also of agro-industrial integration. The dependence of the general rates of economic development and the growth of the standard of living on the maintenance of the proportions between agriculture and industry increases. Simultaneously, the interrelationship between industrial growth rates and those of agriculture became more complex. The agro-industrial complex emerges as an integral structural subdivision of the national economy, dependent for its effective functioning on the relation between the industrial growth rates and the agricultural growth rate and their levels. As was indicated in the documents of the 27th Congress of the CPSU, the importance of factors such as the improvement of the structure and economic mechanism of the agro-industrial complex, the interaction of productive, social and organisation factors of economic growth increased (KPSS, 1986, pp. 30–1, 296, 297).

2. THE URBAN AND RURAL DEVELOPMENTS

The stages of development of industry and agriculture can, in effect, be identified with those of urban and rural environments. Over a long period of time, the backwardness of agricultural development, compared to industry, determined the backwardness of the socio-economic development of the country, compared to towns. The history of the socialist construction in the USSR is also the history of equalising the socioeconomic conditions of work and life of the workers in town and country and the establishment of the social homogeneity of Soviet society.

Under capitalism, the raising of real levels of production and the development of the agro-industrial sphere widen the gap between the production potential of town and country. The faster development of large-scale industrial production in Russia intensified the capitalist activities in the town, the insulation of the country, and the extent to which it lagged behind the towns. The expansion of the production relationships between town and country has taken place on the basis of the greater involvement of the villages in the system of capitalist relations, the proletarisation of the main mass of the rural population, the destruction of the rural structure, and the deepening of antagonistic contradiction.

In liquidating exploitation and in establishing the homogeneity of society which is intrinsic in socialism, socialism eliminates the socioeconomic antithesis between town and country, and consequently their socioeconomic insulation, and promotes the equalisation of their level of development. The means to achieve this are: agro-industrial integration, overcoming the limited opportunities for the application of labour in rural localities especially in agriculture, and combining the functions of town and country to create conditions for the fullest satisfaction of the requirements of members of society. Instead of maintaining insulated urban and rural subsystems characterised by the level of development of their productive and social potentials, and the social position of the urban and rural population in both work and life, socialism gradually establishes a single socioeconomic system, combining town and country in productive economic and social relations (the 'town-country' system).

Though the socioeconomic differences between town and country grew gradually weaker during the development of socialism, the combination of specifically urban and rural peculiarities of work and life were nevertheless continued over a long period of time because of the complexity of the problem.

In conditions of mature socialism the situation begins to change radically. The intensification of the interrelationships between town and country, and the improvement in the level of development of production and life in the countryside in the preceding periods, have now grown into the integration of the various functional and structural urban and rural subdivisions. The integration process has increasingly extended to the structure of production (above all, to the agro-industrial sphere), gradually eliminating the insulation between town and country. This agro-industrial integration creates the basis for the formation of the national economic agro-industrial complex (AIC).

3. THE AGRO-INDUSTRIAL COMPLEX

The national AIC is not the mere sum of agricultural sectors and the industrial sectors associated with it, though these have been developing for decades and their range is gradually widening. The establishment of AIC, as well as other national economic complexes, is the result of the radically new supersectoral or inter-sectoral level of socialisation of production characterised by the emergence and deepening of the processes of inter-sectoral integration, which have led to integral structural subdivisions, which have become the largest blocks of the national economic complex.

The first attempts to combine agricultural and industrial production at primary levels in the forms of agro-industrial combines were made as early as the 1920s and the 1930s, but on the whole they were not a success, primarily because an adequate material and technical foundation did not exist at that time. The inter-sectoral integration in the agro-industrial sphere and the development of the AIC began on a relatively wide scale in the 1960s. In the 1970s it increased in intensity. The main condition for establishing national economic AICs was the industrialisation of agriculture, with the gradual transition from the application of machines to organising agricultural production on the basis of sectoral and regional systems of machines.

It is precisely the industrialisation of major agricultural sectors that ensured the equalisation of their level of development and of those sectors associated with them. This was helped by further specialisation in agriculture, and by the development of related industries such as processing, storing and marketing agricultural output. All this provides the basis for the emergence of integrated agro-industrial production. The requirement is to orientate all interrelated parts in the agro-industrial sphere to effect a transition from a sectoral mechanism to the kind of mechanism which would ensure the strengthening of the unity of the national economic agro-industrial complex, the improvement of proportions and interrelationships within it, the complex approach to the utilisation of technical, biological, socioeconomic and organisational factors of agro-industrial production, with the objective of obtaining the optimum result. At the present time over 30 per cent of the gross social product, of productive fixed assets, and of the number of workers, are accounted for in the sectors which are part of the agro-industrial complex.

4. TASKS FOR THE FUTURE

But, of course, it cannot be said that all problems have already been solved, and that the accelerated formation of the national economic agro-industrial complex as an integral, organic system has been completed, though the management of the agro-industrial complex as an integral structure has, indeed, been ensured.

There are still discrepancies in the development of the major spheres of the complex. The level of industrialisation of agriculture is inadequate, the provision of material and technical services and the storage and marketing output, lag behind and losses of output still take place and resources are not fully utilised.

In accordance with the decisions of the 27th Congress of the CPSU, large-scale measures aimed at improving the systems and effectiveness of the agro-industrial complex are being taken (KPSS, 1986, pp. 296–7). There are three principal interrelated aspects of these measures: (i) solution of the AIC structural problems; (ii) improvement of its economic mechanism; and (iii) improvement of the interaction of production and social development.

The AIC structural problems are varied, but above all, their solution requires the transformation of its asset-producing sectors and the acceleration of the scientific-technical progress in them. The production of assets is oriented largely towards machines, implements, and equipment intended for the mechanisation of individual technological processes and operations. The expanded supply of tools and equipment for labour had led to a fast growth of costs of agricultural enterprises, but the build-up of fixed assets in this case has not been accompanied by a corresponding expansion in production, nor by any substantial reduction in labour intensity. From 1965 to 1984 the fixed assets of the social economy grew 4.2 times, while the number of workers in it decreased by 12 per cent. (By comparison, annual average output from 1961–65 to 1981–85 increased only 1.5 times.)

The complex integrated mechanisation of agriculture is not complete, though the mechanisation of a majority of main operations in its most important sectors is complete or almost so. (Table 11.1 shows figures for some of the later years.) This situation can, in many respects, be explained by the fact that, over a long period, the mechanisation of labour was adapting itself to the technologies established earlier, gradually replacing individual manual work by machine operations. But this way is usually expensive and not always

Table 11.1 Output, productive fixed assets and number employed in the agro-industrial complexes (AICs)

	1970	*1975*	*1980*	*1984*
Volume of output (constant prices, 1000m roubles)	253.1	310.0	347.2	395.2
Fixed assets at end of year (constant prices, 1000m roubles)	162.9	256.7	366.0	476.0
Mean annual number of workers in production sectors of AICs (millions)	43.9	44.2	44.4	45.4

Source: Narodnoe Khozyaistvo SSSR v 1984 g. (The National Economy of the USSR in 1984), Statistika, Moscow, 1985, p. 213.

---

effective. Not all manual work is amenable to mechanisation; sometimes the quality of output declines, as in flax-growing or tea-growing, or the product is damaged, as in potato growing.

Experience shows that the transition in agriculture to production on industrial lines frequently calls for non-traditional solutions, the development and introduction of radical new technologies, the creation of technico-technological systems, such as would combine the possibilities of chemistry, biology and modern technical progress. Industrial technologies have been utilised in many areas of plant breeding, including those for growing maize, rice, sugar-beet, and in all sectors of livestock breeding. But to expand them, technical and resource support are necessary. What is called for here is to combine the efforts of the asset-producing sectors and create their own technical revolution. Table 11.2 shows the contribution of various industrial sectors to the AIC.

To stimulate this process, the impact that the agricultural and processing sectors of the AIC have upon their suppliers should be increased. This is connected with the improvement of the economic mechanism with the interrelationships between the suppliers and consumers within the complex, and with the increased role of economic contracts in determining output and schedules for production and supply.

Another condition for accelerating the technical re-equipment of

Table 11.2 Inputs to the agro-industrial complex: share of industrial sectors in supply of inputs and in fixed assets 1984

Industrial sectors supplying AIC inputs	*Share of supply of inputs (per cent)*	*Share of fixed productive assets (end of year) (per cent)*
Tractor and agricultural machinery industry	47.9	32.0
Food machine-building	5.5	3.1
Maintenance of tractors and agricultural machinery	24.8	24.3
Production of chemical fertilisers and chemical plant protection means	16.7	32.8
Peat production for agriculture	1.1	2.0
Microbiological industry	4.0	5.8
Total	100.0	100.0

Source: Narodnoe Khozyaistvo SSSR v 1984 g. (The National Economy of the USSR in 1984), Statistika, Moscow, 1985, p. 215.

---

agriculture is the improvement of its regional structure. The industrialisation of agriculture is connected with the creation of sectoral and zonal systems of machinery utilisation of technology, and of the system of crop growing as a whole. This implies strengthening the regional specialisation of agriculture, leading to a more complete utilisation of biological potential, to a rise in the productive capacity, and to a fall in costs per unit of output. Here again one can clearly see the interrelationships between solutions of structural problems and improvements of the economic mechanism; between the increased powers of regional and local bodies and agricultural enterprises and associations in the formation of the structure of production; between internal development and the interregional exchange of products remaining after their sale to the centralised all-Union funds.

Another important aspect of the improvement of the AIC structure is a faster development of the material and technical base of the processing industries, the location of output processing and storage at the place of production and, in rural areas, even in the agricultural enterprises themselves. The solution of this task is especially important from the point of view of creating enterprises to link up agricultural and industrial production both technologically and economically. To expedite the implementation of the measures that have been mapped out, aimed at developing the competing sphere of the AIC and at strengthening its unity with agriculture, it was necessary to organise the management of the complex at all levels as a single whole, that is, to effect the transition to the inter-sectoral and territorial management by the creation of the Agroprom of the USSR. Simultaneously, it was necessary to increase the possibilities of local bodies and farms to organise the processing of output and expand storage capacities and improve the relations between agricultural enterprises and procurement organisations. The provision of technical support for the development in this sphere is concerned with the provision of standard, easily-assembled small-size processing enterprises of an industrial type and equipment for them.

The structural shifts within the system of the AIC, the development of industrialisation in agriculture and the agro-industrial integration aggravate the problem of the rational utilisation of the labour resources. The general direction for solving these problems seems to be clear: workers in agriculture are to be reduced in absolute terms on the basis of the industrialisation of agriculture and redistributed between the developing sectors of production services, processing and the social infrastructure. Practically, however, ensuring the national structure of labour resources within the system is complicated by a number of circumstances and contradictory tendencies. Thus, in agriculture, the reduction in the number of workers is far from being an end in itself. It should be a logical result of the introduction of new technologies, organisational forms and the provision of work incentives. It should be compensated by an increased number of highly skilled workers. But in many regions, the outflow of farm workers can be explained by completely different reasons, notably dissatisfaction with the conditions of life and work in the country. As a result, there is an increasing shortage of skilled mechanics and specialists as well as manual labour in such regions and farms. Thus agriculture has ceased to be a potential source of labour resources for other sectors. Indeed, it is itself in need of an inflow of people, but this is impeded by the same causes of the existing shortages.

In some other regions, mainly the Central Asian region, there is a high saturation of agriculture with labour. Within these areas the level of territorial and inter-sectoral migration of population is traditionally high. In such conditions, different variants of population movement from labour-saturated to labour-short areas are offered. Of course, the inter-regional migration of population, including that of various nationalities, is gradually developing. This is a logical process. But it should be the result of development of the requirements of the population itself, including the rise of its vocational and social mobility. With this in view, it is necessary to expand the spheres of labour application, to develop industrial enterprises and sectors of social infrastructure immediately in the rural areas of the labour-saturated regions, the more so as they themselves require a great number of seasonal workers.

As for the areas and farms in which agriculture is not provided with sufficient labour resources, the problem can be partially solved at the expense of a certain interregional, inter-farm and inter-sectoral redistribution of workers. But the main way of achieving a sufficient labour supply lies, not in an additional number of workers, but in saving labour resources by technical re-equipment, improving labour organisation, and mobilising social factors.

In solving the tasks of saving labour resources it is necessary to prevent irrational migration and the outflow of workers from agriculture and rural localities, and to improve the structure of labour supply. To meet the requirements of the industrialisation of agriculture and of the entire agro-industrial sphere for labour involves raising the prestige of agricultural labour, reducing the mobility of labour and engendering a more active attitude to work by refashioning the social structure of the village, including a change in the nature of agrarian work.

The extent of social changes in the countryside are widening quickly. The mean annual volume of capital investment assigned to the social development of the village, according to the item 'Agriculture' in 1981–85 was four times larger than in the first half of the 1960s. From 1965–85 one-third of the currently available dwelling space was constructed, and the major part of existing schools and other community buildings. The earnings of agricultural workers have doubled and the volume of fixed assets per worker has increased 4.5 times.

Still the impact of the social transformation on labour resources, and the growth of the industrial and agro-industrial production, is not as great as one might expect. There are several reasons for this:

1. The social changes are implemented, in many cases, by means of a simple extension of various subdivisions of the social infrastructure, such as the housing fund and the technology in use, without either transforming social conditions or orienting the population towards achieving the social conditions and standard of living enjoyed by the urban population.
2. The implementation of social changes on an intensive basis, that is, by direct transfer of urban forms and conditions of life to the village, is difficult because, on the one hand, it contradicts the functions and peculiar features of village organisation of agricultural production and the use of natural resources; on the other hand, it increases the requirements for resources.
3. Adequate measures for dealing with the complex and closely interlinked aspects of the conditions of life and work which are not easily replaced are not available.
4. The relatively autonomous development of town and country remain. The rural problems were solved, but mainly within the rural communities, and even on the basis of individual farms. However, the real insularity between the urban and rural populations in terms of the structure and level of requirements, education, the degree of information and other characteristics is quickly lessening.

Experience shows that overcoming the socioeconomic differences between town and country does not mean turning the village into a species of town, nor of creating conglomerates of town and country, and consequently the forms of social transformation should take into account the particular features of village life.

The village remains necessary, on the one hand, for the immediate organisation of agriculture, and on the other hand, for maintaining the balance between the present use and future potential of the natural biological resources. The realisation of those functions implies the preservation of a number of structural and functional peculiarities of the production and ecological potential of the village; specific forms of the conditions of work and its organisation, the interaction between man and environment and, of course, of the conditions of life, and the settlements themselves. It should be noted that the same processes of agro-industrial integration, industrialisation of agriculture and urbanisation which place rural and urban conditions of work and life closer to each other simultaneously raise the importance of agriculture in utilising the production, social and ecological possibilities of society.

Given rural and urban characteristics, a number of urban conditions will, inescapably, be preferable to rural conditions, while certain rural conditions will exhibit advantages compared to respective groups of urban conditions. Therefore, in ensuring the social equality of the combinations of conditions of work and life in town and country, certain advantages of the town should be compensated by the advantages of the village, implying the creation of a system of social compensations. It follows that overcoming the socioeconomic distinctions between town and country does not lead to eliminating the differences in rural and urban conditions but, on the contrary, creates the possibility for an all-round development of the positive features of rural and urban societies, including those within their integrated urban-rural structural and functional forms.

The integration of town and country, that is, the development of organic interrelations between them, the establishment of general rural-urban structures, and the creation of the integrated production and social potential of town and country, constitutes a real, and objectively logical, process of drawing together, and gradually achieving, the social equality of the conditions of work and life of the urban and rural populations.

Reference

KPSS (1986) Materially XXVII s''yezdi KPSS (Documents of the XXVIIth Congress of the CPSU) (Moscow: Politizdat).

12 The Relationship between Industry and Agriculture in the Development of the Hungarian Economy

Csaba Csáki
KARL MARX UNIVERSITY OF ECONOMICS, BUDAPEST

More than forty years have passed since the ending of the Second World War in 1945 and these years have considerably changed the Hungarian economy. From being an underdeveloped agricultural country Hungary has become a moderately developed industrial country, and the life of the people in the country has also changed. These past decades were above all years of industrialisation, but Hungarian agriculture also maintained its importance. It can indeed be said, without any agricultural chauvinism, that among the recent achievements of the Hungarian economy it has been the achievements of agriculture that have really attracted wide-ranging international interest and appreciation. In this study we shall examine the development of Hungarian agriculture in comparison with the development of industry, approaching the question from the side of agriculture. We shall try to show the most important characteristics of development of the two sectors and their effects on each other, as well as their future prospects.

1. *Growth in agriculture and in industry*

The development of the two main sectors of the Hungarian economy can be assessed by comparison with international standards and in the light of the country's economic and natural resources.

1.1 Agriculture

The last few decades have been the most successful period of the Hungarian economy. Figure 12.1 shows the growth of gross agricultural production in Hungary over a period of more than thirty years. Table 12.1 gives information about the growth of gross and net agricultural production since 1950. We can see that during the thirty three years from 1950 to 1983 the volume of agricultural production more than doubled, and after the completion of the socialist reorganisation of agriculture in the early 1960s production increased by about 80 per cent. Net agricultural production grew to a much smaller extent: in 1983 it was only about 20 per cent higher than in 1950.

Examining the last five-year period, we regard it as a significant achievement that, in spite of the severe drought in 1983, agriculture managed to increase its production by 12 per cent between 1981 and 1985. The 18 per cent increase in net production is especially remarkable: as a result of this, the contribution of agriculture to the national income also increased as compared with the previous five years. Improved efficiency in agricultural production is unquestionably reflected in this result. The increase in net production can be attributed partly to the economically unjustifiable fact that worn-out means of production were not replaced.

The growth of agricultural production was particularly rapid in Hungary in the 1970s (Table 12.2) but even between 1981 and 1985 the annual growth rate averaged 2.9 per cent. Of course, growth was not uniform within agriculture (see Table 12.3). The fastest development was observed in the branches of cereals, pig farming and poultry husbandry, while the horticultural branches and bulk feedstuffs production lagged behind, relatively speaking.

The growth of Hungarian agricultural production has outstripped that of world food production. According to FAO's figures, in the period between 1971 and 1980 world agricultural production increased by an average of 2.2 per cent annually, and this pace continued in the early 1980s as well (FAO, 1981). The rate of Hungarian agricultural development was more rapid than in the majority of developed countries and one of the highest within the socialist camp (Table 12.4).

According to calculations by the Hungarian Central Statistical Office, in the agricultural production ranking of twenty-three European countries, Hungary was ranked twelfth in 1969–71, whereas ten

Table 12.1 Indices of agricultural products and food industry production (1950=100)

Year	*Gross agricultural production*			*Net agricultural production*			*Gross food industry production*
	Arable and horticultural	*Live animals and animal breeding*	*Total*	*Arable and horticultural*	*Live animals and animal breeding*	*Total*	
1951	134	94	117	145	73	123	114.7
1952	80	95	86	73	74	73	141.0
1953	123	80	105	130	49	105	158.5
1954	107	103	105	103	80	96	167.9
1955	124	109	108	125	88	114	180.6
1956	101	104	102	94	80	90	176.8
1957	124	106	117	121	81	109	182.9
1958	121	125	123	113	104	110	189.8
1959	130	124	128	125	90	114	207.5
1960	121	118	120	112	80	102	217.6
1961	113	125	118	98	90	96	239.3
1962	120	125	122	107	89	101	257.3
1963	130	126	128	117	84	107	278.2
1964	131	138	134	117	95	110	304.8
1965	122	134	127	105	87	99	312.1
1966	139	139	139	123	78	109	321.5
1967	143	145	144	126	75	110	346.6
1968	141	149	145	120	85	109	357.0
1969	161	146	155	144	77	123	368.8

1970	135	162	146	104	86	98	392.8
1971	148	171	157	116	83	105	416.8
1972	156	169	161	123	71	105	439.3
1973	169	177	171	126	78	110	454.7
1974	169	188	177	127	71	108	482.0
1975	177	193	183	128	74	109	494.1
1976	165	198	178	110	80	100	501.0
1977	185	217	198	131	88	116	550.6
1978	182	225	200	118	92	110	552.8
1979	176	225	197	107	91	103	567.2
1980	190	230	206	125	86	111	581.4
1981	193	235	210	121	90	111	598.8
1982	211	247	226	145	96	127	624.4
1983	195	253	220	126	101	118	633.9

Source: Hungarian Statistical Yearbooks, Központi Statisztikai Hivatal, Budapest.

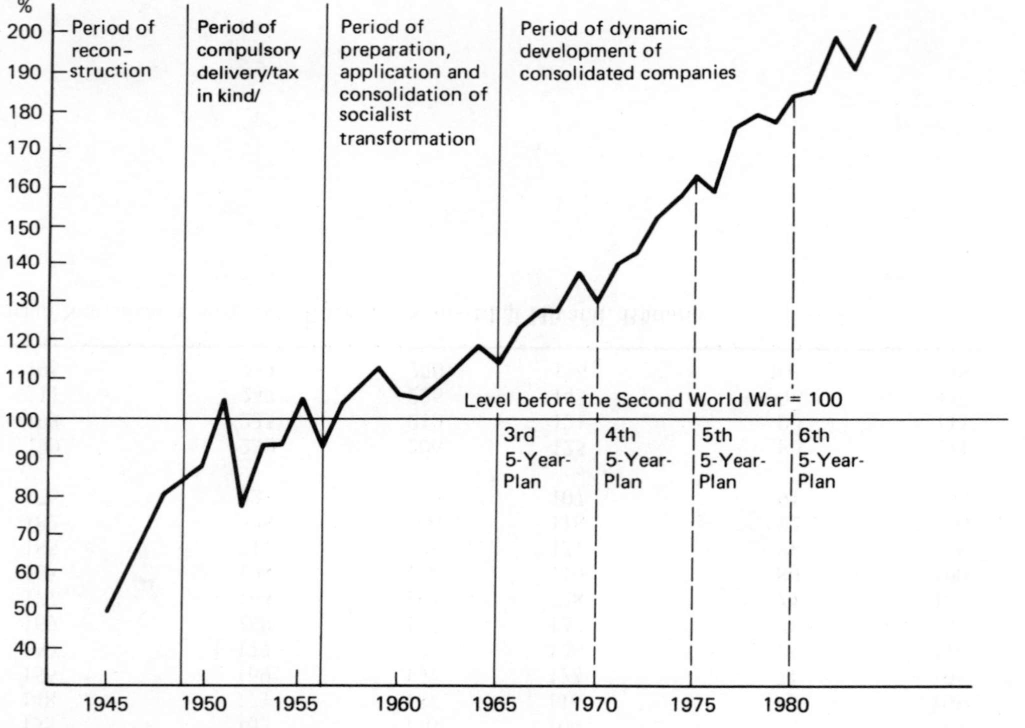

Source: Nyitrai (1985).

Figure 12.1 The development of the gross production of Hungarian agriculture

Table 12.2 The average annual growth of gross production of agricultural products expressed as a percentage of the average for the previous five years

1961–65	*1966–70*	*1971–75 (in real terms)*	*1976–80*	*1981–85*
1.4	3.0	3.5	2.9	2.9

Source: Kovács (1986).

Table 12.3 The average annual growth of agricultural production in the 1970s

	Average annual growth (per cent)	*The ratio of the branch within gross agricultural production (per cent)*
In fast developing branches of which:	6.3	55.5
grain	7.2	12.3
maize	5.7	12.1
pigs	6.1	18.1
poultry	6.5	13.0
In more slowly developing branches of which:	1.4	29.6
vegetables	2.4	3.7
fruit	1.4	5.1
grapes	–0.65	4.8
cattle	1.8	12.1
bulk feedstuffs	1.9	3.9
In other agricultural activities	1.3	14.9
Total for agriculture	3.2	100.0

Source: Agricultural Statistical Yearbooks, Központi Statisztikai Hivatal, Budapest.

Table 12.4 The average annual rate of growth of agricultural production in different countries between 1965–1980 (per cent)

Average for the socialist countries of which:	3.0
Romania	4.4
Hungary	3.5
Czechoslovakia	3.1
Soviet Union	3.1
Bulgaria	2.8
German Democratic Republic	2.5
Poland	2.2
Average for the developed capitalist countries of which:	2.2
West European capitalist countries	2.0
German Federal Republic	2.3
France	1.6
United States	2.3
Canada	1.7

Source: FAO Production Yearbooks, various issues (Rome: FAO).

---

years later, in 1979–81, it ranked ninth. The comparison is even more favourable to Hungary in terms of per capita product volume: in Europe in the early 1980s Hungary was first in poultry meat, second in corn and eggs and third in pork meat. In view of the domestic and internationl achievements in the first half of the 1980s, we can safely say that up to 1984 we continued our advance towards the forefront of European development, even though at the same time food production grew considerably in most European countries (Kovács, 1986).

The relatively favourable natural conditions of agricultural production in Hungary are reflected in the achievements of Hungarian agriculture. As regards the proportion of its land area devoted to agriculture of the total, Hungary ranks among the highest in Europe, 70.6 per cent. In 1984 the agricultural land amounted to 6.6 million hectares and more than half of this is under cultivation. The relatively good supply of arable land is accompanied by soil quality and climatic conditions which are good by international standards. Hungarian economic policy recognised the potential in these relatively favourable natural conditions. The agricultural policy pursued in recent years made possible and encouraged the exploitation of these endowments in the rapid growth of agricultural production in response to domestic and foreign demand.

1.2 The Food Industry

The food industry, the other main component of the Hungarian food economy, developed more slowly than industry as a whole until the mid-1970s (see Table 12.1). In the past decade this branch was more successful in keeping pace with the development of agriculture and its growth was somewhat faster than the industrial average. Food industry production increased by 17 per cent between 1975 and 1980, and by 13.5 per cent in the period from 1981 to 1985. In the past five years the annual growth rate of food processing has exceeded the average growth rate of industry every year; it was over 2 per cent per year and in 1984 actually exceeded 4 per cent.

1.3 Industry

The production of Hungarian industry increased faster than the development of the food economy branch. In 1985 industrial production was 9.2 times higher than its 1950 level (the national income within this period increased fivefold). The growth of industrial production is shown in Table 12.5: the development of industry after the Second World War up to the end of the 1970s was virtually unbroken. Between 1951 and 1968 the growth rate was around 8 per cent and even between 1970 and 1978 it topped 6 per cent annually. Then growth came to a sudden halt. In 1979 the growth rate fell back to 3 per cent and in 1980 the absolute volume of production also dropped. After that, industrial production started to increase again but at a modest rate, averaging about 2 per cent annually. In earlier years the growth rate of industry was always faster than that of agriculture, but in the 1980s agriculture surpassed industry in growth several times.

As regards the structural changes in industry, the importance of metallurgy and mining has decreased in recent years, while at the same time the chemical industry has developed rapidly and the position of the engineering industry has also improved. In the past decade the chemical industry has been the most dynamically developing branch of Hungarian industry.

Despite absolute growth which has been spectacular in itself, the development of Hungarian industry cannot be regarded as an extra-

Table 12.5 Indices of the production of socialist industry (1960=100)

Year	*Gross production value*	*Net production value*
1960	100.0	100.0
1961	108.3	110.6
1962	115.8	118.6
1963	123.6	124.7
1964	132.4	134.3
1965	137.7	139.1
1966	146.8	151.1
1967	154.0	164.1
1968	166.0	171.0
1969	170.8	177.7
1970	184.4	190.1
1971	196.8	203.2
1972	207.2	217.9
1973	221.7	237.6
1974	240.5	258.5
1975	251.9	272.1
1976	263.5	290.4
1977	278.8	306.4
1978	294.1	323.2
1979	303.3	339.4
1980	298.0	329.7
1981	305.1	344.6
1982	312.4	360.0
1983	316.0	364.3
1984	326.3	372.7
1985	328.5	359.4

Source: Hungarian Statistical Yearbooks, Központi Statisztikai Hivatal, Budapest.

---

ordinary achievement by international standards. Till the end of the 1970s the annual growth rate was around the world average or a little above it and higher than in the majority of developed capitalist countries. In these years the rate of development of domestic industry adapted itself well to the tendencies observable in the socialist countries. But even then the development of Hungarian industry fell behind the average of the socialist countries (see Table 12.6).

The stagnation which started in 1979 and is still in effect today deserves special attention. Our economists see as the immediate cause of this our late and inappropriate reaction to the world economic changes that started in the early 1970s. In fact this explanation

Table 12.6 The average annual percentage growth rate of industrial production according to groups of countries between 1956 and 1968

Groups of countries	*1956–60*	*1961–65*	*1956–65*	*1951–68*
World total	7.2	6.6	7.0	7.0
Socialist countries	13.1	7.3	10.3	10.8
of which:				
CMEA countries	10.1	8.3	9.3	10.3
Non-socialist countries	4.3	5.9	5.2	5.5
of which:				
Developed capitalist countries	4.1	6.0	5.0	5.2
EEC countries	6.8	5.2	6.0	6.7
Developing countries	7.8	7.0	7.4	7.0
Hungary	7.8	8.0	7.9	9.5

Source: Nyitrai (1985).

---

is an oversimplification. It is my conviction that the negative features of the industrial policy of the 1950s and 1960s are reflected in our present-day problems. The main factor is that in developing industry we did not build sufficiently on the actual natural and economic resources of the country (see section 2). The shortcomings of our economic management, some of which are having especially unfavourable consequences in the current world economic situation, also play a part in our present problems.

Without going into a detailed examination of the recent problems of the industrial sector, we should like to quote the Polish economist J. Winiecki, who attributes the present problems in industrial production in the small European socialist countries to the disproportionate size of industry and to inappropriate participation in the international division of labour. Semi-autarkic, inward looking development swells abnormally the raw material and semi-finished product producing industries like iron and steel, chemicals, paper-making and textiles, but mainly the extractive branches. The size of these branches far exceeds the level which is normal in countries of the same size that satisfy their demand for semi-finished products within the framework of the international division of labour. The smaller the economy, the greater the problems this sort of economic structure causes. For example, Czechoslovakia produces 5 million kinds of semi-finished and finished products, and Hungary, which is even smaller, produces more than 3 million. (Winiecki, 1986)

2. THE ECONOMIC ROLE OF INDUSTRY AND AGRICULTURE

2.1 Industry

The development of a few decades as outlined briefly in the previous section made industry a crucially important sector of the Hungarian economy. In the early 1980s industry accounted for 46 per cent of the national income. Industrial organisations employ 31 per cent of the working population and take up 25 per cent of the operating fixed assets of the national economy. About one-third of investment is allowed to industry, and industrial products account for 74 per cent of total exports, excluding food industry products. It is evident that the economic development of a country depends first of all on the performance and efficiency of its industry.

2.2 Agriculture and the Food Industry

But the role of agriculture and the food industry in the Hungarian economy should not be underestimated. For many years the importance of the sector showed a downward trend within both gross and net production. It is noteworthy that this decline has stopped in recent years: in the early 1980s the share of agriculture in the production of national income started to increase again slightly, and the number of people employed in agriculture also grew. In 1985 agricultural production in the narrower sense accounted for 16.3 per cent of gross national production and 16.0 per cent of net national production, as compared with 15.1 per cent and 13.9 per cent respectively in 1980. The share of agriculture in the employment of the working population was 20.7 per cent in 1985.

In Hungary the importance of food production is determined by the fact that agricultural land is the only abundant natural resource which if properly used is permanently available, and can even be improved in quality. Hungarian food production has great traditions and its products are world famous. So in this country every possibility is given to produce foodstuffs in excess of domestic demand so as to permit considerable food exports.

In the last few decades Hungarian food production has almost always had a positive balance of trade. A considerable surplus, however, has been produced only since the end of 1960s (see figure 12.2). Hungarian food production provides the country with a steady supply of good quality foodstuffs, and in addition:

- some 30-35 per cent of agricultural produce is sold on foreign markets;
- the food economy accounts for 25 per cent of total foreign trade turnover;
- more than a third of non-rouble-accounting exports consist of agricultural and food products;
- food production in recent years in both rouble-accounting and non-rouble-accounting trade yielded a substantial export surplus approximately 21 billion forints' worth of roubles and US $1 billion respectively).

On the basis of all this, Hungarian agriculture has become a very important stabilising factor in the national economy, a vital source of foreign exchange at a time when industry is not capable of producing a net export surplus, and when paying off debt is a very important task for the country. Hungarian food production makes it possible to import materials and energy carriers which are indispensable in our economic life, and it is one of the most important sources of the foreign exchange we need for industrial development.

Hungarian agriculture, apart from its role in our positive foreign trade balance, contributes to solving the financial difficulties of the national budget and the economy. In recent years the income position of agriculture has become decidedly unfavourable. The balance of subsidies and grants allocated to agriculture became adverse in the early 1980s (see Table 12.7). The position of agriculture further deteriorated as a result of the fact that the cost increase resulting from the rise in price of agricultural means of production was higher than the price increase of agricultural products (between 1981 and 1985 the prices of the means of production used in agriculture rose by 32 per cent, while those of agricultural products rose by only 22 per cent). As a result of all this, the profitability of producing agricultural products deteriorated considerably. Between 1980 and 1985 the

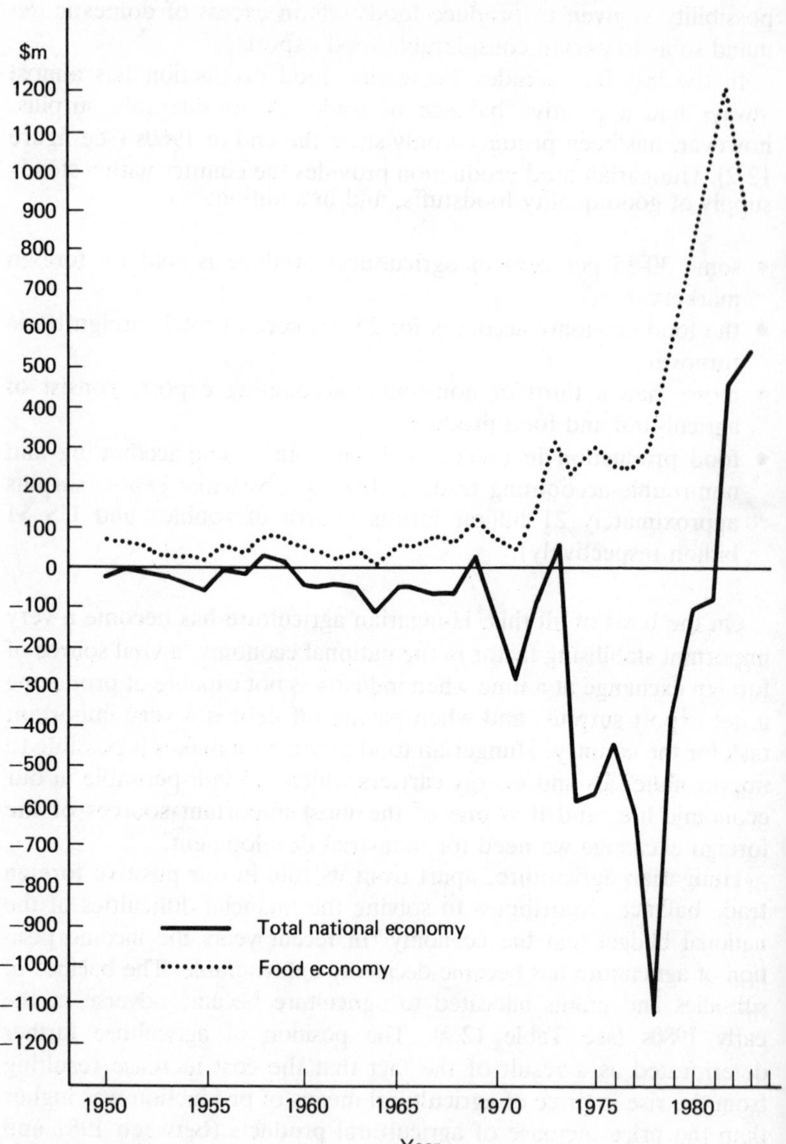

Source: Hungarian Statistical Yearbooks, Központi Statisztikai Hivatal, Budapest.

Figure 12.2 The development of the non-rouble accounting foreign trade balance in the national economy and within this in the food economy

Table 12.7 Subsidies and withdrawals in Hungarian agriculture (billion fts)

Year	*Total subsidy*	*Total withdrawal*	*Balance*
1970	16.2	7.1	+ 9.1
1975	28.8	16.9	+ 11.9
1980	30.3	26.1	+ 4.2
1984	31.2	47.3	− 16.0

Source: Központi Statisztikai Hivatal, Subsidies and withdrawals in agriculture 1975–82.

profits of agricultural companies at current prices rose by only 16 per cent, an increase considerably smaller than that achieved in their production. Profitability decreased especially in the production of animal products, causing a reduction in the volume of products and hence of production.

2.3 The Effects of the World Markets

The situation on the world market for foodstuffs is reducing the favourable influence exerted by Hungarian food production on the development of the country's economy; it is even exerting an unfavourable influence on the whole Hungarian agricultural sector. More than 50 per cent of Hungarian food exports go to non-rouble-accounting markets. This market is greatly distorted by the agricultural protectionism of developed OECD countries but mainly by that of the EEC. It is well known that production stimulated by unrealistically high domestic agricultural producers' and consumers' prices, coupled with a high degree of protectionism in the interest of the home market, leads to the accumulation of enormous surpluses in these countries. Eventually these surpluses pour on to foreign markets; they encourage state budget supported exports, since the capacity of domestic markets is limited. This is how the protectionist agricultural policy of the developed countries is becoming one of the most important factors contributing to the extremely depressed prices on the agricultural world market.

The enormous subsidised surplus food exports of the developed capitalist countries harm the market position of countries which are otherwise producing efficiently, which do not want, or are not able, to take in the price competition financed by state budgets. Instead of a competition in production by agricultural producers in the individual countries, the world market for agricultural products is becoming a competition between state budgets. In this situation there are both winners and losers. Obviously the smaller and poorer exporting countries are on the losing side, while the solvent importers can be found on the winning side.

Hungary is one of the countries most unfavourably affected by the protectionist policies influencing the world market. Like the other small agricultural exporting central European countries, Hungary has been driven out of its historical markets for agricultural products by the developed west European countries and the EEC without any compensation; in other words, its exports directed there are being hard hit by the present discriminatory measures. Furthermore, at the low world market prices caused by the protectionist policy of the developed capitalist countries, the Hungarian food economy is becoming less and less capable of competing with the export-subsidised products of the rich countries. At the CMEA level preference is given to domestic production only in an indirect way; within the CMEA the system of bilateral agreements can only partly give protection against the effects of the export policy of the developed capitalist countries, which is state-subsidised and aims at selling the accumulated surpluses, and which damages Hungary's price position.

It is evident that an agricultural world market free from protectionism, or at least less protectionist, would immediately bring economic advantages for Hungary. Indeed, every step towards agricultural free trade would considerably improve the favourable effects of Hungarian food production on the development of the whole economy, and would provide additional resources which would help us meet our debt payment obligations and lay the financial and technical foundation for structural change in industry.

3. INTERSECTORAL RELATIONS BETWEEN INDUSTRY AND AGRICULTURE

The large-scale development of industry and agriculture has opened up new prospects in the relations between the two sectors. These possibilities are of great importance from the point of view of the development and efficiency of the whole economy.

Table 12.8 shows how much material of agricultural origin has been used in industry. We can see that agriculture supplies a rela-

Table 12.8 Material consumption by Hungarian industry

	1970	*1975*	*1979*	*1981*
Consumption of material of agricultural origin – % of total	18.6	16.8	18.1	23.1
(Index 1970 = 100%)	100.0	148.6	214.2	324.5
Total material consumption (Index 1970 = 100%)	100.0	164.4	220.7	261.1

Source: Hungarian Statistical Yearbooks, Központi Statisztikai Hivatal, Budapest.

tively small proportion of the materials consumed by industry. These figures indicate that industry has only partly exploited the primary production possibilities provided by the economic and natural resources of Hungarian agriculture.

Among the European socialist countries the structure of Hungarian industry is most similar to that of Czechoslovakia, the GDR and Poland. Compared with the developed European capitalist countries, Hungary's lack of mineral resources, the high proportion of metal-using industries and machine production (higher than in Denmark, Norway or Finland), and the importance of the extractive industries, are particularly striking. At the same time, too, compared with the developed capitalist countries light industry based on agricultural resources is relatively insignificant, as are the various branches of the food industry.

In the 1970s the development of the sectoral structure of Hungarian industry was determined by central development programmes. Unfortunately, in these programmes (natural gas, aluminium, public vehicles, computer technology, petrochemicals), and in the big investments associated with them, developments related to agricultural production potential were not emphasised. Programmes initiated in recent years pay more attention to the possibilities of agriculture. The foremost development areas include pesticide production and biotechnology. However, as regards its relation to agriculture, Hungarian industrial policy has not changed substantially and owing to the shortage of capital in the whole economy, adjustment to the needs of agriculture has been much more difficult that it would have been in the 1970s.

The relative backwardness of the food industry, and the fact that it did not benefit from the main developments of past years, and of the last ten years in particular, constitute very serious disadvantages for Hungarian food production. Almost three-quarters of our agricultural exports consist of products which have gone through some sort of processing stage. The level of processing is visible not only in our export products, but to a large extent it determines their prices and saleability. Because of the saturation of markets, the requirements concerning the quality of foodstuffs and the demand for up-to-date packaging have risen considerably. Consequently, the competitiveness of Hungarian agricultural products is decided by these factors for the most part. Can we meet these increasingly rigorous quality and packaging requirements or not? If we cannot, our economical agricultural production and all the efforts made, and partial successes achieved in the earlier stages, have been in vain; we cannot hope to export our food industry products profitably.

The growth of agricultural production mentioned in section 1 caused an increase in demand for means of production which exceeded the domestic supply. In Table 12.9 we show the development of the use by agriculture of materials of industrial origin, examining separately the proportion of imported means of production. Agricultural production in 1983 was about one and a half times that of 1970. However the table 12.9A shows that in that period producers' consumption of the means of production had multiplied three and a half times (Index 358.6), and within that the use of means of production of non-agricultural origin had multiplied nearer four times (Index 382.7). Hungarian agriculture opened up to industry as an outlet in proportion to the increase in the use of means of production of industrial origin. Unfortunately, Hungarian industry only partially exploited the increase in demand resulting from the prosperity of agriculture. It failed to meet demand in terms of quantity, range and quality and its share in agricultural producers' consumption decreased. As is shown in Table 12.9B, in 1983 in the use by Hungarian agriculture of means of production of industrial origin the proportion of imported products was about 20 per cent. In itself this is not a high proportion, but the rate of increase in imports compared with 1970 is significant (see Table 12.9A). Within the increase to 1983 in the use of means of production of non-agricultural origin of less than four times, imports rose to nearly five and a half times their 1970 level.

The share of the machine industry in the supply of the means of production of Hungarian agriculture is particularly modest. In fact, almost the whole stock of basic machinery of Hungarian agriculture

Table 12.9A Indices of usage of means of production in agriculture (1970 = 100)

Usage of means of production	*1970*	*1975*	*1980*	*1983*
Total	100.0	166.9	271.7	358.6
of agricultural origin	100.0	164.4	244.9	339.8
of non-agricultural origin	100.0	170.0	306.1	382.7
of which:				
imports	100.0	337.2	387.1	541.7
of which:				
rouble-accounting	100.0	208.5	213.5	382.7
non-rouble-accounting	100.0	507.3	617.0	752.5

Table 12.9B Distribution of exploitation of means of production in agriculture (percentages)

Exploitation of means of production	*1970*	*1975*	*1980*	*1983*
Total	100.0	100.0	100.0	100.0
of agricultural origin	56.2	55.3	50.6	53.2
of non-agricultural origin	43.8	44.7	49.4	46.8
of which:				
imports (non-agricultural origin)	14.1	27.9	17.8	19.9
of which:				
rouble-accounting	56.9	35.2	31.4	40.2
non-rouble-accounting	43.1	64.8	68.6	59.8

Source: Hungarian Statistical Yearbooks, Központi Statisztikai Hivatal, Budapest.

---

originates from foreign markets and foreign factories. In 1984, of the 3663 new tractors which were put into operation in Hungarian agriculture only 251 were Hungarian-made. In 1984, 1012 new combines and 4242 new trucks were put into operation in Hungarian agriculture. All were imported.

As a consequence of all this, the industrial background of modern agricultural production has only partly developed in Hungary. Apart from a few exceptions, in developed countries, regardless of whether large or small companies dominate agriculture, the renewal of products and means of production in the interests of developing modern industrialised agriculture is provided by the research departments of industrial companies producing agricultural inputs and processing agricultural produce, and by state and private institutions connected with agricultural activities. They generally do this by taking into account the actual factory circumstances and conditions in agriculture. Thus agricultural companies have at their disposal a relatively wide variety of means of production and technologies for the innovative development of production.

In Hungarian agriculture the situation is somewhat different. The majority of inputs come from abroad; the choice is poor and often accidental. In the renewal of technologies the agricultural sector is the decisive one. The partial absence of a domestic industrial background is unfavourable on the whole, even though by means of imports Hungarian agriculture can obtain means of production which Hungarian industry would not be capable of producing. In our domestic circumstances, mainly import-based mechanisation results in a narrower choice of technologies and greatly reduces the possibility of accommodation to specific circumstances by individual farms in the choice of technologies.

We are convinced that the partial failure of Hungarian agriculture to exploit the possibilities is disadvantageous for Hungarian industry too. Domestic food production, with its relatively favourable circumstances, and considerable productive capacity, could be an assured market and an excellent recommendation for domestic agricultural machine production, not to mention the foreign currency that would be saved through import substitution. A more highly developed food industry linked to agricultural raw material sources would probably be a more efficient investment than other areas of industry with less favourable possibilities in the country. Last but not least, this would improve the effectiveness of our agricultural policy and the market position of export products.

4. THE PROSPECTS FOR GROWTH

4.1 The Importance of Industry

The alternatives for the growth of industry and agriculture are often the subject of debate in Hungary these days. The above-mentioned facts also prove correct the thesis formulated in the seventh Five-

Year Plan (1986–90), that the dynamic role in the Hungarian economy should be played by industry. The future of the Hungarian economy, the easing of its international payment problems and the raising of the living standards of the population, all depend, above all, on the extent to which Hungarian industry can step up production, increase the choice and quantity of exportable products and improve efficiency in production.

4.2 Potential and Appropriate Growth of Production

As regards the desirable pace of agricultural development, opinions differ more widely. Owing to the falling world market prices of foodstuffs, our increasing difficulties in selling, and resultant lower efficiency, many ask the question: under the circumstances, is it necessary to maintain the present level of Hungarian food production? Would it not be advisable to redistribute part of the resources tied up in agriculture to other branches of the economy? Some people supplement this argument by referring to the unfavourable environmental effects of intensive agricultural production.

(a) Potential. In spite of the undoubtedly increasing environmental problems from the ecological and biological point of view, there are clear possibilities for increasing production in agriculture. The ecological potential of agricultural production, the possibility of increasing biomass production, have been the subject of two recent national surveys in Hungary.[2] According to these surveys, in 1980, 54 million tons of dry vegetable organic matter were produced in Hungary. The quantity for 1985 is estimated to have been 4–5 million tons greater. Considerably more biomass is produced per year than the aggregate of the annual output of the whole raw material extracting branch (mining and the building material industry) plus the total amount of imported raw materials. The findings of the surveys prove that through the exploitation of ecological possibilities considerable production reserves can be opened up. The natural environment would enable an average 80 per cent increase in agricultural production to be achieved by the end of the century in Hungary.

The possibilities of agricultural growth have also been examined with the help of a simulation model which describes the Hungarian food production system (Hungarian Agricultural Model: HAM).[3] Table 12.10 shows the results of calculations made by the most recent versions of HAM. It is clear from the table that predictions concern-

Table 12.10 The annual rate of agricultural growth in the various versions of the Hungarian food production model

	The whole economy				*Food Production*			
Version[1]	*1–5*	*6–10*	*11–15*	*1–15*	*1–5*	*6–10*	*11–15*	*1–15*
5x	4.0	3.9	3.9	3.9	3.4	3.2	2.8	3.1
6x	3.3	2.5	2.3	2.7	3.3	3.0	1.8	2.5
8x	3.3	2.3	1.9	2.4	3.6	2.7	1.0	2.4
8/a	4.0	3.5	4.1	3.9	3.6	1.9	4.1	3.1
11x	3.1	1.6	1.2	1.9	3.6	1.7	2.5	2.6
14x	3.3	2.8	3.5	3.3	4.2	3.7	6.0	4.7
15x	3.3	2.3	2.8	2.8	4.0	2.1	3.5	3.2
16x	3.3	2.4	3.0	2.9	3.5	2.2	3.6	3.1
18x	3.1	1.8	1.6	2.1	3.3	1.5	1.1	1.9
20/a	3.3	2.4	3.0	2.9	3.5	2.1	3.6	3.1
21x	3.1	2.2	2.8	2.7	3.5	2.2	3.5	3.1
25x	3.1	2.3	2.9	2.8	3.2	2.6	3.9	3.9

Note: 1. All versions are variants of the HAM model
Source: Calculations from HAM

ing agricultural growth do not differ greatly in the different versions. On a fifteen-year average, the two extreme values for annual growth are 4.7 and 1.9 per cent. Broadly speaking, the calculations made according to the HAM model predict the possibility of 3 per cent average growth annually. Over the next 15 years the expected growth rate will be highest in the first five-year period; later it will decrease a little. In our opinion an annual growth rate of 3 per cent represents the upper limit of possible development, which could be achieved only in the event of the most favourable development of conditions. The natural and economic conditions for an annual growth rate of 2 per cent, however, can be regarded as guaranteed. These figures show that Hungarian agriculture has considerable additional reserves and is capable of continuing in the future to produce growth similar to that recorded in the 1960s and 1970s.

To what extent is it advisable to exploit this production capacity? Should the Hungarian economy partly or wholly renounce the exploitation of existing agricultural production possibilities?

(b) Resource considerations. First let us approach this question from the point of view of the resources of agricultural production. It can be stated that, in the case of arable land, it would be positively beneficial

to withdraw from cultivation those poor quality areas which are most at risk from environmental hazards. In these areas the efficiency of production is low and they produce the least marketable export products. A change in methods of cultivation in marginal areas would certainly have a favourable long-term effect on the quality of our arable land.

The problem of the labour force and the means of production that might be released is somewhat more complicated. The possibilities of redeploying agricultural labour within a company have become limited owing to falling demand in auxiliary activities. The relocation of the agricultural population is hardly feasible, and the location of industry in the country is hampered by the state of the infrastructure there.

The utilisation of capital released in the event of a possible limitation of agricultural production is not easy either. Today the food economy accounts for only 13.6 per cent of the total stock of fixed assets. This is only half the size of its share of exports. So in the agricultural sphere the contribution to national production is much greater than the sector's share of the means of production and of investments.

> This means nothing more or less than the fact that in spite of its obviously increasing need for means of production, the food economy still produces unit national income, or unit export income, with much less capital demand than the whole of industry. (Varga, 1986).

So, to limit food economy production in the interests of industry would probably not have a favourable effect on the capital intensity of the national economy; rather the reverse is likely to occur.

(c) Balance of Payments. In the country's present situation, the dilemma of the exploitation of agricultural production potential will not ultimately be decided by the advantages or disadvantages of resource utilisation. Alas, the country's payment obligations tend to simplify this question. Kovács (1986) writes in a recent article:

> If we have a product which can be exported for hard currency in the same quantity, more economically and over a longer period instead of a certain category of foodstuffs, then that is what we should export. In this case export-oriented food production could

be reduced by eliminating the least profitable exports. This question has come up several times in the last few years, most recently during the drawing up of the seventh Five-Year Plan, and the same answer has had to be given every time. There is no product or product line that would meet the above requirements and could replace part of our foodstuffs in our non-rouble-accounting exports.

d) Conclusion. The maximal exploitation of agricultural potential and the export of marginal or barely exportable products are obviously not in the interests of agriculture either. The Hungarian national economy, however, needs these products at present. For this reason the planners expect an annual growth rate of 1.7–2.0 per cent to continue in the coming years in agriculture. The main thing is that marketability and efficiency of production should continue to be stepped up under the circumstances of increased growth. To this end the food industry should receive priority development and certain structural changes should be made in production, concentrating clearly on products of greater exportability in both the production and the marketing spheres.

So the idea of limiting or reducing our overall agricultural production cannot be regarded as either realistic or justifiable, because reducing production would not really result in freeing resources for the national economy that could be converted into something else and used to help exports, and because the replacement of lost agricultural earnings by income from other products seems hardly feasible. On the other hand, a vigorous improvement of the exporting ability of the industrial sector – which has certainly not been restrained by agricultural growth in the past – is of vital importance to the entire national economy. What is more, if the domestic industrial background does not expand and become more competitive, not even the planned development of the food economy at a slower rate than in previous years will be achieved.

* * *

THE FUTURE

As a matter of fact, considerable interdependence between the industrial and agricultural sectors can be observed in Hungary's economic development. Today the agricultural sector, with its export earnings and adequate food supply for the country, is having a favourable effect on the development of the entire economy. The guarantee of the economic development of the country is the performance of industry, and finally in the long run more efficient agriculture cannot exist without the partnership of a successful, competitive industrial sector.

Notes

1. The amount of arable land per capita in Hungary is amongst the highest in Europe.
2. The findings of these two surveys are summarised in an article by Láng (1986).
3. On the Hungarian Agricultural Model see Csáki (1985).

References

Csáki, C. (1985) *Simulations and Systems Analysis in Agriculture* (Amsterdam: Elsevier).

FAO (1981) *Agriculture toward 2000* (Rome: FAO)

Kovács, I. (1986) 'Our Agricultural Development Strategy', *Társadalmi* Szemle, nos 8–9 (in Hungarian).

Láng, I. (1986) 'Prospects for Exploiting Biomass', *Közgazdasági Szemle*, no. 5 (in Hungarian).

Nyitrai, F. (1985) *Forty Years of the Hungarian Economy* (Budapest: Kossuth Könyvkiadó) (in Hungarian).

Vági, F. (1985) 'The New Economic Conditions and Agricultural Policy', *Gazdálkodás*, no. 11 (in Hungarian).

Varga, Gy. (ed.) (1986) *The Competitiveness of our Agricultural Production* (Budapest: Mezögazdaságikiadó) (in Hungarian).

Winiecki, J. (1986) 'The Outsize Dimensions of Industry in the East European Socialist Countries', *Közgazdasági Szemle*, no. 5 (in Hungarian).

13 Problems of an Industrial Economy with Intensive Agricultural Production: the German Democratic Republic

Wolfgang Heinrichs and Klaus Steinitz
CENTRAL INSTITUTE FOR ECONOMICS,
ACADEMY OF SCIENCES OF THE GDR

1. INTRODUCTION

The socialist planned economy is faced with the task of determining the appropriate balance between industry and agriculture in line with the changing internal and external conditions of production. These two substantial sectors of the national economy have to be developed, taking into consideration the role of state socialist ownership for industry, and co-operative socialist ownership for agriculture, so that the driving forces of development will be given the best opportunity to secure a dynamic economic growth and to satisfy the needs of the population. The political dimension of this task is to establish a closer alliance between workers and co-operative farmers by gradually overcoming the essential differences in the working and living conditions in towns and in the countryside.

In the long-term planning of industrial and agricultural production in a developed modern socialist economy with an intensive agricultural production such as the German Democratic Republic (GDR), the balance of the development of industry and agriculture is linked to factors determining the rate, effectiveness and social efficiency of economic growth; it must be considered with reference to a number of complex relationships in society and the economy, and it changes with scientific and technical development. In sections 2 to 4 these matters are considered, and the future tasks are assessed in section 5.

2. RELATION TO ECONOMIC GROWTH

Experience has shown that in all developmental periods of socialism, from the construction of socialism to the shaping of a mature socialist society, the balance of development of industry and agriculture is linked to the factors in the national economy which determine the rate, effectiveness and social efficiency of economic growth. Agriculture, which provides the material basis for nutrition and is the supplier of raw materials for many industries, also plays an important part in highly-developed production, in spite of the fall in the share of agriculture and forestry in the national income and in employment. From 1950 to 1985 their share in the national income fell from 29.2 to 7.8 per cent and of employment from 31.8 to 10.8 per cent. Industry and agriculture together account for 81 per cent of the gross social product, industry having 72.5 per cent and agriculture/forestry 8.5 per cent (1984). Agriculture strongly influences economic growth and its effectiveness. Years of a high growth in the overall economic development were, as a rule, coupled with an increase in agricultural production; whereas the years with a stagnant or falling agricultural output generally had perceptibly adverse effects on the growth rate of the gross social product and on the national income. For instance, within the last period of the Five Year Plan of 1981–85 there was only one year (1982) with a total economic growth of less than 4 per cent (in 1982 the growth of national income was 2.5 per cent). In that year the net product of agriculture and forestry was 2.5 per cent lower than in the previous year.

The importance of agriculture for the living standard of the population is reflected in the share of foodstuffs in retail turnover, which is 31.4 per cent (1985).[1] The share of foodstuffs and luxuries in retail turnover is about 50 per cent (Statistisches Jahrbuch, 1986, p. 234).

3. SOCIAL AND ECONOMIC RELATIONSHIPS

3.1 Relations between Workers and Peasants

The political and social relations between workers and peasants have to be perfected and consolidated in such a way that they will prove effective as foundations of the social system. They influence political stability and the social climate. Their satisfactory functioning is indispensable for economic and social progress as well as for the expansion of democratic co-determination in all social spheres, among them for the relations between urban and rural areas. Some of the many changes needed include (a) the utilisation and development of the specific possibilities and benefits of co-operative ownership in the interest of the members of the co-operative and the entire society; (b) the improvement in the working conditions of co-operative farmers and all those employed in agriculture with the use of industrial methods in agricultural production; (c) the upgrading of the living conditions of co-operative farmers by housing construction; and (d) the creation of better possibilities for education, health care, culture and sports activities in villages, partly by diversified forms of government support from central government funds.

The change in social relationships in the countryside is apparent from the fact that, with the help of capital resources, output per man in agriculture and forestry increased from 23 400 marks in 1960 to 125 500 marks in 1985, that is more than fivefold, bringing it above the average of the national economy of 123 000 marks. Moreover, the proportion of workers with a completed technical training (graduation from universities, colleges, certificate as foreman or skilled worker) has increased from 24 per cent in 1965 to 90 per cent in 1985. The figure of university and college graduates rose from 26 400 in 1965 to more than 72 000 in 1985. Rural living conditions have improved. From 1971 to 1983, for instance, the number of dwellings in villages with interior toilets rose from 18 to 45 per cent, and those with baths or showers from 15 to 61 per cent. The number of rural outpatient clinics has more than tripled since 1950, from 136 to 435.

In the GDR in 1985 3 236 000 persons out of a total of 6 750 000 in all production were employed in industry and 922 000 in agriculture and forestry. Of the latter, 68 per cent are working members in production co-operatives, 31 per cent are workers and employees and 1 per cent are self-employed persons and helpers of affiliated family members (Statistisches Jahrbuch, 1986). There are modern efficient forms of organisation that play an essential role in the development of agricultural production and the conditions of equilibrium between industry and agriculture. The basic units of socialist agriculture comprise 1144 agricultural co-operatives (Landwirtschaftliche Produktionsgenenossenschaft – LPG) for crop production, 2761 LPG for animal production, and 465 nationally-owned farms and their co-operative institutions. These co-operative and nationally-owned production units work together in an ever closer manner in 1190 merged co-operative institutions. It is typical of these large co-operatives that one co-operative partner in crop production, and two to three co-operative partners in animal production, work closely together. On average, they cultivate 4800 hectares of arable and agricultural land and keep some 4500 head of cattle. The co-operative partners have established their own democratically elected councils to ensure close links with all enterprises and institutions involved. Some 43 per cent of the members of all LPGs are in crop production, and 49 per cent in animal production. The remaining 8 per cent belong to other co-operatives and related institutions, such as agro-chemical centres and amelioration co-operatives (Statistisches Jahrbuch, 1986, p. 182).

3.2 Relations between Industrial and Agricultural Production

The priority function of agriculture in the GDR, as in most countries, is to satisfy the needs of the people for foodstuffs on a stable and rising level. Notwithstanding the rapid development of other need complexes, the share of economic resources needed for nutrition has increased; in the 1970s the total cost of labour increased by 13 per cent and of fixed assets by 18 per cent. At the beginning of the 1980s the complex of nutritional needs still retained a share of some 40 per cent in the final product for consumption. To meet the need for nutrition it was necessary to provide an above-average share of resources used for consumption as a whole, some 50 per cent of expenditure on labour and about 48 per cent of expenditure on fixed assets. This shows the economic importance of the most rational deployment of labour, fixed assets and other resources for the various branches of the nutrition complex (Zentralinstitut, 1986).

The results achieved so far in raising the level of supply of foodstuffs are shown in Table 13.1.

Past development was, among other things, characterised by an increase in the quantity of consumption of most foodstuffs. Quantitative demand is now met for certain items. Hence, in the future there is likely to be a shift of demand to high-quality products with a higher user value, especially with regard to healthy nutrition, durability, freshness, taste and degree of prefabrication. The share of vegetable raw materials in the total domestic availability of raw materials was more than 50 per cent. The major share of these raw materials serves

Table 13.1 Consumption of foodstuffs per capita

Foodstuffs	*Unit*	*1960*	*1970*	*1975*	*1980*	*1984*
Meat and meat products	kg	55.0	66.1	77.8	89.5	96.2
Butter product	kg	13.5	14.6	14.7	15.2	15.7
Certified milk	litres	94.5	98.5	100.8	98.7	105.6
Fat and low-fat cheese	kg	3.6	4.6	5.5	7.5	8.7
Vegetables	kg	60.7	84.8	90.0	93.8	104.4
Fruit	kg		55.5	66.6	71.1	79.3
Pure coffee roasted	kg	1.1	2.2	2.4	2.8	3.5
Non-alcoholic beverages	litres		40.8	70.3	81.3	89.3

Source: Statistisches Jahrbuch der DDR 1986, p. 282 (Berlin, 1986).

directly, via processing in the foodstuffs and nutrition industry, or indirectly, as feed for animal production, for the production of foodstuffs. Some 10 per cent are used for seed.

A considerable part of agricultural produce is used as raw material for the production of a number of different goods which are not foodstuffs. Thus agriculture makes an important contribution, in its capacity as a raw material supplier, to the chemical industry, the textile industry, the leather and fur industry, the cellulose and paper industry, and the pharmaceutical industry. Agrarian raw materials are used in the manufacture of 125 production groups, or some three-quarters of all production groups in industry. The scale of the use of agrarian raw materials for industrial production is shown by their share in consumption of materials for leather goods, footwear and furs which amounts to about 50 per cent, more than 20 per cent in pharmaceuticals, some 9 per cent in special chemical products and some 10 per cent in textile raw materials (Schieck and Schmidt, 1984, pp. 6–7). The average share of agricultural products in total industrial material consumption is about 10 per cent. These agricultural raw materials are made available for a very wide spectrum of products; for instance, grain is used for more than 180 different products, potatoes for some 90 and sugar-beet for more than 35 different products. The characteristics of the economic functions of agricultural produce within the macroeconomic cycle are important for the planned formation of balanced reciprocal relations between the growth of industrial and agricultural production. Some 85 per cent of crop production is used as intermediate goods, either for industry or agriculture. More than half of that remains within the cycle of agricultural production, e.g., as feed for cattle, so that its ultimate purpose is food to satisfy consumers' nutritional needs.

In contrast to this, the output of animal production to a larger extent leaves the process of agricultural production. Some 75 per cent of animal products are reprocessed into nutritive goods, and 18 to 20 per cent remain as breeding and productive livestock within their own production process (Schieck and Schmidt, 1984, p. 7).

A significant function of GDR agriculture is, and will be in the future, to make a contribution towards improving the foreign trade balance. A thorough analysis of the change in the domestic production and imports has shown the need for the GDR to cut back on agrarian imports, especially grain, and to increase the share of its own agriculture in meeting the demand for foodstuffs and raw materials for industry. This is not a temporary and transient task, but a strategic objective for the development of agriculture in the 1980s and 1990s. The following factors have been taken into account in formulating this objective:

(a) An increase in the supply of agricultural produce by the GDR is a contribution towards solving the global nutrition problem. It should be noted that with 0.37 hectares of agricultural land per inhabitant, a considerably smaller area is available than in most other European countries. By 1984 it had been possible to raise the contribution of local supplies in foodstuffs for domestic consumption to more than 85 per cent.
(b) Economic expenditure to increase local grain production is less than that required by the national economy to pay for imports.
(c) Most agricultural crops still offer considerable potential to raise yields.
(d) With the increase in the rate of domestic supply there is also a better utilisation of the reproducible natural resources to increase local availability of raw materials.

From 1976 to 1980 some 3.5 million tons of grain were imported each year (Schieck and Schmidt, 1984, p. 68). By raising the annual grain production from an average of 9 million tons in the period

1976–80 to an average 10.4 million tons in 1981–85, and even more than 11 million tons in 1984–85, in the years 1982–85 it was possible to reduce perceptibly the share of grain imports within the total amount of grain consumed in the GDR, which in 1980 had been about 30 per cent. The interrelationship between agriculture and industry has assumed a key importance, especially for the long-term planning of the equilibrium of industrial and agricultural production.

In general, the higher the level of social and productive development, the more comprehensive, versatile and intensive are the interrelations between agriculture and all the other fields of the national economy, especially industry. This applies to the interrelations with the agriculture-subordinate stages of processing and reprocessing of agricultural raw materials (especially nutritive goods and the foodstuffs industry) with transport and trade, catering and works-supply as well as with those fields preceding agriculture. The industries producing agricultural inputs, agriculture itself and the nutritive goods economy (including the foodstuffs industry and trade) form together the complex of nutritive goods. Table 13.2 shows their share in the entire national economy.

The share of agriculture in the nutritive goods complex amounts to 31–33 per cent for production, whereas it is 48 per cent for fixed assets and 40 per cent for labour.

3.3 The Development of Income as a Central Element of Socioeconomic Relations

The socialist state, by resorting to many measures of material and financial support, provided great help for the newly-emerging pro-

Table 13.2 Share in the national economy of the nutritional goods complex and agriculture 1980 (in percentage)

	Nutritive goods complex	*Agriculture*
Gross product	32.4	10.1
Net product	26.3	8.7
Fixed assets	26.7	12.7
Labour	32.5	12.9

Source: Schieck, H. and Schmidt, K. (1984) *Intensivierung der Landwirtschaft de DDR* Berlin: Dietz Verlag, p.8

duction co-operatives, especially in their early years. This help was indispensable in terms of stability and level of income of the members of the co-operatives in the 1950s and 1960s. However, with the consolidation of the co-operatives, the growth of income of the members became more directly dependent upon their own performance. The formulation and application of agrarian price policies have played an important role in increasing the stability and the level of income, and in upgrading the living standard of the co-operative farmers. These policies provide for the sales of agricultural produce at prices which take account of the rising expenses for certain inputs, e.g., energy, fuel and feedstuffs, and ensure the profitability of agricultural enterprises and guarantee the purchase of products by the state. Thus, for the co-operative farmers in the GDR the uncertain existence and the ruin of agricultural holdings have become a thing of the past. Their income obtained from social production and from individual economies has, for a considerable part of LPG members, both a higher level and a higher rate of growth than for workers in industry and other fields of the national economy.

In 1984 an agrarian price reform was carried through. This, taking into consideration the development of production expenses, was aimed at fixing producer prices for agricultural produce that reflect the actual social expenditures and secure the profitability of production for the majority of co-operatives. Thus the producer prices for potatoes rose, for example, by 70 per cent, for milk by 66 per cent. To guarantee stable retail prices for basic foodstuffs despite the rise in producer prices for agrarian produce, substantial funds are used from the state budget as subsidies, and these have increased from 7.8 billion marks in 1980 to 27.6 billion marks in 1985 (Statistisches Jahrbuch, 1986). The price supports for foodstuffs has thus risen to more than three-quarters of the retail turnover of nutritive products.

The price supports for foodstuffs are amongst the largest economic redistributions in the GDR. The low prices for foodstuffs, virtually unchanged in the last few decades, exercise a very important social function with regard to social security and justice. However, they are also the subject of controversy in matters including the use of refined and finished agricultural produce such as bread and milk for feeding commercial animals.

4. SCIENTIFIC AND TECHNOLOGICAL DEVELOPMENT

The conditions of equilibrium between industry and agriculture also change in conjunction with the higher rate of scientific-technical development of the entire national economy.

4.1 Quantitative and Qualitative Change in Agricultural Production

With the development of the 'nutrition' complex there is also a change in its internal equilibrium conditions in quantitative and qualitative terms. The quantitative changes are that the share of agriculture in the services and resources of the complex of nutrition will decrease, whereas the share of the foodstuffs economy, including foodstuff processing and manufacturing, trade, catering, and the works-supply system, and industries supplying inputs which are outside agriculture, will increase. The performance of machine construction, electrical engineering/electronics and chemistry shows a particularly sharp increase in their share. The nutrition complex will be able to cope with the growing demand in future, only if the proportions between the production complexes involved are constantly changed in line with the change in conditions of production. In this context, the idea is to analyse the interrelationships and to plan them, e.g., the ratio of requirements of agricultural machinery, agro-chemicals, crop production, the storage and transport capacities, the processing industry and trade.

Changes in quality in the equilibrium conditions proceed mainly along the lines of raising the level of the user value of goods within all the parts of this complex. Increasingly the entire process of raw material extraction up to the manufacture of highly finished food products is determined by science and technology. This implies that the key technologies of microelectronics and the related information and automation engineering, together with bio-technology, will determine the future development of performance and effectiveness within the entire nutrition complex. At present the national economy of the GDR is gaining useful experience in this field.

The scale of quantitative and qualitative changes has increased, encompassing all major facets of social life and economic activity. Their influence extends from production methods, the use of sophisticated and efficient technologies, the deeper interrelations (manufacture, storage, transport, marketing) within the economic units and among them (co-operation), up to the higher levels of education and

Table 13.3 Growth rate of gross product of productive consumption and of net product (percentage/change)

	1975–80	*1980–85*
Gross product	12	8
Productive consumption	18	4
Net product	1	15

Source: Statistisches Jahrbuch der DDR 1986, p.13, Berlin, 1986.

---

qualification, the enrichment of work content, to the living conditions of working people.

4.2 Changes in Ratios of Broad Economic Measures

In the various stages of development, the type of growth of agricultural production, and accordingly the proportionality between industry and agriculture, is subject to considerable changes. This refers to such proportions as those between (i) gross product, productive consumption and net product; (ii) growth of animal and vegetable production; and (iii) use of resources and production growth.

At the beginning of the 1980s it was possible to alter the trends in proportions of some of these factors. The slower growth of net product compared with gross product was reversed, to the benefit of the satisfaction of needs (see Table 13.3).

This increase in economic efficiency was achieved by a growth in performance of plant and animal production over the period from 1981 to 1985 with a cutback in energy and material. Thus from 1981–85, compared with the period from 1976–80, a growth in the performance of agriculture was accomplished with 15 per cent less cost for mineral fertiliser per unit of plant production, 3 per cent less feed cost per unit of animal production and 14 per cent less energy cost per unit of production. The slower growth rate of plant production than animal production before 1980, and the consequent rising grain imports to safeguard the supply of animal feed, was corrected in the period 1980–85, as is shown in Table 13.4.

The Five Year Plan 1986–90 provides for a continuation of the more rapid development of plant production with an average increase in yields per hectare of 1.7 per cent, and in animal production of 1.4 per cent (SED, 1986). All these processes reflect a new quality

Table 13.4 Growth of production of grain and of animal products (percentage/change)

	1971–75 to 1976–80	*1976–80 to 1981–85*
Grain production	4	15
Animals for slaughter	12	7

Source: Statistisches Jahrbuch der DDR 1986, pp.193, 206, Berlin, 1986.

---

in the equilibrium relations between industry and agriculture, determined primarily by the intensive type of growth which is already shaping the entire economic development in the GDR, including increasingly plant and animal production, and the proportions within agricultural production as well as between agriculture, industry and other fields.

5. FUTURE TASKS

What problems have to be solved in the future in the conditions of the intensive growth of the national economy?

Until 1980 the development of agricultural production was marked by the time-saving variant of intensification. The hours of labour per 100 kg had been reduced for the major products of agriculture from 1961–65 to 1981–84 as follows (1961–5 = 100):

Grain	44.0
Potatoes	46.9
Sugar-beet	45.3
Beef cattle	84.0
Pork	55.2
Milk	54.7

Source: Heinrich and Stegmann (1986) p. 349.

This increase in productivity had in part been reached by a simple substitution of labour by fixed assets, energy and fertilisers. The new quality in intensification in the 1980s is that higher growth rates of labour productivity are being reached on the basis of a simultaneous cutback in labour and other resources.

The scope of the task for the 1980s is illustrated by energy consumption. From 1970 to 1980 in agriculture, forestry and the food economy, the use of service energy rose by about 50 per cent, of which electric energy rose by 86 per cent and gas by 51 per cent (Schieck and Schmidt, 1984, p. 24). Thus the consumption of energy increased far more rapidly than production. By contrast, the higher production rate of agriculture in the years 1981–85, compared with the previous five years of the plan, was achieved with an absolute reduction in the use of service energy. In future, the objective is to lower continuously energy consumption per unit of production by the employment of new technologies with a higher rate of efficiency in the use of energy for soil cultivation and in harvesting and transport, and by a more rational utilisation of the biological potentials of yield and performance of soil and animals. Within the period 1986–90, the average consumption of feed per unit of animal production is scheduled to be cut back by 1 per cent annually. The establishment of new yardsticks for intensification is shown by the plan to achieve an increase in animal production, with the exception of wool, exclusively by an increase in the efficacy of existing animal stocks rather than by further growth of these stocks. Thus, compared with the development in the 1970s, the ratio between the growth of fixed assets and the growth of production is planned to be improved substantially by rationalisation, modernisation and better utilisation, so that the needs for nourishment can be satisfied with a relatively low share of economic resources, thus releasing some resources for developing other need complexes.

This task is closely intertwined with further co-operation between animal and plant production as well as with other branches of the national economy, and is coupled with the increase in the responsibility of co-operatives and nationally-owned farms for a rational development of the agricultural production cycle. A higher rate of processing of food is the objective. The idea is to reach a higher user value per unit of agricultural raw materials produced in the interest of satisfying the nutritive needs of the population. Greater processing is also decisive in securing economy in energy and materials in the long term, and for constantly improving techniques. The goal of higher user value and higher quality relates to all spheres and parts of the nutrition complex:

- industry and civil engineering, which provide inputs and infrastructure as well as increasing the use of agricultural machines, computer-assisted information systems, civil engineering projects, agro-chemical and bio-technological products;

- the vegetable raw materials and animal products sector, e.g., an increase in the protein and energy content of feedstuffs by improved breeding;
- the more efficient utilisation of agrarian produce in processing, with a view to more effective satisfaction of nutritional needs. This includes such things as the manufacture of calorie-reduced foods without loss of quality in terms of consumer taste, substitution of imported ingredients of foodstuffs, substitution of ingredients to reduce energy costs, the increase in the degree of prefabrication of foodstuffs for consumption (this includes instant meals and food products involving an easier preparation of foods in households) and improvement in quality of food products with regard to taste, durability, freshness etc. All processes have to aim at satisfying the nutritive needs with a higher economic effectiveness. They incorporate the use of new scientific-technical knowledge, especially in the field of bio-technology, new auxiliary materials and additives (such as aroma and enzymes) and processes.

Another important task related to the more intensive utilisation of scientific-technical knowledge is to make full use of all agricultural raw materials available, possibly without any losses. This task refers to all stages of production in the nutrition complex. Of special importance is the reduction of losses in the use of energy, feed, agro-chemicals etc., for the production of plant and animal produce, upon harvesting, during transport and during the storage of vegetable produce, in reprocessing agricultural produce (complex utilisation of raw materials and low-loss processing) as well as the circulation and consumption of food products. A cut in losses of agricultural produce and foodstuffs has a special problem when compared with other products because of the nature of agricultural production as a biological process, and the low durability of many agricultural and food products. A cut in losses is tantamount to an increase in output, either directly, whenever harvesting losses are reduced, or indirectly, whenever the demand for any foodstuff declines because of the reduction of losses in consumption. Part of the losses can be saved without much expense, that is, by better organisation of production, a rational combination of qualitative factors of growth, a more effective utilisation of available techniques, and by better efficacy of economic and moral stimuli. The elimination of other losses requires considerable expenditure, such as the expansion of storage and cooling capacities and the introduction of new harvesting and transport technologies. Yet further losses can be limited only on the basis of new scientific and technical knowledge, especially information engineering and bio-technology, as well as their comprehensive use. The scope for avoiding losses is evident from the fact that the annual losses amount to some 7 to 8 per cent of the total harvest of products. By the provision of storage and preservation capacities for fruit and vegetables, e.g., it would be possible to lower the losses during storage for winter and spring by some 20 to 30 per cent (Schieck and Schmidt, 1984, p. 82).

In spite of the considerable expenditure needed to reduce losses, economic computations have shown that the reduction of loss has almost always been more effective for the national economy that the corresponding expansion of production.

Note

1. This share would be still higher if the state subsidies used for maintaining stable retail prices for basic foodstuffs were taken into account.

References

Heinrich, R. and Stegmann, H. (1986) 'Probleme der Entwicklung und der rationellen Nutzung des gesellschaftlichen Arbeitsvermögens in der Landwirtschaft der DDR', *Wirtschaftswissenschaft*, no. 3.

Schieck, H. and Schmidt, K. (1984) (eds) *Intensivierung der Landwirtschaft der DDR* (Berlin: Dietz Verlag).

SED (Sozialistische Einheitspartei Deutschlands) (1986), *Materials on the XI Party Congress of the SED*, Berlin.

Statistisches Jahrbuch der DDR 1986 (Berlin: Staatsverlag 1986).

Studie zur Entwicklung des Ernährungskomplexes, Zentralinstitut für Wirtschaftswissenschaften, Berlin.

Part IV

Unbalanced Productivity Advance, Agriculture-led Development and Rural Activities

14 Rural Employment Linkages through Agricultural Growth: Concepts, Issues, and Questions

John W. Mellor
INTERNATIONAL FOOD POLICY RESEARCH INSTITUTE

1. INTRODUCTION

Contemporary development theory has had little place for agriculture in growth. This is because of a perceived weakness of backward and forward linkages (in Hirschman's, 1958, strong condemnation of agriculture on this count, for example); or, because of an emphasis on capital formation as the primary engine of growth, with agriculture as a consumer goods industry with low savings rates (e.g., see Mahalanobis, 1953 – he was the father of the Indian Second Five Year Plan); or, because of an emphasis on import substitution, with agriculture seen as an export industry, as a producer of nontradable output, or with both inelastic demand and supply (e.g., see Prebisch, 1971).

Of course, in practice a number of economists have a common sense orientation to agricultural growth as a percursor of overall growth (see, for example, an early statement of Kaldor, cited in 1964). But the key interacting relationships between agriculture and other parts of the economy are not specified. In periods of rapidly rising real food prices (for example in the mid-1960s and the mid-1970s) considerations of food security and foreign exchange savings argue for attention to agriculture, but usually in the form of higher prices, which tend to elicit only modest short-run responses in aggregate production. A short-run view of the importance of agriculture will not take supply-shifting technological change into account. But this is the only means by which agriculture helps accelerate economic growth.

2. CONCEPTUALISING THE ROLE OF AGRICULTURE IN ECONOMIC GROWTH

If agriculture is to serve as a primary engine of economic growth, three conditions must be met (Johnston and Mellor, 1961). First, of course, agriculture must be a major sector of the economy so that it can have a significant aggregate effect. That condition is fulfilled in virtually all low-income countries. In the middle-income countries agriculture's share in domestic output has declined to the point that overall growth is determined to a significant extent by autonomous developments in the non-agricultural sector even if agriculture grows vigorously. Even in middle-income countries, if agriculture does not play its optimal role, it is like running a four-cylinder engine on three cylinders.

Secondly, agriculture must grow on the basis of cost-reducing technological change. Such technology, virtually without exception, requires an indigenous research system that can borrow copiously from abroad and can adapt and develop technologies that increase factor productivity. The need for technological change arises from diminishing returns, a classic feature of agriculture. Moving out on a production function characterised by diminishing returns requires increases in the amounts of inputs per unit of output, rising costs and hence, a negative contribution to growth. That is why price policy cannot be the engine of agriculture growth, even though it may be an important complement to an effective technology policy (Mellor and Ahmed, 1987). In contrast to the diminishing returns in agriculture, industrial expansion is seen as subject to constant returns at worst. Perceived positive externalities in industry and increasing capital intensity in agriculture were the key bases for Amartya Sen's early negative view about the potential for a developing country to achieve high employment growth rates and a positive place in such processes for agriculture (Sen, 1968).

Although diminishing returns in agriculture imply that technological change is needed if agriculture is to play a dynamic role in growth, our knowledge of growth processes make an even more powerful case. We know that economic growth has been largely a product of increased factor productivity arising from technological change. Developing countries must turn to agriculture if technological change is to play a major role in growth. If growth driven by agriculture is to fit in that classic mould, then research to increase factor productivity is needed. That this is possible is well demonstrated by the 'Green Revolution' (Mellor, 1976).

Thirdly, the rate of growth of demand for labour must be accelerated as agriculture faces a potentially difficult problem in playing a major role in economic growth because the demand for its goods is inelastic. In developed countries, virtually all growth in factor productivity in agriculture can be realised only by a transfer of resources out of agriculture. That poses economic and social problems that severely limit the extent to which factor productivity can be increased and accelerate overall growth. However, the circumstances for developing countries are different.

Because of exceedingly low incomes, the labouring class in developing countries has a high marginal propensity to spend on food, and labour tends to be substantially underemployed due to misallocations of capital arising from inefficient national and international policies (Mellor and Johnston, 1984). Thus if the rate of growth of the demand for labour can be accelerated, demand for food can grow rapidly, absorbing the increase in supply induced by technological change. If the food supply increases more rapidly than food demand, real food prices will fall. Some reduction in output prices can be absorbed by producers without a decline in output or net income since new technology increases factor productivity.

Increased employment while productivity rises is, of course, desirable, not just because it provides effective domestic demand for rising agricultural production, but because it is essential if participation in the processes of growth is to be broad. That in turn is essential to meeting equity objectives of growth. Of, course, where factor proportions are weighted heavily towards labour, as they are in low-income countries, high growth in employment or, more properly, rapid growth in the demand for labour, is consistent with the tenets of neoclassical economics and comparative advantage.

To achieve accelerated growth in employment with rising productivity requires solutions to two problems. First, since the labour market is comprised of two interacting markets – a food market and a labour market *per se* (Lele and Mellor, 1981), growth in the demand for labour must be accompanied by an increased supply of food. Because low-income people have high marginal propensity to spend on food, an increase in the wage bill increases the demand for food (Mellor, 1978). Because the supply of food is highly inelastic in the short run (and the long-run elasticity) is driven by the forces of technological change, best addressed directly by policy rather than indirectly through market forces an increased wage bill will raise the cost of wage goods. This will slow the growth of employment either through market forces working from higher wages to increased capital-intensity; or, more likely, direct public efforts will be made to combat the obvious acceleration in inflation by employment-decreasing restrictive monetary and fiscal policies. Accelerated growth in food production, of course, takes care of the problem of inflation of the wage goods prices.

At first glance, food imports can provide a perfectly elastic supply of food to back the rapid growth in the wage bill. However, food imports will, in effect, also be inelastic in supply, at least in early stages of economic growth. That is because accelerated growth of wage goods demand will cause food imports to absorb a large proportion of foreign exchange earnings. The supply elasticity of food imports will then be equal to the most limiting of the demand and supply elasticities of exports. That is likely to be very inelastic in early stages of growth because of the nature of the export mix in such countries. The effect will be devaluation of the currency and hence rising real food prices in domestic terms. National policy may constrain food imports long before exchange rate changes have run their course, raising domestic food prices more directly. Of course the problem recedes as development increases the supply elasticity for exports or if the prices for principal exports such as oil increases sharply. The critical point is that if the wage bill grows rapidly, food will make up a sizeable percentage of GNP, of foreign exchange earnings, or of any other economic variable. Thus it is unrealistic not to treat accelerated growth in food production as essential to rapid mobilisation of labour. Of course, rapid growth in food supplies that is achieved by relaxing a wage goods constraint facilitates taking advantage of exports and eventually facilitates additional food imports. This is a major reason why food imports tend to increase rapidly in countries that have achieved rapid growth in domestic food production (Bachman and Paulino, 1979) – the very growth in food production facilitates a pattern of growth in other sectors that favours exports and reduces the main constraint on food imports.

Second, growth in employment and the wage bill requires increases in the effective demand for goods and services and high labour to capital ratios. Otherwise a capital constraint will force wage rates and the wages bill down, reducing effective demand for food and the supply of labour. Under optimal policies, agricultural growth distributes income broadly through small-holder agriculture. We know from empirical studies that small-holder farmers have a high marginal propensity to spend on labour-intensive goods and services. Typically the marginal propensity to spend on such goods and services is 0.40 for locally produced non-agricultural goods and services and another 0.20 or more for agricultural commodities such as horticultural and livestock products that tend to be labour-intensive in their production processes (Hazell and Roell, 1983).

Thus, an acceleration of agricultural growth through technological change facilitates employment growth by relaxing the wage goods constraint on the supply of labour, increasing the demand for labour through its impact on the structure and amount of effective demand, and on the increased effective demand for the increased food production (Mellor and Lele, 1973). In a closed system, technological change in agriculture will reduce real food prices (Lele and Mellor, 1981). Empirically, this mild price depressing effect is easily overbalanced by concurrent autonomous growth in other sectors. A study of developing countries achieving rapid growth in food production shows that such countries rapidly increase their imports of food (Bachman and Paulino, 1979). That implies that the demand for food and hence employment grows more rapidly than the supply of food.

Empirical structure of the relationship between agricultural growth and overall growth show similar multipliers of somewhat less than 1.0 (see; Bell et al., 1982, Rangarajan, 1982). These structures are for countries with policies not particularly favourable to the small- and medium-scale industries that are potentially most responsive.

There seems to be a tendency for agriculturally-led growth to be less effective in Latin America than in Asia. The most important reason for that is the tendency for the distribution of land in Latin America to be highly skewed. Since the addition to net national income from technological change in agriculture is distributed largely to landowners, the benefits are skewed to high-income people, with a consequent tendency for the goods comprising the marginal propensities to consume to have a larger import content and a higher capital intensity. The result is smaller domestic employment and income multipliers and greater dependence on foreign markets – which are innately less stable, more risky, and hence have lower net returns – for growth in food output. The consequence that lower efficiency in the conversion of agricultural growth has for overall growth in Latin America is a lower employment multiplier in the non-agricultural sector and hence lower economic returns to investment in agriculture. The solution lies with a broader distribution of land and a greater relative emphasis on technological change for small-holders. But this may be unsatisfactory because of poor land resources and the high costs of distribution to the small-holder part of bimodal agricultural production systems.

Thus, the logic and the empirical evidence makes it clear that technological change in agriculture can be a powerful engine of overall growth. We can speculate that differences in income distribution and the consequent effects on consumption patterns makes a difference according to country, income distribution, and, no doubt, many other factors. We need to know more about the variations in the strength of basic macroeconomic forces, including the nature of differences in consumption patterns, the differences in capital-labour ratios, and the differences in the factor-share bias and income effect of the types of technological change in agriculture. In particular, we need to understand better how various economic policies determine the efficiency with which accelerated technological change in agriculture accelerates growth in other sectors. As we understand those processes, we can then develop policies to increase the power of the engine of agricultural growth.

In the following section several sets of policy issues are explored as basis for judging from available evidence what policies will be needed to foster an efficient agricultural and employment-oriented strategy of growth. That discussion will also suggest what research is needed to better understand the underlying processes and the policy needs. The broad question posed is 'what policy variables influence the size of the multipliers between agricultural and non-agricultural growth rates?' The question is prompted by the observation that the efficiency of that conversion rate varies immensely among countries and time periods.

3. POLICIES TO ENHANCE RURAL GROWTH LINKAGES

The Broad Problem of Misallocation of Capital

The basic cause of inefficiency in the conversion of accelerated growth in agriculture into accelerated growth for the economy as a whole is the misallocation of capital. The problem is one of bimodalism – the allocation of a large proportion of a society's capital to a small number of workers in capital-intensive industry, and the allocation of a small proportion of society's capital to a large number of employees in the less organised small- and medium-scale manufacturing and service sectors. The latter sectors have high employment content and high average factor productivity. This gross misallocation of capital explains large quantities of unemployed labour. That unemployment is largely explained, in the Harris-Todaro model (1970), by the queuing of labourers for the higher paid jobs in industries with high capital intensities and hence high marginal products of labour and wage rates.

In a neoclassical world, there is, of course, no unemployment. Poverty is purely a problem of low labour productivity. What one needs to do then is to simply increase the capital stock, so that the productivity of labour can be increased – or induce technological change. In the real world, however, the situation is not so simple. Interference with the market may restrict the potential for importing capital intensive goods and services and reduce the use of labour-intensive processes in domestic production for export. Market interferences and asset distribution structures may skew the domestic demand structure toward goods and services produced with high capital intensity.

In the late 1950s and the early 1960s, of course, there was an extensive literature dealing with fixed factor proportions and a debate about whether factor proportions were fixed or not. We can now say, with the benefit of considerable experience, that, in certain industries, processes that operate with a high degree of capital intensity are low cost even with extreme differences in capital cost and labour cost. But in other industries, the choice of low-cost technique is a function of the capital-labour ratios. Given these differences in the range of different capital intensities in the production of commodities, the development problem is one of structuring demand through trade, and through the distribution of income, to favour industries with wide choices of capital intensity. Then, less capital-intensive processes can be chosen, and capital can be evenly spread over the labour force. In practice there will be some distortions and some inequality of capital-labour ratios from one industry to another.

A prime example of bimodalism in capital investment is in the Philippines. The Philippines entered the post-second World War development period with a developed rural infrastructure and a highly-educated work force. It then adopted effective policies for agricultural development, and agriculture consequently grew rapidly. Yet, except for a brief period in the 1970s, that growth was not translated effectively into overall growth (Bautista, 1987).

Taiwan is in sharp contrast (Lee, 1971). Throughout the rural areas of Taiwan, small- and medium-scale manufacturing and service industries grew rapidly, substantially in response to rural demand. Employment also grew rapidly and capital was spread thinly and much more evenly than in the Philippines. One of the important issues in development is why such differences in the allocation of capital occur and what policies are needed to allocate it more efficiently.

Again at the simplest, one might argue for a return to neoclassical, market policies to spread capital more thinly. One might wonder why market-oriented policies were being adopted at such a slow pace, when the international lending institutions and much of the intellectual community is arguing so strongly for a neoclassical, liberal approach.

Presumably, one of the reasons for this slow pace is that we know so little about how to deal with some of society's difficult problems in a neoclassical framework. For example, most democratic political systems are concerned about large fluctuations in food prices, because they automatically bring large fluctuations in the real incomes of the lowest income people least able to absorb them. Measures to deal with this problem may co-opt public resources and impede public investment, which is an essential complement to small-scale, labour-intensive investment. One must also add that a high level of foreign assistance, particularly in countries where it makes up a major portion of GNP, and the bulk of savings and investments, may hinder reform when that aid finances old habits that distort markets.

3.2 Trade to Reduce Capital-Labour Ratios

Trade offers major opportunities for reducing capital-labour ratios. It allows importation of goods such as fertilisers, steel, and plastics that may be essential to high employment growth rates, but that uses processes that are capital-intensive even with the cheapest of labour. In turn, exports allow concentration on labour-intensive goods and services beyond the amount that can be absorbed in the local market. Thus, a development strategy that is oriented to agriculture, high-employment, and production processes with low capital-labour ratios must take place within relatively open trading regimes. Trade will be elastic with respect to the growth rate in such a development process. However, as indicated below, domestic demand is the basic driving engine for effective demand in such a development strategy, so that trade may not make up a particularly large proportion of the total economic activity.

Trade also has an important influence on the effectiveness of an agriculture- and employment-based strategy through its effects on the exchange rate. An overvalued exchange rate is inimical to such a strategy because it discriminates against exports, which tend to be labour-intensive. Since an agriculture-based strategy provides low-cost food through technological change, it hardly seems necessary to further depress the price of food, and, perhaps the incentives to produce food through an overvalued exchange rate. An overvalued exchange rate also encourages substitution and capital-intensive imports for labour-intensive goods and services from the domestic economy.

3.3 Infrastructure and the Commercialisation of Agriculture

It seems likely that the infrastructure requirements for an agriculture-led employment-oriented development strategy will be immense (Olson, 1980). Although the strategy may require less infrastructure in the major metropolitan centres, there must be a proliferation of smaller centres and complete coverage of all parts of the country that have a potential for accelerated growth in agriculture from technological change.

The financial requirements for infrastructure development may be reduced somewhat if agricultural research breakthroughs make infrastructure development cheaper by reducing the cost of production of food, which would thereby make wage goods and labour somewhat cheaper. Nevertheless, the requirements for public finance are sufficiently large that difficult decisions about the geographic allocation of infrastructure investment will have to be made. Similar decisions must be made between rural and urban structures. In general, scarce infrastructure resources will have to be allocated first to those parts of the country which seem to have the best prospects for accelerated technological change in agriculture. Those will tend to be the more prosperous rural areas. Such allocations may have profound implications for regional income distribution and may even have implications for distributional equity across income classes. They certainly have profound political implications that will require somewhat less efficient allocation of resources.

One of the most important means of holding down capital-labour ratios is to take advantage of demand shifters for livestock products and fruits and vegetables. We can make a generalisation that in developing countries, despite the efficiency of labour-intensive production, the real prices of such commodities tend to rise in the face of increasing demand. This is probably due to somewhat inelastic production and severe inelasticities of marketing services and, particularly, failings of the infrastructure system. Since livestock and horticultural products are labour-intensive at the margin and not restrained by land, it is sound economic policy for government to invest to increase the elasticity of their supply. One needs a well-developed infrastructure of roads, electricity, and communications, so that efficiency in the marketing of these labour-intensive commodities can be increased rapidly.

3.4 Education

We hypothesise that the demand for educated people is elastic with respect to the rate of growth of employment. That is, when employment grows slowly, educated unemployment tends to increase rapidly, but when employment increases rapidly, the demand for educated people runs ahead of even a rapid increase in supply. The reason for this is the rapid growth of small- and medium-scale enterprises and of organised service trades. These all seem to use less skilled labour but, in practice, require labour with at least primary, and often secondary, school education. An important research need is to find the extent to which, and under what circumstances, education limits growth and employment, and what kind of education is needed.

3.5 Measures to Increase the Supply of Capital

Obviously, the more the supply of capital can be increased, the more the capital-labour ratio can be raised with a further impact on productivity and growth. The argument for labour-intensive investment patterns is for the even spread of capital, not for low capital-labour ratios *per se*.

The commercialisation of agriculture through technological change and improved rural infrastructure tends to bring about large increases in rural savings rates. That is because the rate of return to capital is increased. The standard view that savings rates are low in agriculture comes from agricultures that are technologically static and with an equilibrium between savings and investment at low rates of return and low rates of savings and investment. However, with technological change, returns to capital increase rapidly and savings rates increase.

Perhaps far more important than the above, the effect of growing agricultural incomes in creating effective demand for production from the non-agricultural sector will, if capital is limiting, result in a shift in the terms of trade toward those commodities. If they are produced in small-scale sectors, the increase in relative prices will increase profitability, which will in turn encourage greater savings and investment, and will also provide the basis for direct investment by increasing retained earnings in the small firms.

Foreign resources may, of course, increase capital availability further. Foreign assistance may have a perverse effect because it may be spent, as in Africa in the last decade, largely for local nontradable services, including education, which are quite inelastic in supply, driving up the prices of those services at the expense of other commodities.

However, as an optimal development process proceeds, the rate of return on investment will tend to be quite high. That should be attractive to foreign investors. That should allow an increase in capital-labour ratios, which will gradually raise productivity and the national income.

Obviously, in this context, the current debt problem is a serious one. Even though that debt may have been generated by inappropriate types of investments, it overhangs the market which reduces the possibilities of net increases in debt. It may in fact, cause domestic capital formation to be considerably lower than the domestic savings rate. That will then require lower capital-labour ratios to take advantage of agricultural growth. That may simply not be possible, given current technology.

3.6 Measures to Raise the Productivity of Agriculture

Of course, the driving engine for the development strategy discussed here is technological change in agriculture. The primary fuel for that engine is, of course, agricultural research. Much of agricultural research can be transferred from one country to another, but enough adaptation is needed that such transfer rarely takes place except when the national system of agricultural research is highly developed. Hence, a high quality national agricultural research system plugged into the farmer through extension services and input supply systems is the first requirement of an agriculture-led strategy of growth.

Next, a host of forces must come into play that facilitate the efficient commercialisation of agriculture. This would include input supply, education, and so on.

Finally, it is important that the technology for agriculture is spread broadly. This should not involve attempts to push technological change where the environment simply is not suitable to it. On the other hand, the more broadly improved agriculture technology can spread, the more powerful will be this basic engine of growth. That requires massive infrastructure investment.

4. RESEARCH NEEDS

The preceding sections were intended to provide a conceptual basis for identifying research needs, particularly in the quantitative fields, for a development strategy within which agriculture plays a major role.

Mathematical models substantiate the strength of the linkages between employment growth and agricultural growth (e.g., see Lele and Mellor, 1981; Mellor and Ranade, forthcoming). There is now a need to develop further these models that incorporate technological change, a food market, and a labour market in order to incorporate capital markets and to elaborate and simulate them.

Measurement of growth relationships in agriculture-led growth is needed. The simplest approach is to note relationships in case studies between agricultural growth and employment. Taiwan is the best documented case (Lee, 1971); the Punjab of India is also well documented (Mellor, 1976). Similarly, note may be made from cross-section analyses of the relation between the demand for wage goods and agricultural growth (Bachman and Paulino, 1979). It would be useful to do cross-section analyses of the key relationships for a range of countries in different stages of development. Despite the usual risk of confusing structural changes with marginal changes, the evidence of broad association would be useful.

Econometric studies for individual countries with differing development strategies and records are useful in measuring the association over time between rates of growth of agriculture, other sectors, and overall growth. Rangarajan (1982) notes an overall multiplier of 0.80 between agricultural growth and overall growth. Studies conducted over a wide range of conditions would allow the forces and the remedial policies that would facilitate raising that multiplier to be gradually pinned down.

Analyses of aggregate input-output and programming models are needed to provide a break down by sectors of the multipliers between agricultural and non-agricultural growth, and hence to provide a basis for policy to improve the efficiency of the processes using such techniques. Bell *et al.* (1982) for Malaysia, and Hazell and Ramasamy (forthcoming) for India, estimate net regional income multipliers close to that found by Rangarajan through econometric analysis.

Key inputs into a programming approach are data for marginal propensities to consume by income class and capital-labour ratios for sources of output by output function. We need both of these for a wide variety of conditions; and the capital-labour ratio data are particularly scarce.

The role of mathematical models, econometrics, and programming models in developing an agriculture-led and employment-based strategy has changed. The replication of past types of work is still important. However, there does not now seem to be a need to prove that there can be large multipliers from agricultural growth. That has been done. There is now a need to show how those multipliers vary to shed light on the policies needed to achieve higher multipliers. Thus, the needed focus is on the variables that affect capital intensity, savings rates, and the effective demand for food.

There are also four areas in which partial analyses are needed. We need to know more about trade relationships and how they can fit with a development strategy in which the development of domestic markets through the growth of income in agriculture can play a major role. Secondly, we need to know how public expenditures can be used most effectively to create the rapid rate of capital formation in the private small- and medium-scale sector that is so necessary. Thirdly, we need to know much more about the scale of infrastructure investment and its geographic dispersion and the composition of the elements of infrastructure needed for an optimal contribution. As part of this, we need to know more about the role that education plays in this kind of structure, and what kinds of education are needed.

It should be clear from the public choice elements in the strategy, that the questions of how to raise tax revenues and how to allocate them are critical. Of course, there needs to be more analysis of how the commercialisation of agriculture, public fiscal policy, and foreign assistance policy can facilitate high rates of capital formation as capital resources are spread evenly.

References

Bachman, K. L. and Paulino, L. (1979) *Rapid Food Production Growth in Selected Developing Countries: A Comparative Analysis of Underlying Trends, 1961–76*, IFPRI Research Report no. 11 (Washington, DC: International Food Policy Research Institute)(October).

Bautista, R. (1987) *Effects of Trade and Exchange Rate Policies on Agricultural Production Incentives: The Philippines*, Research Report no. 59 (Washington, DC: International Food Policy Research Institute).

Bell, C., Hazell, P., and Slade, R. (1982) *Project Evaluation in Regional Perspective* (Baltimore, MD: The Johns Hopkins University Press).

Harris, J. R. and Todaro, M. P. (1970) 'Migration, Unemployment, and Development: A Two Sector Analysis', *American Economic Review*, vol. 60, no. 1 (March) pp. 126–42.

Hazell, P. B. R. and Ramasamy, C. (forthcoming) *Technological Change in Rural Welfare: Impact of the Green Revolution in South India* (Washington, DC: International Food Policy Research Institute).

Hazell, P. B. R. and Roell, A. (1983) *Rural Growth Linkages: Household Expenditure Patterns in Malaysia and Nigeria – Research*, IFPRI Research Report no. 41 (Washington, DC: International Food Policy Research Institute)(September).

Hirschman, A. O. (1958) *A Strategy of Economic Development* (New Haven, Conn.: Yale University Press).

Johnston, B. F. and Mellor, J. W. (1961) 'The Role of Agriculture in Economic Development', *American Economic Review*, vol. 51, no. 4 (September) pp. 566–93.

Kaldor, N. (1964) *Essays on Economic Policy*, vol. 1 (London: Duckworth).

Lee, T. H. (1971) *Intersectoral Capital Flows in the Economic Development of Taiwan, 1895–1960* (Ithaca, New York: Cornell University Press).

Lele, U. and Mellor, J. W. (1981) 'Technological Change, Distributive Bias, and Labor Transfer in a Two Sector Economy', *Oxford Economic Papers*, vol. 33, no. 3 (November) pp. 426– 41.

Mahalanobis, P. C. (1953) 'Some Observations on the Process of Growth of National Income', *Sankhya*, vol. 23 (September).

Mellor, J. W. (1976) *The New Economics of Growth: A Strategy for India and the Developing World*, A Twentieth Century Fund Study (Ithaca, New York: Cornell University Press).

Mellor, J. W. (1978) 'Food Price Policy and Income Distribution in Low-Income Countries', *Economic Development and Cultural Change*, vol. 27, no. 1 (October).

Mellor, J. W. and Ahmed, R. (1987) *Agricultural Price Policy for Developing Countries* (Baltimore, Md.: The Johns Hopkins University Press).

Mellor, J. W. and Johnston, B. F. (1984) 'The World Food Equation: Interrelations Among Development, Employment, and Food Consumption', *Journal of Economic Literature*. vol. 22, no. 2 (June) pp. 531–74.

Mellor, J. W. and Lele, U. (1973) 'Growth Linkages of the New Foodgrain Technologies', *Indian Journal of Agricultural Economics*, vol. 28 (January–March) pp. 35–55.

Mellor, J. W. and Ranade, C. G. (forthcoming) *Technological Change in a Low Labor Productivity, Land Surplus Economy: The African Development Problem* (Washington, DC: International Food Policy Research Institute).

Olson, M. (1982) *The Rise and Decline of Nations* (New Haven, Conn.: Yale University Press).

Prebisch, R. (1971) *Change and Development – Latin America's Great Tasks* (New York: Praeger).

Rangarajan, C. (1982) *Agricultural Growth and Industrial Performance in India* IFPRI Research Report no. 33 (Washington, DC: International Food Policy Research Institute) (October).

Sen, A. K. (1968) *Choice of Technique: An Aspect of the Theory of Planned Development* (New York: Augustus M. Kelly).

15 Agricultural Development-led Industrialisation in a Global Perspective

Irma Adelman
UNIVERSITY OF CALIFORNIA, BERKELEY

Jean-Marc Bourniaux
OECD

Jean Waelbroeck
FREE UNIVERSITY OF BRUSSELS

1. INTRODUCTION

The theme of this conference is the balance between agriculture and industry. The aim of our paper is to explore this issue within the framework of a global agricultural model, which traces out the static and dynamic effects of alternative strategy sequences with respect to the two sectors.

While we believe that in the long run agriculture and industrial expansion must be in balance, several strands of research indicate that an optimal development pattern in the next decade will require unbalanced investment strategies. First, historical research on the early stages of growth of currently developed countries indicates that the Industrial Revolution started in countries that had already experienced substantial increases in agricultural productivity (Bairoch, 1973; Jones, 1967; Adelman and Morris, 1984). Economies that had not done so (e.g., Tsarist Russia) quickly ran into trouble with their industrialisation programmes and were unable to maintain high rates of industrial development. By contrast, developing countries have generally neglected their agricultural sectors and invested the overwhelming bulk of their resources in other branches of activity, thus starting their industrialisation programmes in economies characterised by low productivity agricultural sectors. The World Bank report on agriculture, for example, indicates that in most developing countries the share of investment in agriculture is only between 5 and 10 per cent of total investment (World Bank, 1982).

Secondly, individual country experience with agricultural strategies implemented in open economy trade regimes has been quite favourable. This is evidenced by the recent experience of India and Mainland China, as well as by the earlier experience of Korea and Taiwan in the late 1960s and early 1970s. By contrast, the experience of most developing countries in the 1970s has demonstrated that, in the absence of increases in agricultural productivity, countries quickly find themselves in balance-of-payments problems as they find themselves compelled to import food in order to avoid upsurges in real wages that would jeopardise their industrialisation programmes.

Thirdly, simulations with single-country computable general equilibrium (CGE) models (Adelman, 1984); de Janvry and Sadoulet, 1986) have indicated that, with current initial conditions and in the present low-growth world environment, an Agricultural Development-Led Industrialisation (ADLI) strategy leads to higher rates of economic growth, better income distribution, more rapid industrialisation, and a stronger balance of payments than does continuation of a purely export-led growth strategy. The main reasons for the favourable result of the ADLI strategy are that:

1. the strong domestic linkages of agriculture with manufacturing, through both the demand and the input sides, lead to high domestic demand multipliers for agricultural output;
2. investment in agriculture is less import intensive and more labour intensive than investment in industry and so is agricultural production;
3. the rate of return to investment in agriculture is high, equal to, or exceeding that of investment in industry as indicated in the World Bank study devoted to agriculture (World Bank, 1982);
4. as long as the agricultural sector is poorer than the urban sector, policies that raise the incomes of farmers improve the domestic size distribution of income.

But can the benefits indicated by single-country CGE models be achieved if all less-developed countries (LDCs) were simultaneously to implement ADLI strategies, or would the inelasticity of world demand lead to a sufficiently large decline in world prices for agricultural commodities to more than counterbalance the favourable domestic effects of the agricultural strategy, once domestic import substitution possibilities are exhausted? These are important empirical issues which the current paper is designed to examine.

The structure of this paper is as follows. In the next section, we describe the ADLI strategy. We then sketch the structure of the Rural-Urban North-South (RUNS) model with which the policy experiments relating to alternative agricultural and industrial strategies will be performed. Next we look at the simulation results for the policy alternatives examined. We conclude with a more reflective section in which we interpret the policy import of the simulations.

2. DESCRIPTION OF THE ADLI STRATEGY

The essence of the ADLI strategy lies in shifting a greater share of total investment to the agricultural sector, with a view to improving agricultural productivity. Within the agricultural sector, the emphasis is on food production rather than on export crops, and on medium to small owner-operators rather than on large farms, plantations, and estates. The rationale for these choices is partially in terms of induced growth effects and partially in terms of distributional consequences. It is described in detail in Adelman (1984).

The strategy is implemented in an open-economy trade regime in which incentives are biased neither in favour of exports nor of imports and in which there is no discrimination either way between agricultural imports and exports and manufacturing ones. From the point of view of trade in agricultural commodities, the initial impact of the implementation of the ADLI strategy will be to replace agricultural imports as the ratio of domestic to international prices declines. The next phase will involve becoming an exporter of foodgrains.

As the simulations below will indicate, the ADLI strategy cannot be pushed too far. If not counteracted by policy measures, a very large increase in output of wage goods will result in a drastic drop in the agricultural terms of trade and transfer all the benefits of ADLI to the urban-industrial classes domestically. When the strategy is generalised to all LDCs, the resulting drop in the international terms of trade between agricultural and other products will transfer some of the benefits of ADLI to consumers of imported foods in food-importing countries. Thus, the implementation of the ADLI strategy requires supporting policies partially to counteract the internal terms-of-trade consequences of these policies. We examine some of the options below.

The same reasoning also implies, however, to limitations of purely manufacturing export-led growth strategies that also cannot be pushed to an extent where the development of the agricultural sector is threatened. Here also, the agricultural terms of trade provide the essential link. If pushed too far, such strategies will increase the demand for wage goods without increasing their supply, thus resulting in price increases for the wage good basket, which will generate pressures for increasing urban wages and reduce the international competitiveness of manufacturing exports.

Thus, in the long term, the arguments are for balanced growth of the two sectors. However, in the next decade or so, since there has been a long policy of little investment in agriculture and of surplus transfer from agriculture to the manufacturing sector, and since the decline in the rate of growth of the OECD countries now makes the ADLI option relatively more attractive (see our simulations below), some unbalancing in favour of agriculture is now needed.

3. THE MODEL

The RUNS model used is the Bourniaux (1986) version of the model built in Brussels for use in the World Development Reports of the World Bank. This version trades off a smaller amount of regional disaggregation than in the 'Varuna' version of the system, against a more elaborate description of agriculture.

The regions described are: developing oil-exporting countries; Mediterranean countries (a nondescript group of countries bordering the Mediterranean, including Spain, Greece, the Maghreb, Lebanon, and the island states, with Yemen and Portugal added because this seemed the most logical place to include these countries); Africa south of the Sahara; low-income Asia (i.e., the Indian peninsula plus Burma, East Asia, and other non-oil-exporting countries of Asia); Latin America and the Caribbean; the rest of Western Europe and resource-rich developed countries, including South Africa; and the rest of the world (i.e., the CMEA countries, continental China, and Japan).

For non-agricultural products, the model is slightly more detailed than the other version of the Brussels model, as it adds fertiliser to the capital goods, other manufactures, energy, and services identified in that world model. It includes thirteen agricultural products: wheat, rice, coarse grains, coffee, cocoa, tea, cotton, wool, tobacco, vegetable oils, sugar, meat, and other foods.

Construction of the model was greatly facilitated by the availability of two valuable data bases. One is the world social accounting matrix constructed by the Economic Projections Department of the World Bank, under the direction of P. Miovic. The other is the aggregation of the Food and Agriculture Organization (FAO) trade and utilisation data tape undertaken at the International Institute for Applied Systems Analysis (IIASA), principally by U. Sichra, which we were able to use thanks to the permission of the FAO and of the IIASA Food and Agriculture program (FAP) group. Construction of a world model without this assistance would have been a questionable undertaking for a small university team.

As to coefficients, we followed the conveniently cheap and established procedure of 'picking up coefficients from the literature', except for the agricultural production function. A substantial effort to estimate Extended-Linear-Expenditures-System demand functions was also undertaken. This largely confirmed the indications provided by such studies as those of Lluch, Powell, and Williams (1977) in particular. Another critical source for a model that emphasises agriculture was Mundlak's (1978) study of the determinants of rural-urban migration.

The structure of the model may be sketched as follows. As in the other Brussels models version, the rural and urban sectors are modelled entirely separately so that they may be regarded as separate countries whose trade flows could, if needed, be calculated explicitly. This sharp separation is warranted by the crucial importance of agriculture in the development process.

Price determination is competitive. Market distortions are modelled defining policy-determined wedges between supply and demand prices. For urban goods, demand is defined by a standard Armington (1969) specification, i.e., it is assumed that the goods exported by various countries are imperfect substitutes. For energy, the world price is determined exogenously in terms of the numeraire, the average export price of manufactures by OECD countries, reflecting the market power of OPEC, and the assumption that this organisation seeks by and large to index oil prices to the export prices of manufactures produced by the main developed trading nations. This well-known specification is not adopted for agricultural goods: those produced by the various regions are assumed to be perfect substitutes.

The price system is distorted by various wedges and rigidities. The market power of OPEC is represented by specifying an export tax which absorbs the difference between the cost of production of oil and its exogenous selling price. Each region subjects its imports of non-agricultural goods to fixed *ad valorem* tariffs. The determination of agricultural protection is more complex. The basic assumption is that governments seek to influence the parity between agricultural and non-agricultural prices. These policies are represented by estimated equations that define the regional support price of each agricultural product as a weighted average of the product's world price and of the price index of urban value added. This specification reflects the varying national commitments to maintaining a stable parity between urban and rural prices, and to keeping the average level of agricultural prices in some specific relation to world prices. Variable import and export levies and subsidies implement these policies.

For urban labour incomes also, the model recognises political realities by assuming that unemployment is prevalent both in the developed and in the developing worlds. This reflects wage rigidities. The rigidity is partial, however. The wage rate lies three-quarters of the way between an exogenous target wage and the wage rate that would clear the labour market in each region, subject to the wage behaviour observed in the rest of the world. This is not thought of as a reflection of trade union power, which is often weak in the developing world, but rather of the diffuse, yet potent, resistances which exist in these countries and find expression in sudden political explosions when urban living standards are under pressure.

An important feature of the model is the strict separation between urban and rural consumers. The consumption functions are derived from extended linear expenditures system utility functions and allocated to the production sectors by an input-output scheme. The two subsectors' production functions are also quite separate. In the urban sector, constant elasticity substitution (CES) production functions define the energy and non-energy value addeds as functions of the labour force and of available capital stock. Labour flows freely between urban industries. The energy and non-energy capital stocks are specific, and the allocation of investment between these two sectors is a policy variable. In agriculture, the production system is defined by a multi-input, multi-output production function. This is built up in two steps. Separate CES production functions define 'aggregate resource' variables for animals and for vegetal products. The arguments of those production functions are physical capital inputs (such as bullocks, tractors, and irrigation), labour, and such basic intermediate inputs as fertiliser. Farmers allocate these 'resources' to the various productions by strictly concave transformation functions that take account of decreasing returns in the production of each of the various goods. Calibration of the system ensures that the supply elasticities for the various agricultural products match the values found in the econometric literature.

Saving that is generated in the rural sector finances agricultural investment, while urban saving is used in urban areas — a weak assumption made necessary by the lack of data. The allocation of foreign aid is a policy variable; it may be used in either sector.

In the cities, the allocation of urban investible funds to the energy and non-energy sectors is exogenous. The growth of the available labour force in urban and rural areas depends on demographic factors (United Nations projections) and on migration from country to town. The latter is influenced by the relevant population levels and by per capita income differentials. The calibration of these functions reflects the work of Mundlak (1978), who used time series/cross-section data for different countries, drawn from the World Bank's world tables.

4. KEY MECHANISMS IN THE MODEL

Before describing the simulation results, it is useful to review some key mechanisms through which an increase in agricultural output in a region influences the economy according to the RUNS model.

The region's net exports of agricultural products, of course, rises. This increased supply reduces the ratio of agricultural to non-agricultural prices on world markets. Through a feedback, the region's export prices for non-agricultural goods tend to rise. This is mediated by the strengthening of the region's balance of payments. There is less need to export these goods and, since their demand is not perfectly elastic with respect to price, the export prices of the region's manufactures improve in relation to those of other regions. This increase is large if the initial level of non-agricultural exports is small and if the elasticity of substitution between the region's non-agricultural products and those of its competitors is low.

Within each region, the relative prices of agricultural and non-agricultural products also depend on the government's parity policy. The reader will recall that, in 'non-adjusting' regions, the authorities use variable subsidies and levies to stabilise this ratio while, in 'adjusting' ones, domestic prices reflect changes on world markets. In the first type of region, therefore, the official price policy will prevent world prices from affecting the domestic price ratios. This, of course, is good for farmers, who can maintain their selling prices in spite of the increase in output, and bad for urban workers. Domestic food demand rises less than would be the case if prices were allowed to move freely. Exports of agricultural goods increase even more than would be the case if the region were 'adjusting', while the fall of non-agricultural net exports is also greater.

Finally, there is an important production feedback. Agriculture is the main producer of 'wage goods' — the important commodities that must be provided to urban workers at affordable prices if industrialisation is to proceed successfully. As the ADLI strategy emphasises, any drop in farm-good prices, relative to those of urban goods, provides an indirect boost to industrialisation. In the model, this effect is mediated by the rigidity of the real wage. Lower food prices make possible a reduction of the wage with respect to the product price which encourages firms to raise output. The enhanced competitiveness of urban producers also helps them to raise exports, strengthening the balance of payments and increasing the scope for import-intensive investment. The increase in urban GNP provides, of course, a secondary boost to the agricultural sector, since the increase in urban purchasing power, in turn, raises the demand for agricultural products.

In 'non-adjusting' regions there is also a stimulation, though this is less powerful, and operates differently. The price mechanism described in the last paragraph is blocked by the non-adjusting price policy. Yet, the ADLI strategy does favour industrialisation. Thanks to stable, government-guaranteed prices, the income of farmers rises in proportion to their output, causing an increase in their purchases of urban goods. Any increase in the international terms of trade which the ADLI strategy brings about (the simulations suggest that in most regions the strategy does lead to such a result) raises the region's overall purchasing power and, in particular, its domestic demand for local manufacturers.

5. MODEL PROPERTIES

Any model is a tradeoff between realism, on the one hand, and system transparency and manageability on the other; RUNS is no exception. There is no space in this paper for a critical review of RUNS; but it does seem useful, before presenting the results, to review two of the model's characteristics, an awareness of which will help in interpreting the simulation results.

The first of those is a consequence of the modelling strategy adopted in which the world economy is envisaged as a system of interacting regions. This makes estimation of coefficients almost impossible due to data problems and has led us to use estimates from the literature. These tend to understate differences in behaviour between regions. As a result, the model does not reflect as well as might be desirable the differences in economic flexibility that seem to differentiate various types of countries. Casual observation suggests, for example, that much of the success of the East Asian newly-industrialising countries (NICs) is due to the speed with which their producers manage to adjust to changes in the international environment. This could be captured, for example, by selecting higher trade elasticities for such countries than for their less flexible competitors. This was not done in the RUNS model. On the other hand, the labour market equations in RUNS do assume that wages adjust more flexibly in East Asia than in other regions. It does remain true, however, that regional models like RUNS provide a picture of the world that understates differences among regions.

The second problem stems from the assumption that world markets for manufactures are imperfect so that, when a country reduces its exports, its prices rise in comparison with those of its competitors. As developing countries shift to ADLI strategy, two mechanisms affect the terms of trade of farmers versus the urban population. On world markets, the supply of agricultural products rises, driving down their prices in comparison with those of industrial goods. In addition, the shift to the ADLI strategy slows down the increase of the urban capital stock. This tends to limit the growth of their exports of manufactures to the rest of the world, raising their prices above those of developed competitors. The combination of the two effects implies a substantial deterioriation of the rural-urban terms of trade and may well overstate the deterioration.

6. RUN DESIGN

The run design is meant to highlight the implication of various aspects of ADLI strategies. The model is used to investigate the implications of the three basic ways of increasing agricultural production.

The first of the basic runs investigates the impact on growth of a switch of investment from the urban to the agricultural sector in developing countries. The simulation assumes that the switch brings about 40 per cent increase in agricultural investment in the middle-income developing countries. In the low-income ones in Africa and in Asia, the rural sector still represents a large fraction of output and employment. When such a policy shift is assumed, its impact on the economy is so large that urban growth is depressed to an extent that appears to be unacceptable. Accordingly, investment is assumed to be increased by 30 per cent only in those two regions.

Increasing the productivity of land is the basic way of fostering faster growth of the agricultural sector.There are many policies that may be used to pursue this goal — from the introduction of improved seed varieties, to the improvement of the access of peasants to supplies and to markets for their goods, and the fostering of improved methods of cultivation by extension service. In the run, a 'pure' case is envisaged where the productivity gain is secured costlessly. This covers situations when the required investments have a very short payoff period or where (as in the case of better seeds, for example) what is made available is a new or improved intermediate input.

The third approach involves an increase in the amount of economic aid provided by the rest of the world which is assumed to be earmarked for use in the agricultural sector. Such an allocation of aid has often been advocated as a means of combating hunger or of improving the balance of economies after a period of excessive industrialisation, though in practice the amounts so provided have been small. Such aid would, for example, probably form a substantial fraction of the aid package to Black Africa which many economists feel will be necessary to enable that region to escape from its present stagnation.

In the simulations, adoption of ADLI strategies by all developing regions, as will be seen, does have a large impact on the world prices of agricultural products. In the model, governments are not assumed to respond to world price changes in a passive way; price- setting equations reflect the way in which politicians have in the past tried to strike a satisfactory compromise between a policy of allowing domestic prices to reflect the world ones, and maintaining an appropriate parity between urban and rural per capita incomes. The large changes in the world prices caused by a general shift to ADLI might change the price- setting patterns which the model describes. But, for the present runs, the patterns of response of governments are assumed to remain unaltered.

7. SIMULATION RESULTS

We now comment on the simulation results obtained with the model.

Table 15.1 describes the impact of shifting investment resources from the industrial to the agricultural sector. The impact is positive both for GNP of developing countries and for their real income (nominal GNP deflated by the price index of consumption). Not all regions gain, however. An improved supply of 'wage goods' is more beneficial to semi-industrialised regions with their large urban sectors than to less-industrialised countries in low-income Asia and, especially, in Africa.

In lower income Asia, GDP drops slightly. This region follows non-adjusting prices policies so that the increase in agricultural output does not reduce the cost of wage goods to urban workers and limits the benefits to its substantial industrial sector. Also, it has limited amounts of spare land, and this is reflected in agricultural capital coefficients that are higher than those that characterise urban production.Thanks to a favourable balance of trade in manufactures, its terms of trade improve markedly, as the general shift to the ADLI strategy changes relative prices on world markets in a way that is favourable to manufactures and services. As a result, though GNP falls slightly, the region's international purchasing power actually increases.

In both of the lower income regions, the increase in agricultural investment tends to choke off the supply of capital to industry to an extent that may be excessive, leading to a sharp rise in the prices of urban goods. This, again, reflects the high initial weight of agriculture in these economies.

The strongly favourable impact on the middle-income NICs confirms Adelman's (1984) earlier results for Korea. The result that ADLI approach is unambiguously favourable to outward-oriented countries that are still past the take-off stage appears to be a robust

Table 15.1 Impact on key variables of a shift in investment from the urban to the rural sector in all developing regions[a]

	Real income	*GNP total*	*Rural GNP*	*Urban GNP*	*International terms of trade*
	(per cent differences in comparison with the base run, 1993)				
Poor regions	0.2	–1.0	2.5	–2.4	5.3
Asia	1.2	–0.5	2.4	–1.5	8.2
Africa	–2.0	–2.3	2.7	–4.6	1.1
Middle income					
Non-oil	1.4	1.1	7.4	0.4	1.5
Latin America	2.4	2.2	8.8	1.3	1.6
South-east Asia	1.2	0.5	3.6	0.2	1.1
Mediterranean	0.5	0.3	7.2	–0.4	2.1
Oil	0.0	0.1	3.2	–0.2	0.2
All developing	0.8	0.5	4.8	–0.1	1.3
Developed	0.8	0.6	–1.0	0.7	0.2
Europe	1.3	0.9	0.1	1.0	0.6
North America, Oceania	0.5	0.4	–1.9	0.5	0.6

Note:
[a] A 40 per cent increase in agricultural investment in the middle-income regions; a 30 per cent increase in the poor ones.

one.Latin America is the strongest beneficiary from the shift. This reflects favourable capital coefficients in agriculture, that reflect an abundant supply of available land as well as the strong gain registered by a region where governments have tended to underinvest in agriculture. (That the same is not true of Africa is due to the fact that, partly as a result of poor management, that region's agricultural capital coefficients have tended to be rather high and the domestic supply of manufactures is inelastic.)

The oil-exporting countries are not strongly affected by the policy change. They tend to pursue non-adjusting pricing strategies that insulate their domestic prices from fluctuations on world markets and block much of the feedback from agricultural to industrial growth. The other reason for their lack of sensitivity reflects the inevitably arbitrary assumption that was made about the price of oil. In the model, this is assumed to be indexed to the prices of manufactures exported by developed countries, a huge aggregate that covers much of the goods which oil producers import. As a result, their terms of trade vary very little in response to the policy shift described by the simulation.

Developed countries, finally, are also overall gainers from the policy shift. Their gain may seem small in percentage terms, but a more realistic perspective is obtained perhaps by comparing it to the roughly 0.25 per cent of their GNP that developed countries devote to development assistance. In terms of international diplomacy, there would seem to be grounds for arguing that it would be fair to ask the developed world to support such a policy turnaround in the South by providing additional aid.

Aid to agriculture, the result of which is considered in Table 15.2 is, of course, more beneficial to the developing world than a shift in the allocation of domestic investment between industry and agriculture. (In examining Table 15.2, the reader should bear in mind that the results are not comparable to those in Table 15.1 as the nature of the shocks and the changes in investment that are induced are basically different; we will try to make a more comparable run.) In the run, aid is forthcoming from the 'residual region' which includes Japan and the CMEA countries (this will be changed). Aid raises total investment; reallocating investment funds merely changes its distribution. Aid also strengthens the balance of payments by making it possible to run a deficit. This permits an improvement of the terms of trade, thanks to which the dollar of aid adds more than a dollar to the resources available; it is the 'transfer burden' of trade theory operating in reverse. Both effects account for the very favourable results evidenced by the simulation.

The table reflects the fact that granting additional aid is particularly beneficial to the lower income countries. This is a robust result which is discussed in detail in the basic paper on the Brussels model (see Gunning *et al.*, 1982).

Table 15.3 finally considers the impact of an increase in the productivity of additional capital in agriculture. It is assumed that this is achieved through 'embodied technical progress'. (The result obtained by assuming disembodied technical progress are very similar and are, therefore, not presented in the paper in order to save space. Here again, it is necessary to bear in mind that the *modus operandi* of the boost to output is basically different from that which is described by Table 15.1; hence, caution is necessary in comparing the tables.) This can be thought of as resulting in part from better planning, which can eliminate the sometimes extraordinary errors to which agricultural planners appear to be prone, of which the Aswan Dam is the best-known example. (The dam was meant to make possible a substantial extension of the cultivated area in Egypt. However, elimination of the yearly deposit of silt, from which Egyptian agriculture had benefited for millennia, that reduced output on existing land and the salination of the newly-irrigated areas were so detrimental that there is reason to doubt that the dam has brought about any increase in output.) The 'bottoms up' approach of the ADLI strategy focused at the grass roots level is, of course, designed to facilitate the elimination of this waste. The other source of technical progress that may be borne in mind is further progress in creating improved plant varieties, both through a continuation of the research on plant hybridisation which brought about the Green Revolution, and through more 'exotic' breakthroughs involving plant genetics.

From an economic point of view, the acceleration of technical progress considered here is equivalent to a form of aid that does not pass through the balance of payments. It is as though the country benefited from additional capital resources without the gains which a direct inflow of foreign currency yields.

Once again, the policy has a strongly favourable impact on the world economy. The only 'minuses' in the table are caused by the pressure of falling agricultural prices on the US farm sector. European producers are insulated from this by the very protectionist Common Agricultural Policy implemented by the European Community.

Table 15.2 Impact of increased aid to agriculture[a]

	Real income	*GNP total*	*Rural GNP*	*Urban GNP*	*International terms of trade*
	(per cent differences in comparison with the base run, 1993)				
Poor regions	1.8	2.2	1.3	2.5	1.4
Asia	2.1	2.2	1.1	2.6	2.5
Africa	1.1	2.2	1.8	2.4	–0.1
Middle income					
Non-oil	2.2	2.2	4.1	2.0	0.7
Latin America	3.6	3.0	4.6	2.8	0.7
South-east Asia	2.2	2.1	3.0	2.0	0.3
Mediterranean	1.3	1.4	4.0	1.1	1.3
Oil	0.8	0.8	1.4	0.8	0.2
All developing	1.7	1.8	2.6	1.7	0.6
Developed	0.8	0.6	–0.6	0.6	0.3
Europe	1.2	0.9	0.1	1.0	0.5
North America, Oceania	0.5	0.3	–1.2	0.4	0.0

Note:
[a] Aid to each region increases by $2 billion; the aid is granted by the residual region.

Table 15.3 Impact of an acceleration of technical progress in agriculture[a]

	Real income	*GNP total*	*Rural GNP*	*Urban GNP*	*International terms of trade*
	(per cent differences in comparison with the base run, 1993)				
Poor regions	3.1	2.9	3.0	2.9	1.8
Asia	3.7	3.0	2.5	3.1	3.9
Africa	1.8	2.9	4.0	2.3	–3.3
Middle income					
Non-oil	2.9	3.0	8.2	2.4	0.4
Latin America	4.9	4.8	10.2	4.1	0.6
South-east Asia	1.2	1.4	2.1	1.4	–0.2
Mediterranean	1.6	1.7	8.1	1.1	1.0
Oil	1.1	1.3	5.8	0.8	0.3
All developing	2.3	2.4	5.9	1.9	0.5
Developed	1.1	0.9	–1.2	1.0	0.3
Europe	1.9	1.6	0.1	1.6	0.5
North America, Oceania	0.6	0.4	–2.2	0.5	–0.1

Note:
[a] An 80 per cent increase in the productivity of newly-invested capital, starting in 1983.

In our simulations we have looked at the individual elements of a global ADLI strategy separately. In practice, they would be combined: increases in investment in agriculture would increase technical progress in agriculture. In turn, especially in Africa, the increased investment in agriculture would be financed by increased foreign aid. The results of an actual ADLI strategy would, therefore, be a weighted sum of the individual experiments and, hence, more favourable than indicated in Tables 15.1 to 15.3.

8. CONCLUSION

Several policy conclusions emerge from our simulations. First, the ADLI policy survives quite well its generalisation to all developing countries. It raises world GNP and world real income, with the benefit split almost equally between developed and developing countries. The ADLI strategy is most advantageous for the middle-income NICs that benefit not only from the domestic production, demand, and terms-of-trade effects but also from the international terms-of-trade changes which raise the relative prices of their manufacturing exports. In African low-income countries, the policy requires financing through international assistance. Otherwise, the cost of withdrawing resources from domestic manufacturing is too high. Even the developed grain-exporting countries benefit from ADLI because of the increase in the prices and incomes of urban producers.

Secondly, the productivity of foreign aid directed at investment in agriculture appears to be quite high. This is especially true of investment in agricultural projects in the low-income developing countries. Of course, the productivity of the use of resources by the recipient countries is critical to the results. The simulations assume that the productivity of the marginal aid-financed investment in agriculture is the same as that of the average investment in the past. To the extent that the marginal aid-financed investment is either more or less productive, the results either understate or overstate the favourable impact of aid to agriculture in less-developed countries.

Thirdly, the effects of agricultural policies within the individual groups of countries on other countries are quite significant. The current effort to bring domestic agricultural policies into GATT could, therefore, have major welfare consequences.

Fourthly, in LDCs terms-of-trade policies, which share the potential income benefits of ADLI between urban and rural producers, are desirable for equity and incentive reasons. In our simulations, the balance between stimulating agricultural output and maintaining incentives to farmers appears to be very delicate, for stimulating output has a strong effect on prices. The model does not recognise that the marginal productivity of investment in agriculture is not independent of incentives facing farmers. To the extent that productivity of resource use and price incentives in agriculture are linked, and we have strong evidence from China that they are, the conflict between equity and growth is overstated in our simulations.

Model simulations seem dry; it is interesting to use historical experience to translate simulation assumptions into the concrete terms of historical experience. Historical analysis (Morris and Adelman, forthcoming) indicates that the start of industrialisation was dependent on the existence of a large overall agricultural surplus. In the initial stage that could be accomplished with dualistic agricultural sectors containing both large and subsistence farms. For a time governments could implement policies that substituted for the tenurial reforms and price policies that were required for sustained, widely-distributed agricultural growth. But ultimately, no country succeeded in generating more than narrowly based, temporary industrialisation if they did not have a prosperous small farming sector.

The simulations point to the sequential nature of the role of agricultural development that is also evident in history. ADLI, the model suggests, is most beneficial in growth terms in the middle-income NICs and somewhat less in low-income Asian countries. In African nations, in fact, the creation of an adequate agricultural surplus cannot, it seems, be financed internally; it must be financed by foreign aid.

What are the policies that enabled successful 'growers' to strike an appropriate balance between industrial and agricultural expansion in the nineteenth century? Of all the development paths followed by countries in that century, only two led to widespread, sustained economic growth. These were the balanced growth path followed by the small European economies, and the industrialisation path of the first-comers to the Industrial Revolution. The common characteristics of these two paths that distinguished them from the other paths were (1) high productivity agricultures, (2) tenurial arrangements in agriculture that gave rise to a widely-dispersed agricultural surplus, (3) policies of tenurial arrangements that generated incentives for cultivators, and (4) open-trade policies. The successful countries had undergone their agricultural revolutions prior to industrialisation.

This enabled them to maintain flexible and dynamic agricultural sectors throughout the period of rapid industrialisation and to avoid having growth choked off by rising prices of food that would jeopardise the competitiveness of their industrial sectors. This is precisely what the implementation of ADLI would produce in LDCs.

The largest food-deficit countries – India and China – have, in effect, already implemented the ADLI strategy in the past decade or so. They have shifted investment from industry to agriculture, given incentives to farmers, have benefited from the more rapid technical progress of the Green Revolution, and shifted from being major grain importers to being grain exporters. India has, in addition, benefited from a substantial amount of additional aid. Their growth rates have increased remarkably as a result, just as our simulations would predict. Other Asian countries, such as Thailand and Indonesia, appear to be following in their footsteps. Thus, with appropriate policies, the potential for improving agricultural productivity in all but the African countries is clearly there. But Latin America and Africa have yet to improve the productivity of their agricultural sectors.

At the same time, the OECD countries are all faced with agricultural surpluses and are all responding by increasing their subsidies to domestic agriculture. Our simulations indicate that, with current policies, ADLI would be of benefit to them by raising the prices of their manufacturing exports despite the fact that it would curtail their agricultural exports to LDCs. Counter-intuitive though this result might appear to be, it would, therefore, be in their interest to invest aid dollars in furthering the agricultural development of LDCs.

References

Adelman, I. (1984) 'Beyond Export-Led Growth', *World Development*, vol. 12, no. 5, pp. 937–49.

Adelman, I. and Morris, C. T. (1984) 'Patterns of Economic Growth, 1850–1914, or Chenery-Syrquin in Historical Perspective'. Syrquin M., Taylor, L. and Westphal, L. E. (ed.) in *Economic Structure and Performance* (New York: Academic Press).

Armington, P. (1969) 'A Theory of Demand for Products Distinguished by Place of Production'. IMF Staff Papers, vol. 16, pp. 159–78.

Bairoch, P. (1973) 'Agriculture and the Industrial Revolution, 1700–1914' in Cipolla, C. (ed.) *The Fontana Economic History of Europe*, vol.3 (London: Collins, Fontana Books) pp. 452–506.

Bourniaux, J.-M. (1986) *Analyse des Dilemmes Alimentaires* (Paris:

Economica). Janvry, A. de and Sadoulet, E. (1986) 'Growth and Equity in Agriculture-Led Growth'. Paper presented at International Economic Association Conference, New Delhi, India (1–5 December).

Gunning, J. W., Carrin, G., Waelbroeck, J. *et al.* (1982) 'Growth and Trade of Developing Countries: A General Equilibrium Analysis', Discussion Paper 8210, CEME, University of Brussels, Belgium.

Jones, E. L. (1967) *Agriculture and Economic Growth in England: 1650–1815* (London: Methuen).

Lluch, C., Powell, A. and Williams, R. (1977) *Patterns in Household Demand and Saving* (London: Oxford University Press).

Morris, C. T. and Adelman, I. (forthcoming) *Where Angels Fear to Tread: Quantitative Analysis of Economic History* (Baltimore: Johns Hopkins Press).

Mundlak, Y. (1978) 'Occupational Migration Out of Agriculture – A Cross-Country Analysis', *Review of Economics and Statistics*, vol. 60, no. 3, pp. 392–98.

World Bank (1982) *World Development Report* (New York: Oxford University Press).

16 The Role of Nonfarm Activities in the Rural Economy*

Carl Liedholm
MICHIGAN STATE UNIVERSITY

Peter Kilby
WESLEYAN UNIVERSITY, MIDDLETOWN, USA

1. INTRODUCTION

Until quite recently it has been conventional to equate, in a rough way, the rural economy with the agricultural economy. Rural households, containing anywhere from 30 to 70 per cent of the nation's population, were envisaged as having as their primary function the production of food and fibre for the home market and one or more crops for the export market. In addition to farm production, household members might as secondary activities be engaged in a certain amount of agricultural processing, transporting and marketing.

This view has begun to change in the past few years.[1] There is a growing recognition that the nonfarm sector plays an important welfare-augmenting role in providing simple consumer goods and services to poorer rural households; from the other side, the provision of these goods and services provides a humble but critical income to landless labour. But for most policy-makers the image remains that of a passive sector – passive in so far as its size is seen as being wholly dependent upon the level of farm income, and passive in that it makes no independent contribution to economic growth.

In this paper, drawing upon recent research, we attempt to present a clearer delineation of the nonfarm rural economy – its magnitude, its anatomy, and how it changes over time. We present evidence to show that nonfarm activities not only make a major welfare contribution with respect to equity and income-smoothing, but that many of these activities add more to GDP than the substitute goods and services supplied by technically-advanced capital-intensive producers. Finally, we argue that the sector is no more or less passive than any other sector in the economy, and that it can make substantial contributions to agricultural growth.

2. EXTENT OF NONFARM ACTIVITIES

Given that conventional statistical measures of employment and output do not exist for most nonfarm activities, how can we measure the sector's size? There are three ways. First, frequently there is information on occupational classification of the rural population that is collected during the decennial population census. Secondly, there are especially-designed establishment surveys within a given sample area. Finally, there are rural household income-and-expenditure surveys undertaken within the context of a national sampling design.

Table 16.1A presents mainly census-based figures on the share of the rural labour force whose primary occupation lies outside of farming. Although the range runs from 14 to 49 per cent, in over three-quarters of the countries the nonfarm share is between 19 to 28 per cent. While this itself is a very large magnitude, it is nevertheless an underestimate (e.g., larger rural towns are excluded, women's nonfarm work is undercounted, secondary occupations – which net out heavily in favour of nonfarm activities – are omitted).

Table 16.1B, showing the composition of nonfarm activities, is also derived mainly from census data. While there is considerable variation between the nine countries, the three major components are manufacturing (including agricultural processing and repair activities), trading and services. Since trading is the most common secondary occupation, it is likely that this category is understated.

A second source of information on the rural nonfarm sector is the specially-designed establishment survey. These are generally limited to manufacturing units which, because of their relative fixity of location, are easier to count than concerns engaged in, say, transportation, construction or petty trade. Table 16.2, which reports the percentage of total manufacturing employment that occurs in the rural areas, is primarily derived from this type of sample survey. These percentages are usually built up as follows: formal urban employment (plus some large-scale processing employment in rural areas) obtained from the standard statistical series, to which are

Table 16.1A Percentage of rural labour force with primary employment in rural nonfarm activities

Country	*Year*	*Coverage*	*Percentage of rural labour force*
Guatemala	1964	All rural	14
Thailand	1970	All rural	18
Sierra Leone	1976	Male-rural	19
South Korea	1970	All rural	19
Pakistan	1970	Punjab only	19
Nigeria	1966	Male - 3 dist. W. State	19
India	1966	All rural	20
Uganda	1967	Four rural villages	20
Afghanistan	1971	Male - Paktia Region	22
Mexico	1970	All - Sinaloa State	23
Colombia	1970	All rural	23
Indonesia	1971	All rural	24
Venezuela	1969	All rural	27
Kenya	1970	All rural	28
Philippines	1971	All rural	28
W. Malaysia	1970	All rural	32
Iran	1972	All rural	33
Taiwan	1966	All rural	49

Table 16.1B Sectoral composition of rural nonfarm employment in selected countries (Percentage)

	Afghanistan (1970)	*India (1966)*	*Indonesia (1971)*	*Sierra Leone (1975)*	*Philippines (1970)*	*Korea (1970)*	*Colombia (1970)*	*Malaysia (1970)*	*Taiwan (1966)*
Manufacturing	46	39	29	40	34	30	33	22	27
Construction	9	14	5	2	11	10	8	5	4
Trade and commerce	11	14	34	35	15	24	19	22	13
Services	10	24	27	23	30	29	33	41	50
Other[1]	24	9	5	—	10	7	7	10	6
	100	100	100	100	100	100	100	100	100

Note: 1. Includes utilities, transport, and miscellaneous; omits 'other and unknown'.
Source: Chuta and Liedholm (1979).

Table 16.2 Percentage of manufacturing (large and small scale) employment in rural areas

Country	*Year*	*Percentage of manufacturing employment*
Sierra Leone	1976	86
Indonesia	1976	80
Sri Lanka	1971	75
Jamaica	1980	74
Ghana	1973	72
Bangladesh	1974	68
Zambia	1985	64
Philippines	1976	61
India	1967	57
Pakistan	1975	52
Taiwan	1976	49
Malaysia	1970	46
Korea	1975	30

Note: Rural defined as all localities under 20 000 inhabitants.
Source: Liedholm and Mead (1986).

added employment estimates for fabricating activities in the urban informal sector with the final component being provided by the rural establishment survey.

Are the reported facts of Table 16.2 to be believed – that in ten of the thirteen countries rural areas account for over half of manufacturing employment? Like census data, establishment surveys are not entirely reliable with respect to aggregate measurement; but unlike census data, we cannot say whether the result is an overestimate or an underestimate. On the side of producing too low a figure, this type of survey does not capture non-commercial production (for own consumption) and it most surely overlooks some out-of-the-way small producers.[2] This source of undercounting can be magnified or reversed, first, by the particular point in the agricultural cycle that the survey took place, since part-time work constitutes a large share of nonfarm activities, and secondly, by the geographical areas of the country that happen to be sampled, since the volume of nonfarm activities typically exhibits substantial regional variation. Hence, there is no obvious bias in the estimates reported in Table 16.2 – the likelihood that they are too low is equal to the probability that they are too high.

Specially-designed establishment surveys also provide a great deal of information about the nature and functioning of the nonfarm sector. While firm size may range above 20 employees, the great preponderance of these rural nonfarm firms are very small.[3] Liedholm and Mead's (1986) review of evidence from over a dozen countries reveals that 85 per cent of the small rural manufacturing firms employed fewer than five employees with the one-person firm generally dominating. Larger units engage both unpaid family workers and wage-paid employees. Combined fixed and working capital per person is typically modest. Unlike the enumerated wage labour force, women constitute a large fraction – 40 per cent or more – of those engaged in the sector and frequently account for the majority of the small-scale entrepreneurs. Acquisition of skills takes place through apprenticeship and other forms of learning-by-doing.

This brings us to the final source of statistics on the size of the nonfarm sector, the rural household income survey. Based upon a carefully-drawn random sample of several thousand rural households and entailing weekly data collection over the course of a year covering household receipts by source, expenditures, labour allocation and a host of supplementary variables, these surveys – if constructed for the purpose – provide the most accurate measurement of both employment and output.[4] Problems of part-time work, seasonality, overlooked enterprises, secondary occupations – all vanish. The bad news is that such surveys are extremely expensive and require great organising abilities from the statistical agency in charge. As a consequence this desirable source of information is not often available.

Table 16.3 presents the nonfarm income share for five countries. Comparison with Table 16.1 reveals that, in four out of the five cases, the income share is substantially larger than the 'primary occupation' share. The one exception, Taiwan, is almost certainly the result of the decade discrepancy between the two measurements. If these few figures are indicative, we may tentatively conclude that the nonfarm sector ranges from one-half to three-quarters the size of the agricultural sector. Thus it constitutes a major sector in all low- and middle-income economies.

3. INCOME FROM NONFARM ACTIVITIES

Are rural nonfarm activities a major source of income for the poorest rural households? If so, do they serve to reduce income inequality in rural areas? Do they also contribute to stabilising income among

Table 16.3 Share of nonfarm income in total rural household income

Country	*Year*	*Percentage of nonfarm income*
Northern Nigeria (3 villages)	1974	28
Korea	1980	34
Sierra Leone	1974	36
Taiwan	1975	43
Thailand	1978	43

Sources: Northern Nigeria: Matlon (1977).
Korea: Korea (1981).
Sierra Leone: Unpublished results from Sierra Leone African Rural Employment Project reported in Chuta and Liedholm (1979) (includes households in rural towns plus in villages).
Taiwan: Taiwan (1976).
Thailand: World Bank (1983).

---

poorer households over the course of the year? Answers to these questions should provide us with a reasonably comprehensive assessment of the equity issue.

Given that land is the farmer's principal productive asset, size of holdings has commonly been used as a variable to stratify rural households into income classes. How important is rural nonfarm income for those with little or no land? Not surprisingly, an examination of data from five countries in Asia and Africa (see Table 16.4) reveals an inverse relationship between size of landholding and the share of nonfarm income in total rural household income. For the smallest landholding categories in each country, nonfarm income sources account for over 50 per cent of household income.

Is the income derived from these nonfarm sources sufficient to reduce income inequalities within the rural areas of these economies? For the two African cases as well as Thailand (see Table 16.4), the nonfarm income sources cause the total income of rural households with the smallest amounts of land to exceed the incomes of those with somewhat larger farms. This 'vertical J'-shaped relationship between total rural household income and landholdings is perhaps not unexpected in Africa, where land is not a limiting factor. It also appears to hold in some parts of Asia, such as in Thailand and Japan, but it is not ubiquitous (see Korea and Taiwan in Table 16.4).

These general findings, however, do call into question the notion that farm size is a consistently good proxy for total rural household income or a good indicator of who are the rural poor. Indeed, a complex set of factors bearing on farming, nonfarm enterprises and off-farm trading and employment opportunities determine rural household income levels. Although this heterogeneity complicates the task facing policy-makers in dealing with the rural poor, it also means that there is a much wider set of opportunities that can be developed.

A better indicator of whether or not rural nonfarm income reduces income inequality, however, can be obtained by relating the total nonfarm income share to total rural household income. Although information on this relationship is sparse, data are available for Sierra Leone, Nigeria, and Thailand. An examination of Table 16.5, in which rural nonfarm income shares are related to total rural household income quintiles or terciles (from low to high), reveals the 'vertical J'- shaped relationship again. Rural nonfarm income is thus relatively important at both ends of the income distribution spectrum. Differing types of nonfarm income are important at the low and high income ends of the distribution. For the low-income rural household, wages from working on other farms and service-type activities are the predominant sources, while for the high-income households salaries from administrative and manufacturing activities tend to predominate. These latter activities tend to have higher entry barriers and yield higher returns than agriculture or the other types of rural nonfarm activities.[5]

What is the net effect of these various nonfarm income sources on overall income inequality in rural areas? The result from two African studies as well as from Thailand indicate that including nonfarm income with farm income reduces the rural Gini coefficients in each case. Gini coefficients calculated on per capita farm income alone were .43 in Sierra Leone and .32 in Nigeria, compared with coefficients on combined farm and nonfarm incomes (rural) of .38 and .28 respectively (Matlon *et al.*, 1979). In rural Thailand, the Gini declines from .58 when only farm income is considered to .38 when all the sources of the rural households' income are included.[6] The available evidence, albeit limited, does suggest that rural nonfarm income reduces rural income inequalities in several countries.

Rural nonfarm activities also contribute to the smoothing of household income over the year. An analysis of the monthly income fluctuations of 424 rural households in Thailand reveals, for example, that the variability of total household income was substantially less than the variability of net farm income over the year.[7] Studies from

Table 16.4 Size of land holding and relative importance of nonfarm income in total household income

Country and date	*Size of holding (acres)*	*Nonfarm income share in total household income (%)*	*Total household income ($)*
Korea – 1980	0.00 – 1.23	74	3 005
	1.24 – 2.47	39	3 450
	2.48 – 3.70	28	4 321
	3.71 – 4.94	23	5 472
	4.95 +	16	7 401
Taiwan – 1975	0.00 – 1.23	70	2 768
	1.24 – 2.47	52	3 442
	2.48 – 3.71	44	3 701
	3.72 – 4.94	39	4 570
	4.95 +	26	5 566
Thailand – 1980–81 (4 regions)	0.00 – 4.10	88	1 362
	4.20 – 10.20	72	974
	10.30 – 41.00	56	1 613
	41.00 +	45	1 654

Sierra Leone – 1974	0.00 – 1.00	50	587
	1.01 – 5.00	23	404
	5.01 – 10.00	14	546
	10.01 – 15.00	12	770
	15.00 +	15	927
Northern Nigeria – 1974	0.00 – 2.46	57	479
	2.47 – 4.93	31	377
	4.94 – 7.40	26	569
	7.41 – 9.87	15	769
	9.88 +	24	868

Sources: Korea: Korea (1981).

Taiwan: Taiwan (1976).

Northern Nigeria: Matlon (1977).

Sierra Leone: Matlon *et al.* (1979) – includes data from 550 rural households in *villages only* nation-wide. Thus, the average nonfarm share is lower than that reported in Table 16.3

Thailand: Figures derived from primary data generated by the survey of 424 rural households (village) in four regions conducted by the Thai Rural Off-Farm Employment Project (for details, see Narongchai *et al.*, 1983). Although households were chosen at random within villages, some of the villages were chosen because of their varieties of nonfarm activities. Thus, they are not "representative" of the entire country. The average nonfarm income share is 65% in this example compared with 43% for farm households reported for the entire country (World Bank, 1983).

Table 16.5 Percentage of rural household income earned from farm and nonfarm sources by income class

Country	*Income class*	*Farm (%)*	*Nonfarm (%)*
Sierra Leone			
	Lowest tercile	80.3	19.7
	Middle tercile	81.2	18.8
	Highest tercile	80.0	20.0
Northern Nigeria			
	Lowest quintile	76.6	23.4
	Middle quintile	78.0	22.0
	Highest quintile	61.4	38.6
Thailand			
	Lowest quintile	37.5	62.5
	Middle quintile	44.0	56.0
	Highest quintile	34.9	65.1

Sources: As in Table 16.4

Northern Nigeria and Sierra Leone point to similar findings (Matlon *et al.*, 1979). Farm and nonfarm activities tend to move in opposite directions over the year and income earned from nonfarm sources complements the pattern of net farm income received.[8] Overall, nonfarm activities are thus seen to make an important welfare contribution with respect to both equity and income stability in rural areas.

4. EFFICIENCY OF NONFARM ACTIVITIES

Are these rural nonfarm enterprises efficient users of economic resources? Although seen to possess equity virtues with respect to the distribution of income, they are frequently thought to be inefficient and thus confront policy-makers with a potentially vexing tradeoff. If, however, some categories of rural nonfarm enterprises are found to generate more real output per unit of resources expended than their larger scale urban counterparts, then agricultural and other policies that enhance these activities can increase both output and employment.[9]

The evidence on the economic efficiency of rural nonfarm activity has been rather meagre. Comparisons of small-and large-scale enterprises using partial efficiency measures, particularly the output – capital ratio have been made, but these have yielded at best a mixed picture of the relationship between capital productivity and size.[10] Moreover, only rarely are rural and non-industrial enterprises specifically examined in these analyses. These studies also suffer from the limitations that surround all partial efficiency measures; if some resource other than the one included in the measure is scarce, and thus has a non-zero opportunity cost, then it may yield incorrect results.

Comprehensive economic efficiency measures, such as total factor productivity and social benefit-cost analysis, overcome the limitations of the partial ones.[11] Ideally, all scarce resources are explicitly included in the analysis and are evaluated at their 'shadow' or 'social' prices that reflect their scarcity values in the economy. Unfortunately, only a few such studies exist (Ho, 1980; Cortes *et al.*, 1985) and none consider rural nonfarm enterprises explicitly.

Liedholm and Mead (1986), however, recently used a social benefit-cost measure to compare the relative efficiency of small rural manufacturing enterprises with their larger-scale urban counterparts in Sierra Leone, Honduras, and Jamaica. Following the approach suggested in Cortes *et al.* (1985), the ratio of the enterprise's value added to the cost of its capital and labour, both valued at their shadow or 'social' prices, was used to measure economic efficiency.[12]

The primary data used to derive the social benefit-cost ratios were generated from the detailed small-scale industry surveys that Michigan State University and host country researchers had conducted.[13] Hundreds of rural firms in each country were interviewed twice weekly over a twelve-month period to obtain daily information on revenues and costs. The information on the large-scale enterprises was obtained from the worksheets used to construct the Industrial Censuses in Sierra Leone and Honduras and from the National Planning Agency's Industrial Survey in Jamaica.[14] In calculating the social benefit cost ratios, the 'shadow' social price of capital was assumed to be 20 per cent, while unpaid family labour was valued at the average price for *skilled* labour in small-scale industry.[15] Since world prices for outputs and material inputs were not available for the Honduras and Jamaican studies, domestic prices were used; this means efficiency comparisons had to be limited to large and small rural enterprises operating in the same product group with reasonably similar mixes of output and purchased inputs.

The key finding from this three-country analysis is that small manufacturing enterprises are found to use fewer resources per unit of output than their larger-scale counterparts in a majority of the industry groups considered. A glance at Table 16.6 reveals that the social benefit-cost ratios are higher for rural small-scale enterprises in eight of the twelve cases examined. Only in the wearing apparel industries of Jamaica and Honduras, and the shoe and furniture industries of Sierra Leone, do the larger-sized enterprises prevail. Moreover, the social benefit-cost ratios for small rural nonfarm enterprises exceed one in all but two industries. Such findings provide at least limited support for the contention that some small rural nonfarm activities in developing countries are economically efficient.[16]

One weakness of this analysis is that output and purchased inputs were valued using domestic rather than world prices. Fortunately, sufficient data were available from Sierra Leone to enable a computation of enterprise social benefit-cost ratios at *world prices* to be made.

Table 16.6 Social benefit-cost ratios (domestic prices)[1] for various large and rural small-scale industry groups in Africa and Latin America

Country/enterprise group	*Rural[2] small-scale*	*Large[3] scale*
Africa:		
Sierra Leone (1974–75)		
Bakery	1.86	1.03
Wearing apparel	1.78	0.53
Shoes	1.65	2.00
Furniture	0.81	0.87
Metal products	1.63	1.61
Latin America:		
Honduras (1979)		
Wearing apparel	0.82	0.89
Shoes	1.27	0.54
Furniture	1.44	0.84
Metal products	1.21	0.74
Jamaica (1979)		
Wearing apparel	1.00	1.79
Furniture	2.51	1.36
Metal products	1.87	1.58

Notes to Table 16.6

1. Gross output and purchased input values used to compute value added (numerator) are evaluated at actual (domestic) prices; hired labour evaluated at actual wages paid for small and at 0.8 of actual wages for large. Unpaid family (including proprietor) valued at skilled wage rate for small-scale industry in each country (Le. 0.16 per hour in Sierra Leone; La. 0.71 per hour in Honduras; J$ 1.50 per hour in Jamaica). Capital was evaluated at a shadow interest rate of 20 per cent in each country. For a rationale for these particular shadow rates, see Haggblade, Liedholm, and Mead (1986).
2. Small-scale firms employ less than fifty persons.
3. Large-scale firms employ fifty persons or more. With one exception, these firms are located in large urban areas.

Source: *Sierra Leone* small-scale enterprise data collected in 1974–75 – survey reported in Chuta and Liedholm (1985); large-scale data from worksheets used to generate Census of Manufacturing figures of Central Planning Unit, Government of Sierra Leone, 1974–75.

Honduras small-scale enterprises data collected in 1979 enterprise survey in four regions reported in Stallman (1983); large-scale industry data obtained from worksheets used to construct the 1975 Census of Industry.

Jamaica: small-scale enterprises data collected in 1979 survey reported in Fisseha (1982); large-scale data collected form worksheets used by the National Planning Agency for their 1977 industrial survey.

The results of this analysis, summarised in Table 16.7, reveal that at world ('social') prices, small-scale manufacturing enterprises in Sierra Leone are more efficient than their larger-scale counterparts in all the enterprise groups considered except for shoes. The aggregate social benefit-cost ratio for rural small-scale industries is +1.57, indicating that small industries overall are economically efficient and have a positive effect on the total output of the Sierra Leone economy. Moreover, except for furniture, the ratios for the individual industries all exceed one, indicating their positive contributions to the economy as well. By contrast, the social benefit-cost ratios for large-scale industry is 0.49 overall, and exceeds one in only a single industry group, shoes.[17] The large-scale activities, consequently, have a negative effect on the Sierra Leone economy. A shift of resources to rural small industry would thus appear to make economic sense.

Table 16.7 Social benefit-cost ratios large[1] and rural small-scale[2] manufacturing enterprises in Sierra Leone – 1974–75

Industry	*Social benefit-cost ratio Domestic Prices*[3]		*Social benefit-cost ratio World Prices*[4]	
	Rural small-scale	*Large-scale*	*Rural small-scale*	*Large-scale*
Food				
Bakeries	1.86	1.03	1.80	0.68
Beverages	–	1.79	–	0.89
Others	–	4.41	–	–2.46
Textiles				
Wearing apparel	1.76	0.53	1.38	–0.30
Gara cloth	4.82	–	3.68	–
Shoes	1.65	2.00	1.14	1.40
Wood				
Furniture	0.81	0.87	0.52	0.48
Metal				
Metal products	1.63	1.61	1.16	0.90
Repairs	4.78	–	4.78	–
All	1.94	1.74	1.57	0.49

Notes:

1. Large firms employ fifty or more persons.
2. Small firms employ less than fifty persons.
3. For the social benefit-cost ratio (domestic prices), the gross output and purchased input values used to compute value added (numerator) are evaluated at actual prices in Sierra Leone; hired labour is evaluated at the market wage for small and at 0.8 of actual wage for the large; apprentice labour is evaluated at Le. 0.06 per hour and family labour at Le.0.16 per hour; capital is evaluated at 20 per cent using the capital recovery factor for the fixed component. For the rationale of these shadow prices estimates, see Chuta and Liedholm (1985).
4. For the social benefit-cost ratio (world prices), the gross output and purchased input values at domestic prices were adjusted for the 'nominal tariffs' on imported elements. Where quantitative restrictions applied, such as for flour, the difference between cif import prices and domestic prices were used.

Sources: Small-scale enterprise data collected in 1974–5 survey reported in Chuta and Liedholm (1985); large-scale enterprise data obtained from Census of Manufacturing data collected by Central Planning Unit, Government of Sierra Leone 1974–5. Data were obtained from fifteen of the twenty-eight large industries; these fifteen firms accounted for over 90 per cent of the large industry value added. Customs data obtained from the Government. Specific tariffs converted to *ad valorem* rates based on current fob prices.

---

5. SIZE AND GROWTH OF NONFARM ACTIVITIES

What determines how large the rural nonfarm economy is in any given country and what are its likely growth prospects? This can be approached by examining, on the one hand, expenditure patterns for those goods and services that could potentially be supplied by this sector and, on the other hand, by the supply response of rural nonfarm enterprises.

We begin with the best documented and largest class of expenditures, namely consumer goods and services. Although rural household expenditure studies are not uncommon, they typically do not distinguish the source of various consumption goods, e.g., whether the shoes purchased were made overseas, in a major urban area or in the rural economy. Investigations which do draw this distinction have been carried out in Sierra Leone (King and Byerlee, 1978) and Nigeria and Malaysia (Hazell and Roell, 1983).

In Table 16.8 the combined budget share of food expenditures (including alcohol and tobacco) ranges from two-thirds to four-fifths of household spending. This, of course, reflects modest levels of per capita income in all rural economies. The lesser reliance on home-produced food in the Muda area of Malaysia and the greater reliance on food imported from outside the region are the joint effect of higher income level and more specialised agriculture.

Among the goods and services that make up the 'local non-food' category are tailor-made clothing, footwear, hats, wooden furniture, pottery and mats; firewood; schooling and medical care; domestic servants, laundering and hairdressing; films, eating and drinking out; repairs, improvement and construction of homes; public transport and the operation of own transport.

In all three countries it is this 'local non-food' category that has the highest expenditure elasticity. This means that a 10 per cent increase in household income in Sierra Leone will lead to a jump in spending

Table 16.8 Rural expenditure elasticities in three countries

	Average budget share			*Expenditure elasticities*		
	Sierra Leone	*Nigeria*	*Malaysia*	*Sierra Leone*	*Nigeria*	*Malaysia*
Own food	47	56	27	.87	.88	.37
Local food	21	19	19	1.06	1.09	.76
Imported food	NA	5	21	—	1.07	.65
Local non-food	9	9	18	1.40	1.34	2.05
Imported non-food	NA	11	15	—	1.16	1.66

Notes: *Sierra Leone*: A national sample 1974, N = 203.
Nigeria: The Gusau region 1977, N = 321.
Malaysia: The Muda region 1973, N = 839.

Sources: *Sierra Leone*: King and Byerlee (1978) p. 204.
Nigeria and Malaysia: Hazell and Roell (1983) p. 28.

on local nonfarm goods and services equal to 14 per cent, to a 13 per cent increase in the Gusau region of Nigeria and to a 20 per cent increase in Muda. Thus we have strong evidence that rural non-farm goods and services are not 'inferior', but rather have the potential to grow more rapidly than agriculture itself, providing an *expenditure share* of all rural employment.

Individual components of the non-farm category have sharply differing expenditure elasticities. The highest elasticities are associated with services. Thus, in the case of Sierra Leone the figure for transport is 1.38 and for personal services and ceremonial outlays 2.38. By contrast, the elasticity for manufactured products originating from small-scale producers is 0.76. In Gusau and Muda the figures for housing construction and repair are 1.40 and 3.02, and for transportation 1.67 and 1.48.

Elasticities for specific manufactured goods for Sierra Leone and Bangladesh are shown in Table 16.9. The Bangladeshi households, at a per capita income of about £100, are the poorest of the four countries and, presumably, have the smallest budget shares devoted to non-food items. Particularly impressive in both countries are the higher income elasticities of demand for rural-based production relative to the products of large-scale urban industry

The actual growth in farm and non-farm rural employment has, in the aggregate, followed the pattern predicted by these expenditure elasticities. However, it is likely that the *composition* of nonfarm activities will be different than that suggested by the elasticity coefficients. Specifically, expenditures on rural manufacturers will be somewhat lower and expenditures on services (particularly trade and transportation) will be higher than predicted.

Beginning with manufacturers, the initial range of rurally-supplied goods will be larger or smaller depending upon craft traditions and the entrepreneurial endowment (e.g., it tends to be larger in Asia than in Africa). But in all countries as per capita income rises there is a shift in location from village to regional town and metropolitan area. Although the rural producer has an advantage in less expensive labour and premises, improving rural roads progressively diminishes the natural protection he enjoys against urban competitors. At the same time the more gifted rural entrepreneurs are attracted to the towns where the larger markets promise higher entrepreneurial returns; economies of agglomeration yield further advantages in the availability of more skilled labour and of cheaper, more diverse raw materials. Production in the towns, while carried out in units four or

Table 16.9 Expenditure elasticities of rural households for various small and large enterprise products

Products		*Sierra Leone*[1] *(1974)*	*Bangladesh*[2] *(1980)*
Food:	Bread – small	+0.69	+1.14*[3]
Clothing:	Dresses and pants (tailoring) – small	+0.72*	+0.96**
	Dresses and pants (clothing) – large	+0.59	—
	Dresses and pants (imported)	+1.49	+0.29
	Lungi – small	—	+1.61*
	Lungi – large	—	+1.00*
	Sari – small	—	+2.00*
	Sari – large	—	+0.63**
	Sari (synthetic) – large	—	+1.74*
Wood:	Furniture – small	+1.61*	+2.00*
Metal:	Agricultural tools and utensils – small	+0.50	+1.06*
	Agricultural tools and utensils – large	+0.89	+1.29*
All small-scale industry[4]		*+0.76**	—
All large-scale industry[4]		*+0.33**	—

Notes:
1. In Sierra Leone, data from 203 rural households were fitted into a modified form of a ratio semilog inverse expenditure function.
2. In Bangladesh, data from 444 rural households were fitted into a semilog expenditure function with the values in table estimated at mean expenditure levels.
3. * estimated coefficients significant at 1 per cent level.
 ** estimated coefficients significant at 5 per cent level.
4. From King and Byerlee (1978).

Sources: *Sierra Leone*: King and Byerlee (1977); *Bangladesh*: BIDS (1981).

---

five times the size of the rural producer, is still comparatively small scale and labour intensive.

To the extent large-scale public investment is made in building up the infrastructure of regional towns, many entrepreneurs will locate here and the output will not be lost to the larger rural economy. But to the extent entrepreneurs do migrate to the urban areas and to the extent urban-based substitute goods – plastic utensils, synthetic textiles – replace traditional products, the demand for rurally-produced manufactured goods will fall. Because these changes – along with other shifts in taste and relative prices – occur over time, they are not picked up in cross-section expenditure surveys and hence the latter's expenditure coefficients are an overestimate.

Expenditure studies may also be deficient with respect to nonfarm transport and trading activities, since most of these are embedded in the price of the consumer good. In so far as there is a shifting away from village-produced goods to more distant sources, the share of these marketing services will rise. Hence, inferences from household expenditure patterns are likely to underestimate the growth in aggregate rural nonfarm services.

The two remaining, smaller catergories of expenditures pertaining to nonfarm activities are production outlays on farm inputs (backward linkage) and expenditures on processing and marketing of agricultural output once it leaves the farm (forward linkage).[18] In the case of production inputs, cement for irrigation works, fertiliser (typically the largest single input expenditure), and other agricultural chemicals, do not originate in the rural economy; the same is true for four-wheel tractors. Equally, some portion of agricultural processing takes place in urban areas. One of the few studies that has attempted to net out intersectoral purchases is that of Bell, Hazell and Slade (1982) for Muda (Malaysia); they found that one-third of the incremental income was due to backward and forward linkage, whereas two-thirds was attributable to consumption expenditures.

While localised forward linkages give rise to considerably more value added than the comparable agricultural inputs,[19] the latter – particularly farm equipment – play a unique role in their potential impact on agricultural productivity. Other nonfarm activities such as trading and transport stimulate farm output by reducing marketing costs, which leads to an outward shift in demand at the farmgate. Farm equipment inputs, on the other hand, can act directly on the yield par acre and output per person.

There are two components to the nonfarm sector's 'productivity contribution' to agriculture. The first is related to the rural farm equipment industry's capacity for idiosyncratic design adaptation. In the animal draft farming sector of many Asian, African and Latin American countries there are only three or four types of ploughs in use, both for breaking the soil and for secondary tillage. In Taiwan, local blacksmiths have provided farmers with a wide array of cheap, highly specialised implements. Primary tillage aside, of eight secondary tillage implements one is the harrow. There are eleven kinds of harrows: the comb harrow, three knife-tooth harrows (standard, bent frame, flexible tooth), two spike harrows, the bamboo harrows, the pulverising roller, the stone roller, the tyned tiller, and disc harrow. A single one of these harrows, the standard knife tooth, has twelve regional variants. Width, length, material, number of teeth, shape of tooth blade and method of affixing teeth are adapted to local topography, field size, soil structure and available construction materials.

The results of idiosyncratic design adaptation is that the task – in this case secondary tillage – is done more quickly (higher labour productivity) and it is done more effectively (higher land productivity). More dramatic, better-known examples of idiosyncratic design adaptation include India's portable irrigation pump based on vertical high-speed diesel engines made in small engineering workshops, and Thailand's Prapradaeng power tiller.[20]

These last two examples also illustrate the second way that rural farm equipment producers raise agricultural productivity. This second component is the supplying of inexpensive partial-mechanisation inputs which break labour bottlenecks and thereby pave the way to higher cropping intensity. Additional examples here include small electric or gasoline pumps, small motors attached to threshers and winnowers, as well as backpack sprayers. The result is higher output per acre per annum, and increased labour income through higher utilisation of manpower over the course of the entire year. In summary, the rural nonfarm sector stimulates agricultural output in three ways: through substantial income effects on food expenditures, through reduction of marketing costs, and through the productivity contribution of localised farm equipment manufacturers.

The extent to which the increase in demand described above will translate into an expansion in rural nonfarm output depends importantly on the supply response. With respect to the short-run supply elasticity, available evidence from the MSU and other field surveys indicates there is a substantial amount of 'excess capacity' – in the range of 20 to 40 per cent – among rural nonfarm enterprises in most developing countries.[21] While there may be periodic shortages of imported raw material and working capital in certain activities, the overwhelming cause of excess capacity is insufficient demand. Hence, the short-run supply curve may be considered highly elastic.

With respect to the long-run elasticity, the potential constraint is blocked entry, with the primary candidates being capital and specialised skills. For the great majority of activities initial capital requirements fall in the range of £50 to £900, which are usually financed from personal savings and loans from family and friends. Hire-purchase credit is available for rice amd maize mills and other fee-for-service processing equipment. Even where capital requirements are far higher, if the past record is any indicator, there is usually sufficient access to finance among select individuals in the rural economy so that no bottleneck will long remain. A similarly optimistic conclusion emerges with respect to skill barriers: formal educational requirements are low; specific skills are formed by apprenticeships and learning-by-doing in a gestation period ranging from six months to three years.

6. CHANGE OVER TIME

Has rural nonfarm activity, in fact, been increasing over time? Aggregate statistics indicate that it generally has. Anderson and Leiserson (1980), using secondary ILO data, have shown that the employed rural labour force increased faster between 1959 and 1970 than the agricultural labour force in all regions except for Latin America. Specific data for nine countries reported by Chuta and Liedholm (1979) reveal that the percentage of the labour force engaged in nonfarm work has risen in all of them. They also report the following annual growth rates in nonfarm rural employment: Korea (1960–74) at 3.2 per cent, Taiwan (1955–66) at 9.4 per cent, Kenya (1969–75) at 8.8 per cent, Mexico (1960–70) at 5.6 per cent, Iran (1956–72) at 4.8 per cent and Indonesia (1961–71) at 5.5 per cent.

There are important variations in the growth rates by type and size of enterprise. By firm size, for example, time series data on differential rural growth rates are sparse, but some limited information on rural industrial growth rates are now available for firms employing from one to fifty persons in India (1961–71) and Sierra Leone (1974–80) (Liedholm and Mead, 1986). These data indicate that a direct relationship exists between the growth rates and firm size. In both these countries, the growth in the number of rural industrial firms is highest in the 10 to 49 employee size category, for example, and lowest in the one-person firm category. Indeed, in Sierra Leone, the number of one-person rural industrial firms actually declined during the time period covered by the study. Such findings tend to reinforce Dennis Anderson's (1982) conclusion, that 'household' manufacturing for the country as a whole 'tends to decline first in relative and then in absolute terms as industrialization proceeds. Moreover, the growth rates were higher the larger the size of locality, and thus reflect the shift to provincial towns noted above.

7. CONCLUSIONS

To sum up our main points. Nonfarm activities productively absorb a large quantity of rural labour and provide a major source of income to a majority of rural households. Because they are the source of a particularly large share of sustenance to the rural poor, they have a substantial impact on reducing income inequality. An exclusive focus on land reform as a solution to rural poverty is mistaken. Finally, nonfarm activities are not only efficient contributors to GDP, but they stimulate agricultural growth through effects on income, farm productivity, and marketing costs.

Differing public policies will result in a larger or smaller rural nonfarm economy. The redirection of large-scale public expenditures towards the development of infrastructure in rural towns is one potent intervention available, and is, of course, highly to be desired on other grounds. A second area is the creation of a general policy environment that is at least neutral with respect to the size of enterprises (Haggblade, Liedholm and Mead, 1986); for instance, implicit tariffs on tools and equipment, raw materials, and spare parts should not be higher for smaller firms than for larger firms as is true in many countries. In addition, it should not be overlooked that, given the strong linkages, policies aimed at increasing agricultural output are relevant to raising nonfarm output and employment. At the project level, the new lending modalities for channelling working capital to microenterprises should be pursued (Kilby and D'Zmura, 1985). Finally, the strength of the nonfarm sector depends upon the infusion of new technical knowledge. Research and development expenditures need to be aimed at design upgrading of farm equipment, transportation vehicles, and traditional consumer products; use needs to be made of best-practice surveys and adaptive research to improve existing artisan production processes. But these steps will in all probability only be taken when those in power are more fully informed of the size and potential contribution of the rural nonfarm sector, and then are willing to commit themselves to the potentially hazardous task of mobilising new constituencies and placating the old.

Notes

*This paper reports on work supported by US AID's Employment and Enterprises Policy Analysis Project (EED, Bureau of Science and Technology). Dr Kilby's contribution was made during his stay at the Woodrow Wilson International Center for Scholars.

1. See, for example, the contributions of Johnston and Kilby (1975), Mellor (1976), Chuta and Liedholm (1979), and Anderson and Leiserson (1980).
2. Comparisons of the street by street, village by village enterprise censuses conducted by MSU and host country scholars with 'official' censuses find that the latter not infrequently undercount the number of small enterprises by a factor of two or more (see Liedholm and Mead, 1986).
3. Small scale is defined for the purposes of this paper as firms employing less than fifty persons. Rural is defined, unless otherwise specified, as localities with 20 000 inhabitants or less.
4. Similar types of 'enterprise surveys' using a cost-route method to collect weekly data from small firms have been conducted by Michigan State University and host country scholars in Sierra Leone, Bangladesh, Jamaica, Honduras, Thailand and Egypt (see Liedholm and Mead, 1986, for details).
5. The high return nonfarm activities, however, still generally yield a lower return on average than their urban conterparts (see Chuta and Liedholm, 1979, for details).
6. Calculated from data on 424 rural households collected by the Thai Rural Off-Farm Employment Project (see Narongchai *et al*, 1983, for details).
7. The coefficient of variation computed for net farm income was 2.07, but was only 0.64 for total household income, which includes nonfarm income sources (computed from monthly data generated by the Thai Rural Off-Farm Employment Project – see Narongchai *et al*, 1983).
8. See below for a more detailed examination of the complementary nature of farm and nonfarm inputs.
9. Employment would increase if the labour capital ratio of smaller firms exceeded those of the larger ones. Virtually all empirical studies find that small rural enterprises are more labour intensive (usually measured in terms of the labour-capital ratio) than their larger scale counterparts in the aggregate. At the industry-specific level, the same results generally hold, although a few exceptions exist (such as in Korea) (see Liedholm and Mead, 1986, for details).
10. See, for example, Page and Steel (1984) and Liedholm and Mead (1986) for a review of the evidence.
11. For a detailed discussion of these measures, see Biggs (1986).
12. More specifically, the social benefit cost ratio (*SBC*) is calculated on the basis of the following formula:

$$SBC = \frac{VA}{r_s K + w_s L}$$

where:

VA = valued added;
r_s = shadow or 'social' price (interest rate) of capital;
K = total fixed and working capital
w_s = shadow or 'social' price of labour;
L = total labour hours, including family and apprentice hours.

A ratio greater than one means that the activity or enterprise has a positive effect on the *total output* of the economy, while a ratio less than one means it has a negative effect. If actual (e.g., domestic) rather than 'social' (e.g., world) prices are used to evaluate value added, however, the *SBC* can only be used to compare the producivity of enterprises in the same sector.

13. Approximately 495 rural manufacturing firms were surveyed in Honduras (see Stallmann, 1983, for details), 200 in Sierra Leone (see Chuta and Liedholm, 1985, for details), and 150 in Jamaica (see Fisseha, 1982, for details). Small scale refers to firms employing fifty persons or less, while rural refers to localities with 20 000 inhabitants or less.
14. The dates of the large and small industry surveys differed slightly in Jamaica and Honduras. Although the small enterprise surveys were both conducted in 1979, the large-scale surveys covered 1977 in Jamaica and 1975 in Honduras. The economic conditions in these countries did not differ markedly between these periods, however, so the validity of comparisons should be not seriously vitiated.
15. The actual wages paid to all workers in large-scale enterprises were included at 80 per cent. For a justification of these adjustments, see Haggblade, Liedholm and Mead, 1986.
16. Ho (1980) for Korea, and Cortes *et al.* (1985) for Colombia, find that large-scale enterprises tend to be more efficient than their smaller-scale counterparts using comprehensive efficiency measures. They do not consider rural activities explicitly, however.
17. The ratio did exceed one for several individual large firms and industries, but because of confidentiality rules their individual figures had to be combined with others.
18. These forward linkages with respect to agricultural output are similar to the marketing services for nonfarm products described in the preceding paragraph.
19. A good overview of specific production inputs and processing activities is available for Thailand in World Bank, 1983. The share of all manufacturing value added deriving from rice milling, rubber processing, cassava chipping, tobacco curing and fruit-canning that takes place in rural areas is many times larger than that of farm equipment and animal feed. For a more general treatment of the relative size of forward and backward linkages over the course of economic development, see A. Simantov (1967).
20. The case of the power tiller in Thailand is instructive. Japanese power tillers for paddy cultivation had not been widely adopted owing to high

purchase price. A low-cost adaptation, developed by IRRI in the Philippines, was introduced in the late 1960s; it did not succeed. The Prapradaeng tiller was developed locally and improved through a prolonged iteration between local farm users and the equipment producers – the forcing house of successful appropriate technology – and is now manufactured by more than forty small firms.

21. For a review of field studies that bear on both short- and long-run supply elasticities in this sector, see Liedholm and Mead (1986).

References

Anderson, D. (1982) 'Small Industry in Developing Countries: A Discussion of Issues' *World Development*, vol. 10, no. 1.

Anderson, D. and Leiserson, M. (1980) 'Rural Non-Farm Employment in Developing Countries' *Economic Development and Cultural Change*, vol. 28, no. 2 (January).

Bell, C., Hazell, P. and Slade, R. (1982) *Project Evaluation in Regional Perspective* (Baltimore: The John Hopkins University Press).

Biggs, T. (1986) 'On Measuring Relative Efficiency in a Size Distribution of Firms', *Employment and Enterprise Policy Analysis Discussion Paper*, no. 2 (Cambridge, Mass.: Harvard Institute for International Development).

BIDS (Bangladesh Institute of Development Studies) (1981) 'Rural Industries Study Project – Final Report,' (Dacca, Bangladesh: BIDS).

Chuta, E. and Liedholm, C. (1979) 'Rural Non-Farm Employment: A Review of the State of the Art,' *Michigan State University Rural Development Paper*, no. 4, East Lansing, Michigan.

Chuta, E. and Liedholm,C. (1985) *Employment and Growth in Small Industry: Empirical Evidence and Assessment from Sierra Leone* (London: Macmillan).

Cortes, M., Berry, M.A. and Ishag, A. (1985) 'What Makes for Success in Small and Medium Scale Enterprises: The Evidence from Colombia', *The World Bank* (Washington, DC).

Fisseha, Y. (1982) 'Management Characteristics, Practices, and Performance in the Small-Scale Manufacturing Enterprises: Jamaica Milieu', PhD dissertation, Department of Agricultural Economics, Michigan State University.

Haggblade, S., Liedholm C. and Mead, D. (1986) 'The Effect of Policy and Policy Reforms on Non-Agricultural Enterprises and Employment in Developing Countries: A Review of Past Experiences', *Employment and Enterprise Policy Analysis Discussion Paper*, no. 1 (Cambridge, Mass.: Harvard Institute of International Development).

Hazell, P. and Roell, A. (1983) 'Rural Growth Linkages: Household Expenditure Patterns in Malaysia and Nigeria', *International Food Policy Research Report*, no. 41 (Washington, DC).

Ho, S. (1980) 'Small Scale Enterprises in Korea and Taiwan', *World Bank Staff Paper*, no. 384 (Washington, DC: The World Bank).

Johnston, B.F. and Kilby, P. (1975) *Agriculture and Structural Transfor-*

mation: Economic Strategies for Late-Developing Countries (London: Oxford University Press).

Kilby, P. and D'Zmura, D. (1985) 'Searching for Benefits', *USAID Evaluation Special Study*, no. 28 (Washington, DC: USAID).

King, R.P. and Byerlee, D. (1978) 'Factor Intensities and Locational Linkages of Rural Consumption Patterns in Sierra Leone', *American Journal of Agricultural Economics*, vol. 60, pp. 197–206.

Korea, Government of (1981) *Korea Statistical Yearbook*, Economic Planning Board.

Liedholm, C. and Mead, D. (1986) 'Small Scale Enterprises in Developing Countries: A Review of the State of the Art', *Michigan State University International Development Working Paper*, East Lansing, Michigan.

Matlon, P. (1977) 'Size Distribution, Structure, and Determinants of Personal Income Among Farmers in the North of Nigeria', PhD dissertation, Cornell University.

Matlon, P. *et al.* (1979) 'Poor Rural Households, Technical Change and Income Distribution in Developing Countries: Two Case Studies from West Africa', *African Rural Economy Working Paper*, no. 29, Michigan State University.

Mellor, J. (1976) *The New Economies of Growth* (Ithaca, New York: Cornell University Press).

Narongchai, A., Onchan, T., Chalamwong, Y., Charsombut, P., Tambunletchai, S. and Atikul, C. (1983) *Rural Off-Farm Employment in Thailand* (Bangkok: Industrial Management Company Ltd.).

Page, J. and Steel, W. (1984) 'Small Enterprise Development: Economic Issues from African Experience', *World Bank Technical Paper*, no. 26 (Washington, DC, The World Bank).

Simantov, A. (1967) 'The Dynamics of Growth of Agriculture', *Zeitschrift für Nationalokonomie*, vol. 27, no. 3.

Stallmann, J. (1983) 'Rural Manufacturing in Three Regions of Honduras', mimeo., Michigan State University.

Taiwan, Government of (1976) *Taiwan Farm Income Survey of 1975* (Taipei: Joint Commission for Rural Reconstruction).

World Bank (Hans Binswanger *et al.*) (1983) *Thailand: Rural Growth and Employment* (Washington, DC: The World Bank).

17 Rural Resource Mobility and Intersectoral Balance in Early Modern Growth

Hiromitsu Kaneda
UNIVERSITY OF CALIFORNIA, DAVIS

1. INTRODUCTION

This paper studies the nature of relationships between intersectoral resource mobility and intersectoral balance in early modern economic growth and, on the agricultural side in particular, the crucial role played by transmission of the best-practice technology to less developed areas. The primary focus is, first, on migration of labour as it is reflected in the shifting of the economy's centre of gravity from agriculture to industry and, secondly, on the so-called eastward-movement of rice cultivation. The present point of departure is the 'concurrent growth' thesis by Kazushi Ohkawa which emphasises the development of agriculture side by side with urban sectors in the early modern growth of Japan (Ohkawa, 1964).

Given the rising labour force and intersectoral mobility of labour, major analytical issues to be addressed in this paper are as follows:

1. Identifying the proximate factors that are responsible for the concurrent development of agriculture and the non-agricultural sector with particular reference to the experience of Japan.
2. Establishing a relationship in concurrent growth among the sectoral variables on the demand as well as on the production side and clarifying, in particular, the implied growth patterns of sectoral per capita earnings.
3. Analysing the sources and rates of agricultural growth across regions to shed light on the agricultural development under conditions of concurrent growth.

From a larger perspective these issues relate to the question of primacy in economic development of industrialisation and/or agricultural development and to the question of transition in income distribution during early modern economic growth. The experience of Japan after the Second World War has often been cited as an important example demonstrating both a high rate of economic growth and a favourable performance in income distribution. Lacking data, however, it has not been possible to confirm that such was the case with respect to the Japanese experience in the early phases of her modern economic growth. It is appropriate then to search for additional evidence to those that have been advanced by students of Japan's modern economic growth.

2. CONCURRENT GROWTH IN JAPAN'S EARLY MODERN ECONOMIC GROWTH

Ohkawa has emphasised that the population pressure of Meiji Japan (from 1867) arose not from the natural increase of the total population, but rather from that concentrated in the urban/modern sector. Comparing with the known European experience, Ohkawa emphasised the following main factors in Japan's modern economic growth. First, the labour force engaged in agriculture remained virtually unchanged indicating that the net increase in the total labour force was absorbed entirely by the non-agricultural sector. Secondly, the growth rate of Japan's total population was moderate (about equal to the European growth rate) during the initial phases of modern economic growth. Although the data are less reliable, the growth rate of the total labour force is taken to be more or less the same. Thirdly, relative to European countries on the eve of industrialisation, in Meiji Japan the proportion of the labour force engaged in agriculture was overwhelmingly large. According to Ohkawa then, other things being equal, industry in Japan had to grow fast enough to absorb much more labour from agriculture in order to keep the farm labour force constant. Therefore, the crucial question of intersectoral interest was how Japan could accommodate this requirement, thereby avoiding a much more aggravated population pressure on the land.

Salient features of the intersectoral relations in this early modern period were the developments in both agriculture and the non-agricultural sector that enabled agriculture, with its overhwelming majority of the labour force, to develop along with the nascent, rapidly growing industry. Indeed, according to a survey of empirical evidence relating to income inequality in the Meiji era by Akira Ono and Tsunehiko Watanabe, the differential of per capita income between urban and rural areas remained almost stable before 1915, and then increased sharply (Ono and Watanabe, 1976). Moreover, according to Ohkawa, both the productivity of labour in agriculture and industrial real wage rates did increase *pari passu*, at least until 1905–10 at the annual rates averaging 1.0 to 1.5 per cent (Ohkawa, 1964). In focusing attention on this process I find Ohkawa's concurrent growth thesis quite useful.

The thesis states that the agriculture of Japan developed not before (as with typical European counterparts), but side by side with the process of rapid industrialisation – a concurrent growth of industry and agriculture. Agriculture that had already been over-populated did none the less grow at a reasonable pace without displacing its labour force. Industry developed rapidly by absorbing labour, the majority of which originated from the rural agricultural sector. First, on the non-agricultural side, rapid absorption of labour was promoted by the confluence of the speedy growth of the sector itself, the sector's output mix favouring labour-intensive products, and the relatively labour-intensive techniques of production which complemented it. The rate of absorption of labour into industry consistently exceeded the rate of growth of population. The upshot of all this as argued, for example, by Gustav Ranis and John C. H. Fei, that during Japan's initial industrialisation phase the maximum use was made of her abundant factor by adopting labour-using (or at least, not very labour-saving) innovations and that clever local adaptation of imported technology was being carried out leading to capital-shallowing (Ranis and Fei, 1963). It is pertinent to note that the rate of increase in non-agricultural employment amounted to two-thirds of its rate of output growth during this period (Ohkawa, 1964).[1]

In agriculture, secondly, it is to be noted that as utilisation of land intensified (i.e., the ratio of cropped area to arable land increased) during 1880 to 1900, and remained at high levels until about 1915, not only did input of man-hours of labour per arable land grow, but also labour input per unit of cropped area increased significantly. According to Saburo Yamada's examination of major statistical sources relating to labour use in Japanese agriculture, the estimated hours of labour per hectare of cropped area rose steadily from 3500 in 1880 to 3670 in 1915 (Yamada, 1982). Given that the rural wage rate (as distinguished from rural per capita earnings) remained virtually constant, this suggests that the nature of technological change was labour-using during this period in the agriculture of Japan.[2]

Quite consistent with this are empirical findings by Yujiro Hayami and Masao Kikuchi. Between 1880 and 1935 total inputs in agriculture (the aggregate of all conventional inputs using factor share weights) increased at a rate of 0.4 per cent per annum, and total productivity (the ratio of total output to total input) grew at 1.2 per cent. This implies only one-quarter of total output growth in Japan's agriculture was accounted for by increases in land, labour, capital, and current input; the rest was from technological progress broadly defined (Hayami and Kikuchi, 1985). Given that farmland was by far the most binding constraint in production, it is reasonable to postulate that substitution was induced of current inputs, especially fertilisers, for land, thereby increasing the input of plant nutrients per unit of farmland area. Hayami and Kikuchi hypothesise that the process involved the technological change biased toward land-saving and fertiliser-using in Hicksian terms. Underlying this process were the development and diffusion of fertiliser-responsive and high-yielding rice varieties and improvements in land infrastructure such as irrigation and drainage.

It is then quite sensible to argue that the increase in labour inputs (flow) observed by Yamada reflected in part farmers' efforts to prevent soil exhaustion resulting from the enhanced intensity of land use by larger inputs of self-supplied fertilisers (organic fertilisers such as compost, manure, green manure, night soil, etc., that are quite labour intensive). Farmers did not hesitate to expend rising amounts of labour in order to coax out the maximum possible yields from their fields.[3] It is worth noting also that the subsequent expansion of less land-dependent, but significantly labour-intensive, farm enterprises such as sericulture (and processing of farm waste products, e.g., rice straw) was also in line with this type of induced innovation.

3. A TWO-SECTOR MODEL OF RESOURCE MOBILITY

In a well-known 1947 paper, Herbert Simon provided a theorem stating that if two sectors in an economy have the same rate of technical progress, labour will migrate towards the sector in which the demand for the product is more income-elastic (Simon, 1947). William Baumol showed in 1967 that in a model of unbalanced growth there is a tendency for the output of the 'nonprogressive sector' – the demand for whose product is not too highly price-inelastic – to decline and perhaps vanish (Baumol, 1967). This case of Baumol's was later recast in a form compatible with the question of labour migration by Artle, Humes, and Varaiya. In this version, in the case of unbalanced growth of two sectors, labour migrates towards the progressive (non-progressive) sector if the demand for its output is elastic (inelastic) to its own price (Artle, Humes and Varaiya, 1977).

More recently Vislie has examined these results. The Vislie version further specifies the conditions under which the Simon and the Baumol conclusions hold (Vislie, 1979). With a model that characterises the production functions of the two sectors with labour as the only one variable factor, neutral technical progress in both sectors, and the Slutsky relation that specifies a substitution effect and an income effect between the two goods. Vislie has shown that, not surprisingly perhaps, the own price elasticity condition is not sufficient to guarantee that labour migrates towards the progressive sector. In this section I extend his model in such a way that highlights differential rates of annual per-capita earnings growth as well as relative rates of neutral technical progress in the two sectors.

I believe that this model is appropriate for studying early phases of modern economic growth and that it is rich in analytical content despite its relative simplicity. Let us start with a production function for each sector where capital is fixed and only labour is variable. Let X_i here designate output in sector i and N_i here its labour force. In logarithms of the variables:

$$X_i = a_i + N_i + \beta_i t, \qquad i = 1.2 \tag{1}$$

where the total labour force is $N = N_1 + N_2$, the rate of technical progress is β_i, and $a_i = 1\text{n}\, A_i$ where A_i is a constant. The underlying assumption is that there is virtually no open unemployment, with the dominant modes of production allowing both income and work sharing. This formulation implies that the productivity of labour in sector i rises at the rate of β_i here per unit time period. The second relation is the per-capita demand function for the output of each sector, whose argument consists of per capita income (Y/N) and the two product prices as follows:[4]

$$(X_i - N) = b_i + E_i(Y - N) + e_{i1}P_1 + e_{i2}P_2, \tag{2}$$

where b_i is constant, E_i is the income elasticity of demand for i-th goods such that $E_i = Y/X_i(\partial X_i/\partial Y_i)$ and e_{ij} is the price elasticity of demand for i-th goods with respect to the price of j-th goods, i.e., $e_{ij} = P_j/X_i(\partial X_i/\partial P_j)$, $(i, j = 1, 2)$. Assume in addition that none of the goods are inferior, i.e., $E_i > 0$, and that these demand functions are homogenous of degree zero in Y/N, P_1 and P_2, i.e., $e_{i1} + e_{i2} + E_i = 0$ $(i = 1, 2)$. We can rewrite (2) as follows:

$$X_i = b_i + E_iY + (1 - E_iN + e_{i1}P_1 + e_{i2}P_2 \qquad (2')$$

From (1) and (2′), we have

$$a_i + N_i + \beta_i t = b_i + E_iY + (1 - E_i)N + e_{i1}\,P_1 + e_{i2}P_2.$$

Differentiating with respect to time and letting R stand for growth rate of the subscript variable, we have

$$R_{Ni} + \beta_\iota = E_iR_Y + (1 - E_i)R_N + e_{i1}R_{P1} + e_{i2}R_{P2}. \qquad (3)$$

Let per capita earnings in each sector be designated by w_i. Then, total income

$$Y = w_1N_1 + w_2N_2,$$

and therefore

$$R_Y = \mu_1(R_{N1} + R_{w1}) + \mu_2(R_{N2} + R_{w2}), \text{ where } \mu_\iota = (w_i\ N_i/Y.)$$

We postulate that the price of the i-th products, P_i respond to the sectoral wage earnings and the rate of technical change in the i-th sector.

$$P_i = (w_i - a_i) - \beta_i t. \qquad (4)$$

Then, it is easy to see that

$$P_{Pi} = R_{wi} - \beta_i \qquad i = 1, 2$$

and, therefore, that

$$R_{Ni} + \beta i = E_i[\mu_1(R_{N1} + R_{w1}) + \mu_2(R_{N2} + R_{w2})] + (1 - E_i)R_N$$

$$+ e_{i1}(R_{w1} - \beta_1) + e_{i2}(R_{w2} - \beta_2).$$

Putting $i = 1$, by letting the labour-recipient (say, non-agricultural) sector the first sector, the relationship becomes

$$R_{N1} = E_1[\mu_1(R_{N1} + R_{w1}) + \mu_2(R_{N2} + R_{w2})] + (1 - E_1)R_N - \beta_1 + e_{11}(R_{w1} - \beta_1) + e_{12}(R_{w2} - \beta_2). \quad (5)$$

4. CONCURRENT GROWTH OF THE TWO SECTORS

Suppose we define the concept of concurrent growth of the two sectors in terms of the equal rates of technical change and also of the equal rates of per capita wages growth. In the context of equation (5) concurrent growth then entails the following assumptions:

$$\beta_1 = \beta_2 = \beta > 0, \text{ and } R_{w1} = R_{w2} + R_w.$$

Then, (5) can be simplified to:

$$R_{N1} = R_N + R_w + \beta(E_1 - 1). \quad (6)$$

Thus, if per capita incomes in the two sectors grow *pari passu*, as hypothesised by students of Japan's early modern economic growth, then the rate of labour migration to sector one responds to the growth rate of total labour force, the common rate of per capita earnings growth, and the product of the rate of technical change, as well as to the income elasticity of demand for the sector one products. Simon's theorem in this context is then as follows:

Given that population is fixed and that wages are controlled (i.e., $R_N = R_w = 0$), if $E_1 > 1$, then $E_2 < 1$ for all pairs of $(\mu_1, \mu_2,)$ and therefore R_{N1} is greater, the greater is E_1.

Suppose next that the first (non-agricultural) sector is progressive in the sense that, while labour productivity in the other (agricultural) sector stagnates, the first sector enjoys labour productivity growth at a positive rate. Therefore, we assume $\beta_1 > 0$, $\beta_2 = 0$, and $R_{wl} = R_{w2} = R_w$. Then, (5) is rewritten

$$R_{N1} = R_N + R_w + \beta_1(-1 - e_{11}). \quad (7)$$

Given once again that the population is fixed, and that the wages are entirely controlled by the authorities, the sign of R_{N1} is positive, that is, labour will migrate towards the first sector, if and only if $e_{11} < -1$. This is Baumol's result put in terms of equation (5).

If, however, $\beta_1 > 0$, $\beta_2 = 0$ and $R_{w1} \neq R_{w2}$, the results are not as simple nor as clean-cut. Thus,

$$R_{N1} = E_1(\mu_1 R_{w1} + \mu_2 R_{w2}) + R_N - \beta_1(1 + e_{11}) + e_{11}R_{w1} + e_{12}R_{w2}. \quad (8)$$

It is evident that in this case it is not sufficient to have $e_{11} < -1$ to guarantee that labour moves towards the progressive sector 1.

4.1 Application to Japan's Early Modern Economic Growth

The fundamental equation (5) is given by growth rates of sectoral variables, N_i, and w_i, and national total variable N, plus some parameters designating income and price elasticities, E_1, e_{11}, e_{12}, and the sectoral rates of productivity growth, β_i, i = 1, 2. Except for certain simplified cases as discussed above, all these parameters are required for studying aspects of modern economic growth.

I estimated β_i with the use of a statistical version of equation (1) and the elasticity parameters, E_1, e_{11} and e_{12} with the first differences of the variables in logarithms in a statistical specification of equation (2). The estimates of the rates of technical change are quite satisfactory, while those of the elasticity parameters are not. The results are as follows, where the asterisk signs signify that the estimates are not significantly different from zero and the figures in parenthesis are standard errors:[5]

Parameters	*1885–1900*	*1901–1915*	*1885–1915*	*1916–1930*	*1916–1940*
β_1	0.8	1.4	0.9	1.2	1.9
β_2	0.8	1.9	1.6	1.2*	0.6

	E_1	e_{11}	e_{12}
Coefficients	1.057	–0.025*	0.045*
(Standard errors)	(0.078)	(0.035)	(0.032)

It is clear that per capita output (real NDP per gainful worker) grew much more rapidly in non-agriculture (sector 1) than in agriculture (sector 2) during the latter period between 1916 and 1940, and that the reverse was the case during the earlier period between 1885 and 1915. Indeed the concurrent growth thesis, in the sense that the productivity of labour in the two sectors grew *pari passu*, is confirmed only for the period between 1885 and 1900, when one of the conditions for Simon's postulates, $\beta_1 = \beta_2$ obtains. However. given the growth rate of total labour force and that of the gainfully employed in non-agriculture, equation (6) can be used to estimate the common rate of growth of annual wages in the two sectors for this period (Simon's other condition being the equal growth rates of wages). Rearranging the terms in equation (6), we get

$$R_w = R_{N1} - R_N - \beta(E_1 - 1) = \\ 1.9 - 0.7 - 0.8(1.06 - 1) = 1.2 \qquad (6')$$

that is to say, the implied growth rate of per capita annual earnings is about 1.2 per cent per annum during the period between 1885 and 1900.

Understandably, data on sectoral annual wage-earnings in early phases of modern economic growth are hard to come by. I use the only wage rate series available from the Long-Term Economic Statistics (LTES) sources, given as the index of annual wage rate for manual and office workers.[6] The growth rate of this series in real terms (R_{w1}) together with those (from the (LTES) of gainfully employed workers in total (R_N) and in non-agriculture (R_{N1}), calculated by taking quinquennial average values, are as follows:

Years	R_{w1}	R_N	R_{N1}
1885–1915	1.5	0.7	1.9
1916–1940	2.6	0.9	1.8

The growth rate of annual real wages in non-agriculture between 1885 and 1915 turns out to be 1.5 per cent per annum, which is quite close to the implied common rate above. In fact, the real annual wage index increased at about the same rate during the first half (1.5 per cent) and the second half (1.6 per cent) of this period. It is reasonable, therefore, to accept this computed rate to be the approximate growth rate of per capita earnings in agriculture during 1885 to 1900, when the two sectors were postulated to have grown concurrently.

Among the rest of the periods, particularly interesting is the period between 1916 and 1930, when the labour productivity growth in agriculture decelerated and estimated β_2 was not significantly different from zero. Given the estimated elasticities and growth rates, equation (8) that assumes $\beta_1 > 0$ and $\beta_2 = 0$ can be rearranged as in the following:

$$R_{w2} = (E_1\mu_2 + e_{12})^{-1} [R_{N1} - R_N - (E_1\mu_1 + e_{11})R_{w1} + \beta_1(1 + e_{11})] \quad (8')$$

and (8) can be given specific parametric values for 1916–30. Serious data problems here concern estimating the absolute levels of wages in the two sectors, which are required for estimating μ_i for the period in question. In the absence of necessary data two simple (extreme) alternatives are taken here. Computation utilises (i) the proportions of NDP contributed by each sector ($\mu_1 = 0.82$ and and $\mu_2 = 0.18$[7] and (ii) those of the labour force in each sector at the mid-point in the fifteen-year period. Then the first estimate is:

$$R_{w2} = (1.06^* \ 0.18 + 0.04)^{-1}[1.8 - 0.9 - 2.6(1.06^* \ 0.82 - 0.03) + 1.2(1 - 0.03)] = 0.5.$$

With almost exactly half of the labour force in agriculture, the second estimate yields $R_{w2} = 1.3$. The growth rate of per capita annual earnings in agriculture, therefore, is estimated to have ranged between a negative half per cent to about half the growth rate of annual wage rate in non-agriculture. Agriculture during the period between 1915 and 1930 was clearly stagnant in productivity growth and in per capita earnings. As equation (8) shows, contributing to this phenomenon prominently were the factors on the non-agricultural side, such as the increase of the total labour force, though moderate, the relatively higher income elasticity of demand for the output of non-agriculture as well as the relatively high growth rate of per capita earnings of that sector, and the increased weight of non-agriculture in the economy as a whole, despite the relatively high rate of labour absorption and that of technological change in the non-agricultural sector.

What were the sources of growth for agriculture during these early periods? The farm labour force for all practical purposes remained virtually constant. Had it been just to maintain the average level of income constant over time, the sector would have had either to renew enough capital to maintain the capital/labour ratio, or to effect technical change that was not overly biased against labour. If living standards had to rise, then it would have been necessary to ensure either that the average capital intensity of production was actually rising or that output augmenting technical change continued to take place to give rising output and consumption per head. It is common ground that the development of Meiji agriculture was achieved with remarkably small demands on the nation's scarce resources. I now turn to analyse the sources and rates of this growth.

5. SOURCES AND RATES OF OUTPUT GROWTH PER FARM HOUSEHOLD[8]

Ideally, it is desirable to have agricultural value added data on the cross-section basis for studying the sources of agricultural growth during the early phases of Japan's modern economic growth. As there are no such data available, I focus on rice production and yield of land in terms of yield of rice in *koku* per *tan*.[9] I hyphothesise here that the crucial source of growth was yield increase and that the rate of growth of Japan's agriculture varied enormously depending on the gains in land yield. I attempt to show in this section that the expansion of cultivated area contributed in the early decades of moderate growth to absorbing the rising number of farm households and to easing the severe land constraints, and that in these early modern decades the source of growth in rice production was the spread of rice cultivation to less favourable areas mainly in eastern regions but also within the most developed regions of western Japan, impacting adversely on the average yields at first. I show further that rice yield began to rise from around the turn of the century and subsequently it did become a significant source of growth in western Japan and then later on in eastern Japan. It is seen that marked growth in rice yields began only with the coming of the commercial (organic) fertiliser industry, and the diffusion of its use in the more advanced west first, and then in the less developed eastern prefectures. From another point of view, from the early phase of modern economic growth the less-developed east began the process of catching up with the traditionally advanced western part of Japan.[10]

On the basis of weighted averages Table 17.1 shows that the number of farm households per unit of cultivated area for all Japan remained virtually constant (with a small decline between 1886–88 and 1896–98 and an equally small rise between 1926–28 and 1936–38 being the only exceptions). In the Western Prefectures the decline in the number of farm households per cultivated area was pronounced between 1886–88 and 1896–98, after which the ratio remained virtually the same. In the East the decline in the number of farm households per *cho* continued gradually from 1886–88 until about

Table 17.1 Farm households, land yield and weights by prefectural output of rice, 1886–88 to 1936–38, Japan, excluding Hokkaido, Eastern and Western Prefectures

Year	*Farm households (L)*[1]			*Yield, koku per tan (y)*[2]			*Weights (v)*[3]	
	Japan	*East*	*West*	*Japan*	*East*	*West*	*East*	*West*
1886–88	1.2	1.0	1.4	1.49	1.39	1.56	40.3	59.7
1896–98	1.1	1.0	1.2	1.46	1.35	1.53	39.3	60.7
1906–08	1.1	1.0	1.2	1.82	1.56	1.97	36.4	63.6
1916–18	1.1	0.9	1.2	1.91	1.81	1.98	40.3	59.7
1926–28	1.1	0.9	1.2	1.98	1.94	2.02	42.2	57.8
1936–38	1.1	0.9	1.2	2.18	2.12	2.22	43.1	56.9
				(in relation to Japan = 1.00)				
1886–88	1.00	.83	1.11	1.00	.93	1.05	.40	.60
1896–98	1.00	.88	1.08	1.00	.92	1.05	.39	.61
1906–08	1.00	.84	1.09	1.00	.86	1.08	.36	.64
1916–18	1.00	.84	1.11	1.00	.95	1.04	.40	.60
1926–28	1.00	.85	1.11	1.00	.98	1.02	.42	.58
1936–38	1.00	.83	1.12	1.00	.97	1.02	.43	.57

Notes:
1. Number of farm households per *cho* (= 2.45 acres) of cultivated area.
2. In *koku* (=6.37 cubic ft, = 150 kg for rice) per *tan* (= 0.245 acres).
3. Weighted by the share of prefectural rice output in the total for Japan.

Source: Prefectural tables in Kayo, N., (ed.) (1958), *Nihoh Nogyo Kiso Tokei* (Basic Statistics of Japanese Agriculture), pp. 608–52.

1926–28, when, during the next decade or so, the return inflows of labour into agriculture came to be reflected in the slight rise in the ratio of farm households per unit area cultivated both in the East and in the West.

In addition to the variations in weighted average yield of rice over time, the change in the weights of the two halves of Japan in rice output is interesting. The West's share increased by some 4 percentage points from 1886–88 to 1906–08 at the expense of the East's. However, since the first decade of the twentieth century the weight of the East has grown steadily. A more impressive regional contrast, nevertheless, was the rise in the East's rice yield, which after an initial decline started to climb once again and eventually caught up with those of the West's. This is nothing but what is known as the eastward movement of rice cultivation. The less-developed prefectures of eastern Japan started the process of catching up with and eventually even surpassing the traditionally advanced western prefectures. The technology of rice cultivation was transmitted to less favoured areas not only in the East's less-developed prefectures but in the more-developed prefectures of western Japan and, as a result, often invited a decline in average rice yield. As such, the eastward movement was by no means a simple straight-forward progression.[11]

As output per farm household is given by the product of acreage per household and yield per unit of land, one can decompose an increase in household productivity into a change in (reciprocal of) the number of farm households per hectare and that in yield per hectare. The former fraction changes as labour use varies due to labour intensity of output mix as well as to substitution of capital for labour by, say adoption of machinery. The latter fraction is rice yield per hectare in this case, although generally output per hectare (e.g., per annum) varies as yields of individual crops rise, cropping intensities increase, and output mix shifts from low to high value enterprises. Estimates of the productivity of different farm types can be weighted by their respective output shares to give the national average for each period. Changes in farm household productivity are then those of the weighted national averages between any two dates. In this framework, therefore, the changes in rice output per farm household we observe after the Meiji Restoration reflect combined effects of (i) technical innovations and changes in inputs that enhanced rice yield per hectare; (ii) technical innovations and changes in inputs that decreased labour requirement per rice hectare; and (iii) changes in the share of low-productivity farms versus high-productivity farms in total rice output between any two dates.

Thus, to analyse the effect of (i) regional shifts in the centre of gravity of Japan's rice production over time, represented by the changes in the weights of agricultural regions with differential productivity (V), (ii) uses of labour, represented by the number of farm households per hectare (L), and (iii) yields of rice per *tan* (Y), we generate a family of eight indices with changes in the period values of the variables V, L, and Y. Then, the independent effect of a factor is taken as the mean difference between the index where it appears in a period 2 value and those where it appears in a period 1 value. Similarly, the independent effect of a group of factors is obtained by linear combination of the indices. Thus, on the assumption that each variable, and each grouping of variables is a factor with measurable independent effect, we obtain estimates indicating the relative importance of each factor in question.[12]

Table 17.2 presents the estimates of the independent effects of the three key variables. The weighted national average output of rice per farm household increased by only 8 per cent between 1886–88 and 1896–98. A fall in yield of land contributed adversely to producing this moderate overall increase in rice output per farm. On the positive side, however, were the improvements in cultivated area per farm (though quite minute) and the gains in the relative weights of high-productivity prefectures. Of these, the contribution of the increase in cultivated area per farm household was the most important factor. Fully three-quarters of the change in rice output per farm was accounted for by this factor, outweighing the share contributed by the shift in the share of output from low-yielding prefectures to high-yielding ones. The estimates imply that during this period the gains in the national average rice output per farm in Japan were due primarily to the growth of cultivated area per farm and secondarily to the rising shares of output from western, more developed prefectures.

The characteristics of the sources of output growth per farm during this early decade or so are more sharply represented by the estimates in the lower panels of Table 17.2. During this period the Eastern Prefectures achieved only a meagre 1 per cent growth in output per farm, to which, by the order of significance, the gains in the expansion of cultivated area per farm and those in the share of the high-yielding prefectures (within Eastern Japan) contributed. These factors together managed to overcome a large negative impact of the

Table 17.2 Farm productivity as affected by land yields, farm household number, and interregional shifts in Japanese agriculture, 1886/88–1936/38

Year	*Productivity index in terminal year*[1]	*Changes attributable to*[2] *Change in weights*	*Change in farms*	*Change in yields*
	All Japan (excluding Hokkaido)			
1886/88–1896/98	108	27.2	76.2	–15.8
1896/98–1906/08	126	4.3	9.9	80.8
1906/08–1916/18	110	–2.7	27.7	67.7
1916/18–1926/28	105	13.3	12.0	69.7
1926/28–1936/38	109	10.2	–22.3	109.6
	Eastern Japan			
1886/88–1896/98	101	34.5	252.6	–204.8
1896/98–1906/08	119	9.7	21.5	59.8
1906/08–1916/18	121	5.2	11.6	77.7
1916/18–1926/28	109	8.0	6.4	83.3
1926/28–1936/38	109	15.6	–21.5	104.2
	Western Japan			
1886/88–1896/98	113	27.8	63.0	–3.2
1896/98–1906/08	130	0.2	3.8	94.5
1906/08–1916/18	104	–26.1	76.4	44.6
1916/18–1926/28	102	16.0	24.4	50.9
1926/28–1936/38	108	4.2	–24.3	116.9

Notes:
1. Productivity index in the initial year equals 100.
2. These may not sum to 100 per cent because the effects of groupings of factors are not included.

Sources: Method adapted from Yates, F. (1937) *The Design and Analysis of Factorial Experiments* (Harpenden, England). Data from Table 17.1.

---

fall in the weighted average rice output per hectare. In contrast, in the Western Prefectures the gains in rice output per farm household were much more significant at 13 per cent. The fall in yield per hectare had an almost insignificant (negative) effect on the farm household productivity increase. Almost two-thirds of the gains were due to the growth of cultivated area per farm. It is noted that the relative magnitude of the West's weights enabled the Western pattern of growth to dominate the Eastern one in affecting the overall growth pattern for Japan as a whole.

During the next ten years or so around the turn of the century rice output per farm household grew spectacularly, for Japan as a whole the increase was 26 per cent, for the Eastern Prefectures 19 per cent and for the West 30 per cent. In contrast to the case in the previous decade, of the 26 per cent growth in output per farm for the whole country, some four-fifths were attributable to the gains in yield per hectare. The contribution of the regional shift in weights was almost negligible this time; and the effect of the expansion of cultivated area per farm accounted only for a tenth of the growth of farm household productivity. Both in the East and the West each of the three independent factors contibuted positively to the growth in rice output per farm. In the East some 60 per cent of its total was accounted for by the effect of growth in yield per hectare, about 10 per cent by the greater weights of high-productivity prefectures, and some 20 per cent by the (minute) expansion in cultivated area per farm. In the West, where the productivity growth per farm was larger, the 25 per cent growth was almost fully attributable to the gains in yield per rice area. It is to be noted that in the period between 1896-98 and 1906-08 yield increasing innovations were solely responsible for the growth in rice output per farm in the West, while in the East the farm productivity growth was derived in addition from the (slightly) enlarged cultivated area per farm as well as the augmented share of high performance prefectures.

For Japan as a whole, gains in yield per hectare were unquestionably the primary factor contributing to growth in rice output per farm after the first decade of the century, from 1906-08 to 1936-38, in each of which the yield gains were responsible for more than two-thirds (and, in the last decade before the Second World War, more than the full increment) of the growth of productivity per farm household. Given the environment of the agriculture of Japan in which the scope was limited for rapidly expanding cultivated area per farm (or for reducing the number of farm households) and for further enhancing the relative output share of the high performance prefectures, growth in output per farm household had to rely primarily on the gains in yield per hectare. This is what the present empirical results show and consistent with the results of previous studies by Ohkawa, Hayami and Yamada, etc. These results demonstrate, at least since around the turn of the century, the crucial role that the land productivity augmentation, and by implication the so-called biological-chemical innovations, played in the growth of rice output per farm household in Japan.

The national averages mask significant regional differences, however, particularly for two decades or so after 1906–08. The rates of farm productivity growth differed markedly between the East and the West while the sources of growth deviated only somewhat between the two parts of Japan. Table 17.2 indicates that the yield factor ceased to be the dominant one in the Western Prefectures once the period of spectacular growth had ended. During the next ten years or so,when the per farm productivity growth fell to only 4 per cent in the West, by far the greatest sources of output growth were the rise in area per farm due to the fall in the number farm households (on account of the war-time upswing in non-agriculture). Moreover, during the period between 1916–18 and 1926–28, when the West's output growth per farm fell further to a bare 2 per cent (over the decade), the yield factor was responsible for only about a half of this growth. On the other hand, in the East where a large growth in farm productivity continued a decade after the magnificent period between 1896–98 and 1906–08, the yield factor continued to be predominant and accounted for more than three-quarters of the still large 21 per cent growth in rice output per farm. During the period between 1916–18 and 1926–28 the farm productivity growth in the East declined to a moderate gain of 9 per cent only. None the less, it is to be noted that the eastern part of Japan kept on outperforming the West during these decades. Although the share of the East was on the rise, the drastically poorer performance of the West exerted a drag on the overall national growth performance.

During these decades, the share of rice in the real farm value of agricultural output fell. Rice had occupied 56 per cent of the total in 1986–88, but the share dropped gradually to 43 per cent by 1936–38. In contrast, sericulture's share rose rapidly from 5 per cent in 1886–88 to 20 per cent in 1926–28, as did the share of livestock and poultry from about 1 per cent to 6 per cent. Farm enterprises such as sericulture were basically less land-dependent, but labour-intensive. In addition, sericulture was relatively remunerative while the prices were favourable. Cocoon prices had peaked in the early 1920s (with the index of 225, 1904–6=100) and then plummeted within a period of about ten years to the levels of 1904–06. However, the sericulture's share declined and that of crops other than rice came to rebound by 1936–38. Despite the rather drastic fluctuations in the relative importance of sericulture, agriculture's diversification progressed steadily throughout the period from 1886–88 to 1936–38.[13]

Of course, agriculture as a whole came to occupy an ever smaller share of Japan's net domestic product (in 1934–36 market prices). Starting with one-third share of the total in 1886–88, the contribution of agriculture decreased steadily to 25 per cent in 1906–08 and to 13 per cent in 1936–38. In contrast, manufacturing and mining which had started with the share of 7 per cent in 1886–88 came to claim some 30 per cent in 1936–38. Equally conspicuous were the changes in the farm households themselves. William Lockwood estimates that half of Japan's 5.5 million farm families had some non-agricultural employment in the 1930s and that for about one-fourth of these farm families the income from non-agricultural activities exceeded that derived from farming. Indeed, in Japan's modern economic growth, it was not large city-located industries that were mainly responsible for growth; it was 'the expansion of Japan's basic economy – agriculture and small-scale industry built on traditional foundations – which accounted for most of the growth of national productivity and income during this period' (Lockwood, 1954, pp. 25, 491). Small-scale rural activities, including para-agricultural activities as well as rural industries, not only utilised labour on the farm in slack seasons, but also channelled into use other farm resources such as family savings and local raw materials that otherwise would have remained idle.

6. INTERSECTORAL BALANCE IN CONCURRENT GROWTH

In Meiji Japan the population and the labour force were rising and, given the prevailing technology, agriculture had already reached a land frontier. Future agricultural growth depended in large measure upon more intensive cultivation. Moreover, the great majority of its labour force was engaged in agriculture (say, 80 per cent) where labour productivity and average labour income were markedly lower than in the non-agricultural sector.

In this paper I have examined the characteristics of intersectoral relationships in this economy. It is appropriate now to review explicitly the nature of its intersectoral problem. Had it been desirable to raise the low incomes of those in agriculture, how could this have been accomplished without making large demands on scarce resources and also minimising adverse consequences? Income equalisation could have led to lower aggregate output in the short run, say as a result of negative incentives arising from current income transfer, or as a consequence of shifting investment funds from industry with a higher rate of technological change and a higher rate of demand growth to agriculture with lower rates. Had these adverse results ensued, such short-run redistributive measures would clearly have been inefficient in a longer-run perspective of aggregate output growth. The dilemma between output growth and intersectoral equity in this case could have arisen, had the low productivity sector been favoured at the expense of the high productivity sector.

Suppose that an alternative had been to rapidly industrialise and raise the incomes of agricultural workers by employing more of them in the non-agricultural sector. In addition to the enormous demands that this would have made on the nation's scarce resources of capital, administrative skills and foreign exchange, the gains in income resulting from rising high-income sector employment would have raised current aggregate consumption at the expense of future growth.[14] The dilemma in this case, therefore, would have been between the impact of enlarged higher-income employment on the current distribution of consumption and on its distribution over time.

It appears that Meiji Japan could avoid these types of predicament by promoting satisfactory growth of the low productivity sector by relying mainly on those measures that made only small demands on the economy's scarce resources. Agriculture's 'technology shelves' were exploited to select (i) indigenous best-practice technology embodied in *rono* (veteran farmers) consisting mainly in improved farm practices, better farm management, and pre-Mendelian methods of plant selection, (ii) techniques of land-infrastructure improvements focused on changing land plots and small-scale irrigation and drainage projects, and (iii) development of scientific biological-chemical technology based on modern agronomy and soil science. More or less these became major sources of technical change in Meiji agricultural growth. Conspicuous in its absence, indeed, was (iv) modern mechanical-engineering technology concerned with the application of mechanical power to field operations, which came into prominence only after the 1950s. The Japanese choice impacted uniquely on her long-term patterns of demand and aggregate output growth. Beside restraining costs of output augmentation by enhancing the use of on-farm resources, the technological options selected were divisible and neutral to the operating scale of the farms. Institutional reforms of Meiji and newly-organised farmers' associations contributed substantially to facilitating communications among formerly isolated farmers. The expanded range of communication enhanced accessibility of farmers to new technology packages. The characteristics of these technological options thus determined the intrasectoral distribution of benefits within the agricultural sector itself. Presumption is yet strong that the agricultural stagnation came during the interwar years, given the traditional structure of small-scale household farming with limited supply of capital on the farm, as the relatively inexpensive sources of technical change had been exploited fully, and as the terms of trade and the income position of the farmers turned unfavourable when imports of colonial rice increased.

Notes

1. In contrast to this, the rates of growth of total employment in the manufacturing sector for all of Latin America were substantially less than half the growth rate of output in 1945–60 (Baer and Hervé, 1966).
2. The virtual constancy of real wage rates is taken as one of the essential characteristics of the initial phase of Japan's modern economic growth. Minami stated that 'our first proposition is that the real wages of unskilled labor in both farm and non-farm sectors were kept almost unchanged (or, if increased, very slightly) from the early years of Meiji until around mid-1910s (Minami, 1970). We ought, of course, to distinguish between wage rates and wage earnings in confronting income distribution questions. During the initial phase of Japan's modern development, despite the unchanged wage rates, workers' earnings increased considerably due to the fuller utilisation of idle labour. In farming activities technical progress of labour-intensive type allowed increasing labour input. The development of rural industries provided the labour force in the traditional farm households with more employment opportunities (Tussing, 1966). There is no overemphasising the importance of the developments in sericulture and tea cultivation, which utilised farm labour and which enabled the expansion of primary product exports. In particular, silk manufacturing (largely for the domestic market) and sericulture enterprises (for raw silk for export) on farms developed synergistic relations, impacting favourably upon each other in the prewar years.
3. The impact of the commercial organic fertiliser industry (soybean cake, rapeseed cake, fish meal, etc.) became appreciable across Japan towards the turn of century, and that of chemical fertilisers significant only after around 1920 (Hayami, 1967).
4. Vislie points out that there is no preference function from which such demand functions for all commodities can be derived. The budget constraint is $Y = P_1X_1 + P_2X_2 = w_1N_1 + w_2N_2$ where w_1 stands for the wage rate in sector i, and is incorporated into equation (2).
5. The data sources are as follows:
   (i) Ohkawa, K. *et al.* (eds) (1974) *Long-Term Economic Statistics*, vols. 1, 9 (Tokyo: Toyokeizai Shimpo-sha);

(ii) Bank of Japan, Statistics Deparment (1966) *Hundred-Year Statistics of the Japanese Economy* (Tokyo: Bank of Japan).

6. The data series are given as the 'Index of Per-Capita Annual Wage Earnings (1930=100), for Manual Workers and Office Workers'. I deflated the series with the deflator for Net Domestic Product (NDP) in manufacturing and mining. The source is in the text discussion of the estimation procedure in volume 1 of LTES (Ohkawa *et al.*, 1974, Table 8-7, p. 140).
7. μ_1 in this case is evaluated by the proportion in the total of NDP originating in non-agriculture (NDP total - NDP agriculture) at the quinquennial mid-point average for each period.
8. The primary source of data for the study of this section is a set of prefectural tables in *Nihon Nogyo Kiso Tokei* (The Basic Statistics of the Japanese Agriculture) (Kayo, 1958, pp. 608–52). The choice of 1886–88 as the starting set of years was dictated by the fact that those are the first set of years for which the data needed for empirical analysis in this section were available with reasonable sufficiency.
9. One *kọku* is a common unit of measuring volume in Japan. It is equivalent to 6.37 cubic ft and equals 150 kg in weight in the case of rice (Kayo, 1958, p. 654). One *tan* is one-tenth of a *cho* (a *cho* is 2.45 acres). The average yield of rice for all Japan was 1.49 *koku* (224 kg) per *tan* (0.245 acre) in 1886–88, in the East 1.39 *koku* and in the West 1.56 *koku* per *tan*.
10. The *Eastern Prefectures* include: From *Tohoku Region*, Aomori, Iwate, Miyagi, Akita, Yamagata and Fukushima; from *Kanto Region*, Ibaragi, Tochigi, Gumma, Saitama, Chiba, Tokyo and Kanagawa; from *Tozan* and *Tokai Regions*, Yamanashi, Nagano, Gifu, Shizuoka, Aichi and Mie; and, in addition, Niigata Prefecture from *Hokuriku Region*. The *Western Prefectures* include all the remaining prefectures in Japan, excluding the northernmost Hokkaido: From *Kinki Region*, Shiga, Kyoto, Osaka, Hyogo, Nara and Wakayama; from *Chugoku Region*, Tottori, Shimane, Okayama, Hiroshima and Yamaguchi; from *Shikoku Region*, Tokushima, Kagawa, Ehime and Kochi; from *Kyushu Region*, Fukuoka, Saga, Nagasaki, Kumamoto, Oita, Miyazaki and Kagoshima; and the remaining prefectures from *Hokuriku Region*, Toyama, Ishikawa and Fukui. In 1886–88 these two parts of Japan shared about an equal proportion of the total rice area, with a slight edge to the East (50.5 per cent). Due to the higher rice yield in the West, however, the weight in the total rice output was in favour of the West (59.7 per cent) instead of the East (40.3 per cent).
11. It is commonly accepted that during the early decades of the Meiji era, the most important contribution to agricultural growth was derived from the nation-wide dissemination of indigenous, *known* techniques and know-how, rather than the introduction or adoption of any *new* technique or implement. When the imported (Anglo-American) techniques based on horse-mechanisation failed and the research in the agricultural experiment stations (the agro-chemical methods of Germany relating to soil analysis and fertilisers) was in its infancy, it was the knowledge and techniques embodied in the veteran farmers which provided the basis for

the initial progress. Veteran farmers travelled throughout the country after 1868, often under government auspices, teaching improved methods of cultivation that were based initially on their own experiences rather than on scientific experimentation. Farmers' associations contributed substantially to facilitate communications between formerly isolated districts.

12. The method of analysis was adapted from Yates (1937), Parker and Klein (1966) and Kaneda (1980). The data were from Table 1.
13. Focusing on four product groups, i.e., rice, other crops, sericulture, and livestock and poultry the index of diversity was defined as: $D = 1 / \Sigma r^2$, where r is the share of each product group in the total value of farm output. The computed index was as follows: 2.28 for 1886–88, 2.51 for 1896–98, 2.62 for 1906–08, 2.91 for 1916–18, 3.15 for 1926–28, and 3.18 for 1936–38.
14. In an earlier paper that focused on food consumption patterns I argued that slow changes in food consumption patterns in Japan (and small food imports) materially 'contributed' to high rates of savings and to raising the demand for products of the domestic industrial sector (Kaneda, 1968).

References

Artle, R, Humes, C. jr. and Varaiya P. (1977) 'Division of Labor – Simon Revisited', *Regional Science and Urban Economics*, vol. 7, pp. 185–96.

Baer, W. and Hervé, M. E. A. (1966) 'Employment and Industrialization in developing Countries', *Quarterly Journal of Economics*, vol. 80, pp. 88–107.

Bank of Japan (1966) *Hundred-Year Statistics of the Japanese Economy* (Tokyo: Bank of Japan)

Baumol, W. J. (1967) 'Macroeconomics of Unbalanced Growth: The Anatomy of the Urban Crisis', *American Economic Review*, vol. 57 (June) pp. 185–96.

Hayami, Y. (1967) 'Innovations in the Fertilizer Industry and Agricultural Development: The Japanese Experience', *Journal of Farm Economics*, vol. 49 (May) pp. 403–12.

Hayami, Y. and Kikuchi, M. (1985) 'Agricultural Technology and Income Distribution: Two Indonesian Villages viewed from the Japanese Experience', in Ohkawa, K. and Ranis, G. (eds) *Japan and the Developing Countries: A Comparative Analysis* (Oxford: Basil Blackwell) pp. 91–109.

Kaneda, H. (1986) 'Long-term Changes in Food Consumption Patterns in Japan, 1878–1964', *Food Research Institute Studies in Agricultural Economics, Trade, and Development*, vol. 8, no. 1, pp. 1–32.

Kaneda, H. (1980) 'Structural Change and Policy Response in Japanese Agriculture After the Land Reform', *Economic Development and Cultural Change*, vol. 28, no. 3 (April) pp. 469–86.

Kayo, N. (ed) (1958) *Nihon Nogyo Kiso Tokei* (The Basic Statistics of the Japanese Agriculture) (Tokyo: Norinsuisangyo Seisansei Kojo Kaigi) pp. 608–52.

Lockwood, W. W. (1954) *The Economic Development of Japan: Growth and Structural change 1868–1935* (Princeton: Princeton University Press).

Minami, R. (1970) *Nihon Keizai no Tenkanten* (The Turning Point in the Japanese Economy) (Tokyo: Sobunsha).

Ohkawa, K. (1964) 'Concurrent Growth of Agriculture with Industry: A study of the Japanese Case', in Dixey R. N. (ed.) *International Exploration of Agricultural Economics* (Ames: Iowa State University Press) pp. 201–12.

Ohkawa, K. *et al.* (eds) (1974) *Estimates of Long-Term Economic Statistics of Japan since 1878, Vol. 1, National Income* (Tokyo: Toyokeizai Shimposha).

Ono, A. and Watanabe, T. (1976) 'Changes in Income Inequality in the Japanese Economy', in Patrick, H. (ed.) *Japanese Industrialization and its Social Consequences* (Berkeley: University of California Press).

Parker, W. N. and Klein J. L. V. (1966) 'Productivity Growth in Grain Production in the United States, 1840–60 and 1900–1910', in National Bureau of Economic Research, *Output, Employment, and Productivity in the United States after 1800* (New York: Columbia University Press) pp. 523–80.

Ranis, G. and Fei, J. C. H. (1963) 'Innovation, Capital Accumulation, and Economic Development', *American Economic Review*, vol. 53 (June) pp. 283–313.

Simon, H. (1947) 'Effects of Increased Productivity upon the Ratio of Urban to Rural Population', *Econometrica*, vol. 15 (January) pp. 31–42.

Tussing, A. R. (1966) 'The Labor Force in Meiji Economic Growth: A Quantitative Study of Yamanashi Prefecture', *Journal of Economic History*, vol. 26 (March) pp. 59–92.

Vislie, J. (1979) 'Division of Labor – Simon Revisited: A Comment', *Regional Science and Urban Economics*, vol. 9, pp. 61–70.

Yamada, S. (1982) 'Labour Absorption in Japanese Agriculture: A Statistical Examination', in Ishikawa, S., Yamada, S. and Hirashima, S. *Labor Absorption and Growth in Agriculture, China, and Japan* (Bangkok: International Labour Office, Asian Employment Programme) pp. 27–58.

Yates, F. (1937) *The Design and Analysis of Factorial Experiments* (Harpenden: Commonwealth Bureau of Soil Science) Technical Communications no. 35.

18 Productivity Change and Growth in Industry and Agriculture: An International Comparison*

Mieko Nishimizu and John M. Page, Jr
THE WORLD BANK**

1. INTRODUCTION

On the supply side, the growth of an economy, an industry, or a firm is determined by the rate of expansion of its productive resources and by improvements in their efficiency; that is, by total factor productivity (TFP) growth. In the short run, where there are limits to how fast employable resources can grow, achieving high rates of productivity change offers an important avenue for accelerating economic growth under resource constraints, while in the medium to long term, differential rates of productivity growth among economic activities have a major impact on resource allocation and structural change in an economy.

Analyses of structural transformation during the process of economic development have tended to emphasise the contributions to growth of resource shifts out of traditional sectors with low productivity to modern sectors of the economy with high productivity. Conventionally, the low productivity sectors were associated with industry. Development strategies tended, therefore, to emphasise implicit or explicit taxation of the agricultural sector to provide a basis for more rapid expansion of industrial output. Trade, exchange rate, and agricultural pricing policies in many countries combined to subsidise import substitution industries at the expense of agricultural producers. The whole industrial sector was treated as an infant industry, and relatively little attention was given to the costs of generalised protection. Policy-makers accepted the view that although the initial costs of production in many industries would exceed the cost of competing imports, industrial productivity change would ensure that the differential remained modest and that its duration was not excessive.

The introduction of new seed-fertiliser technologies created the potential for increased agricultural productivity in a relatively wide range of foodgrains, and established a basis for rapid expansion of agricultural output and incomes in many developing economies. This 'seed-fertiliser revolution' prompted renewed consideration of more balanced approaches to economic growth which emphasised the complementarity between agricultural and industrial development. At the same time, in the middle-income countries the size and degree of differentiation of the industrial sector made it difficult to apply the infant industry argument to all industrial production, and the resource costs of subsidising the industrial sector had become excessive. Thus, export-led and agriculturally-based development strategies were advocated as alternatives to the import substitution strategies of the 1960s and 1970s.

There are three questions concerning productivity growth that are relevant for assessing possible development strategies. First, what is the range of productivity growth rates one can expect? Secondly, do patterns of productivity change differ among countries and between industry and agriculture. And thirdly, are there any identifiable relationships between agricultural and industrial output growth and productivity growth. This paper addresses these questions by examining the historical experience at a sectoral level of a wide range of countries.

2. THE ANALYTICAL FRAMEWORK OF TFP MEASUREMENT

There are essentially two methodological approaches to TFP measurement, which are complementary to each other. One approach is to estimate TFP by econometric estimation of production or cost functions. Advances in this econometric approach benefited from application of mathematical functional forms which are less restrictive *a priori* in specifying economic relationships than Cobb-Douglas or CES forms.[1] The other methodological approach is to measure TFP in terms of index numbers. The basic measurement problem of the index number approach centres on how to aggregate various outputs and inputs in defining appropriate quantity and price indices of output, input, and TFP.[2]

Indices of TFP change are usually given in terms of output per unit of total factor inputs, and are functions of scale elasticities, output and input elasticities, and quantities (or prices) of outputs and inputs. It is usually assumed that product and factor markets are competitive, and that firms maximise profits (or minimise costs) subject to a constant returns to scale production function and market prices which are taken as parameters. Under these assumptions, input and output elasticities are equivalent to the observed cost shares of each input used and revenue shares of each output produced. The index of TFP change can then be computed as the difference between the revenue share weighted output growth rate (at constant prices) and the cost share weighted real input growth rates.

3. TFP GROWTH IN INDUSTRY AND AGRICULTURE – AN INTERNATIONAL COMPARISON

Over the years, the empirical productivity literature has accumulated a substantial body of information at the macro level about rates of TFP growth and about the relative contribution of TFP and factor input growth to economic growth in a number of countries. Recently Chenery (1986) pulled together existing macroeconomic estimates of TFP growth rates (on a value added basis) for thirty economies, eighteen developed and twelve developing countries.[3] He found that there are three groups of these countries, distinguished by the mean rates of TFP growth and by the degree of contribution to economic growth made by TFP growth.

The first group identified by Chenery is developed countries other than Japan. This group is characterised by relatively low factor (capital and labour) growth and TFP growth, but with the latter accounting for about a half or more of GDP growth. The second group is referred to as 'typical developing economies' by Chenery. This group has somewhat faster growth rates in both factor inputs and TFP compared with the first, but with TFP growth accounting for less than 20 per cent of GDP growth. The third group consists of Hong Kong, Israel, Korea, Spain, Taiwan and Japan. They are distinguished by the most rapid rates of factor input, TFP and GDP growth of all countries examined, but the contribution of TFP to GDP growth falls somewhere between the first and the second group.

Decomposition by Chenery of the sources of growth by sectors for a country in the per capita income range of $560–1120 indicates that TFP growth accounts for approximately 24 per cent of output growth in agriculture, 31 per cent in light manufacturing, and 25 per cent in heavy manufacturing. Average rates of productivity growth are 0.86, 2.11 and 2.19 per cent per annum for agriculture, light manufacturing and heavy manufacturing, respectively.

There is now a large enough collection of estimates of TFP growth to allow us to begin looking for similar 'stylisable' facts at the sectoral and subsectoral level for industry and agriculture. Our industrial sector data base consists of the eighteen countries listed in Table 18.1.[4] The estimation period differs from country to country, but the measurement methodologies are reasonably comparable among them. The coverage of industrial activities, and the level of disaggregation, also differ among countries, but we were able to maintain twenty-one industries (roughly at the ISIC two-digit level) as the common denominator among the different estimates. The basic data on industrial output growth, TFP growth and the industrial classification are provided in the Appendix.

Estimates of aggregate total factor productivity growth for the agricultural sector are not as widely available as those for industry. A recent study by Kawagoe and Hayami (1985), however, provides sufficient information to calculate average annual rates of productivity change between 1960 and 1980 for the agricultural sector as a whole for thirteen of the countries for which comparable TFP series exist in industry.

In estimating TFP growth rates, capital (including land in agriculture), labour, and intermediate inputs are included in the underlying production function. Therefore, output is defined as gross output – the total value of production at producer's price – as opposed to value added. When intermediate input is recognised explicitly, it implies that the estimated TFP growth rates are lower than those based on value added and the capital-labour raltionship are lower by exactly an amount proportional to the share of intermediate input in gross output.[5]

The size of our industry data base – TFP estimates of twenty-one industries for each of the eighteen countries – makes it difficult to summarise it neatly. For the sample of industrial TFP growth rates, pooling countries and industries together, the (unweighted) mean annual growth rate is about 0.7 per cent per annum and the standard deviation is about 1.5 per cent. Looking across different countries,

Table 18.1 Countries in the industry and agriculture productivity data bases

Country	*Income per capita*	*Industry period*	*Agriculture period*
Sweden	14 870	1963–80	1960–80
USA	12 820	1960–79	1960–80
Finland	10 679	1960–80	1960–80
Japan	10 085	1960–79	1960–80
Yugoslavia	2 794	1965–78	1960–80
Argentina	2 560	1956–73	1960–80
Chile	2 559	1960–81	1960–80
Portugal	2 518	1973–78	1960–80
Mexico	2 250	1970–80	1960–80
Hungary	2 100	1976–83	
Korea	1 700	1960–77	
Turkey	1 542	1963–76	1960–80
Philippines	787	1956–80	1960–80
Thailand	679	1963–77	
Egypt	651	1973–79	1960–80
Zambia	597	1965–80	
Indonesia	524	1975–82	
India	256	1959–79	1960–80

Notes: Per capita income figures in 1981 US$ from World Bank (1985) *World Development Report*.

---

there is one pattern that is worth noting at this point because it raises some interesting questions. The distribution of TFP growth rates appears to become more dispersed as per capita income declines. There are in fact significant rank correlations (by Kendall's tau) between the standard deviation (significant at 95 per cent level) and the coefficient of variation (significant at the 90 per cent level) of the TFP growth distributions and the per capita income level of different countries. There is no correlation between the mean TFP growth rate and the per capita income level.

The relationship between the variance of TFP growth rates and the level of incomes emerges quite dramatically in a contrast of the four most developed countries (Sweden, USA, Finland, and Japan) with all others. Theoretically, long-run equilibrium in a market economy should allow no differences in TFP growth rates (or levels) among different industries. Because high rates of TFP growth permit more rapid increases in the ability of producers to compensate factors, resources should be pulled towards high productivity growth industries, and this resource shift should continue until differentials in rates of productivity growth are eliminated.

The observed pattern of increasing variance in TFP growth rates with declining per capita incomes may, therefore, reflect impediments to the mobility of factors in lower-income countries, arising from structural deficiencies in factor markets. It is of interest, however, that there are four countries among the ranks of developing economies in our sample that show distinctly smaller variations among industrial TFP growth rates than other developing countries. They are Yugoslavia, Mexico, Hungary, and India. These countries, as a group, also have relatively low rates of average TFP growth in comparison to other developing countries.

Figure 18.1 depicts the mean TFP growth rates for each of four manufacturing subsectors – consumer goods, light intermediate goods, heavy intermediate goods, and capital goods – and for agriculture (for the subset of thirteen countries for which comparable TFP estimates exist).[6] The heavy intermediate goods and capital goods subsectors, particularly capital goods, tend to have higher mean TFP growth rates compared with the other two industrial subsectors. Nishimizu and Robinson (1984) found a similar pattern in their comparison of industrial TFP growth among Japan, Korea, Turkey, and Yugoslavia, and Chenery's (1986) archetype middle-income economy has marginally higher rates of TFP growth in heavy than in light industry.

Agricultural TFP growth rates provide an interesting contrast with those in industry. There is considerably less variation in TFP growth rates across countries for agriculture than for industry. The agriculture rates lie in a band of approximately -0.5 to 2.5 per cent per annum. Agriculture is the leading sector in terms of productivity change in three of the four high-income countries in the sample – Sweden, USA, Finland – with rates of TFP change in the vicinity of 2 per cent per annum. In Japan, which is the fourth high-income country, agriculture ranks second only to capital goods with a rate of TFP change of approximately 1 per cent per annum. Japan, however, began the period 1960–80 with many of the characteristics of a middle-income country. Agricultural TFP growth rates are negative in three middle- and low-income countries – Turkey, Egypt, and India – and are substantially below those for the high-income countries in all of the remaining developing countries except Yugoslavia, Chile and the Philippines. This finding is consistent with the tendency

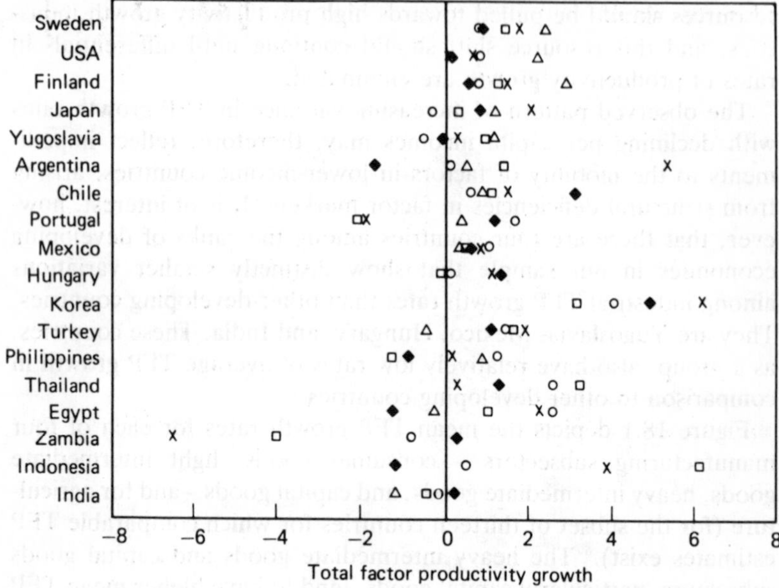

Figure 18.1 Total factor productivity growth

reported by Kawagoe and Hayami (1985) for agricultural productivity levels and growth rates to increase with per capita income, and for a widening gap to emerge between developed and developing countries during the two decades between 1960 and 1980.

There is a sharp shift in the relative position of industrial and agricultural TFP growth rates with the level of economic development. Among the high-income countries contained in the sample, rates of agricultural productivity change are high relative to those in industry, while in the majority of middle- and low-income countries rates of TFP change in agriculture tend to lie in the middle of the range of industrial TFP growth rates. India, Egypt and Turkey are particularly striking for the differential which exists between the rate

Table 18.2 Output and TFP growth in industry and agriculture

	Intercept	*LDC*	*Output growth*	*R*
ATFP	1.568***	−2.165***	.293*	.377
	(.416)	(.449)	(.164)	
CTFP	−.591	−.024	.187***	.281
	(.443)	(.497)	(.031)	
LTFP	−.378	−.621	.189***	.231
	(.599)	(.605)	(.046)	
HTFP	−.831	−.384	.175***	.251
	(1.533)	(.320)	(.032)	
KTFP	−.287	−1.220	.226***	.282
	(.742)	(.775)	(.044)	

Note: Dependent variables are TFP growth rates for:

ATFP = Agriculture
CTFP = Consumer goods
LTFP = Light intermediates
HTFP = Heavy intermediates
KTFP = Capital goods
Significance Levels
*** 99%
** 95%
* 90%

---

of TFP change in agriculture and the rates for industry. In each of these three cases industrial productivity change substantially exceeds productivity change in agriculture.

The regressions presented in Table 18.2 provide a test for differentials in the rate of TFP growth between developed and developing countries and of the Verdoorn-Kaldor relationship between output growth and TFP growth.[7] In agriculture a significant difference exists between the productivity performance of developed and developing countries, controlling for output growth.[8] TFP growth in agriculture is, on average, 2 per cent greater for developed than developing countries. There is no significant difference in TFP growth rates between developed and developing countries in any of the industrial sector groupings. The relationship between output growth and TFP growth is positive and statistically significant (at the .01 level) in all of the regressions. However, in contrast to Kaldor's (1967) argument that the association between output expansion and TFP growth is observed most prominently in manufacturing and other secondary activities, the magnitude of the coefficient on output expansion is greatest for the agricultural sector.

The strong association between TFP growth and output growth observed in agriculture may arise from increases in output made possible by the introduction of new fertiliser, responsive foodgrain varieties, and other technological improvements. In contrast, Kaldor's hypotheses regarding the relationship between output growth and productivity growth in industry centre on the role of economies of scale in generating measured TFP growth.[9] This view is given some support by the relatively high magnitude of the coefficient on output growth in capital goods manufacturing, in which there is some engineering evidence of economies of scale. Across consumer goods, light and heavy intermediates, however, there is no evidence of significant differences in the coefficient on output expansion, despite quite different production functions.

The relative rates of productivity change in industry and agriculture between developed and developing countries provide some insight into changing patterns of comparative advantage. Nishimizu and Page (1986) have recently demonstrated the role of relative rates of productivity change in determining dynamic comparative advantage. The implications of the relative patterns of productivity change in light of their work are that, while on average LDCs have been able to maintain their levels of comparative advantage in industry, they have been losing ground to the industrialised countries in terms of their comparative advantage in agriculture.

Figures 18.2 and 18.3 summarise our data base differently. TFP growth rates and total factor input growth rates always sum to output growth rates. These figures show the relationship between TFP growth and total factor input growth, in a graph where 45-degree lines of constant output growth are drawn in. Figure 18.2 plots mean rates for the industrial sector and the aggregate rate for the agricultural sector of each country, and Figure 18.3 shows the mean 'global' rates for each industry and for agriculture.

The efficiency performance of a country or an industry cannot be judged by simply examining the size of TFP growth rates. Rapid TFP growth may be contributing very little to faster output growth, or slow TFP growth may be more than doubling the growth potential of a slower rate of input growth. The share of the TFP growth rate in output growth is a useful indicator of efficiency that normalises, in a sense, for differences in the rates of economic growth. In Figures 18.2

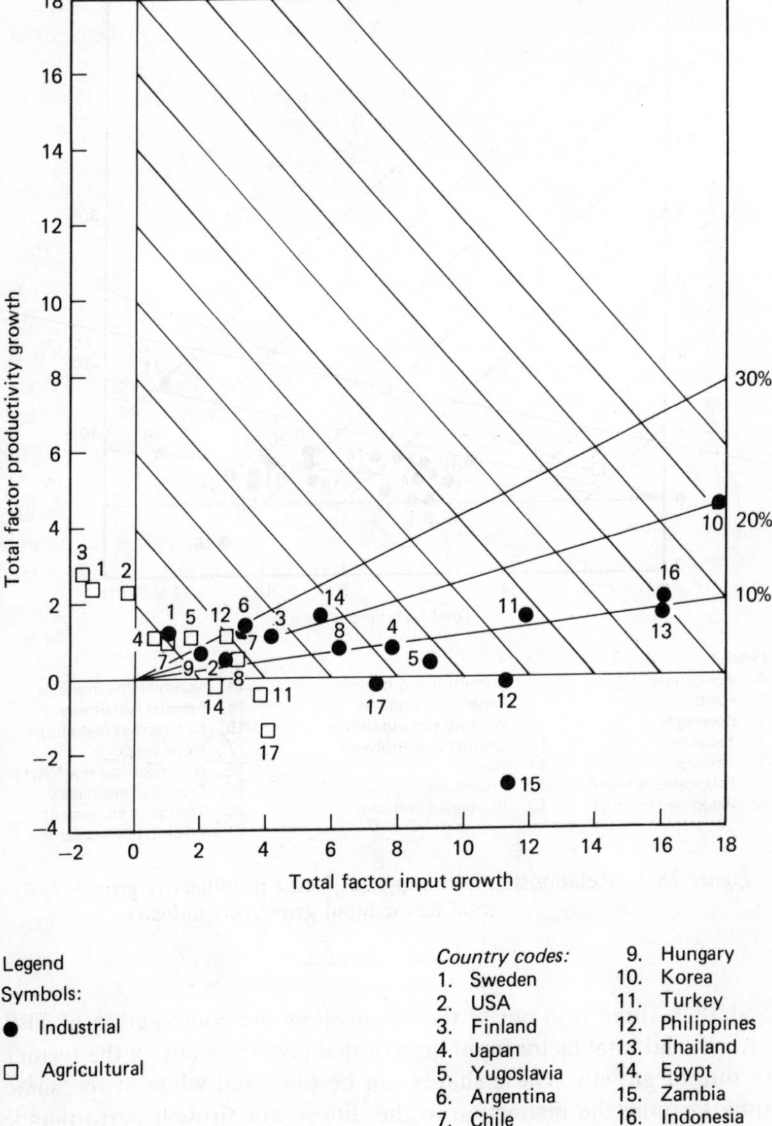

Figure 18.2 Relationship between total factor productivity growth and total factor input growth by country

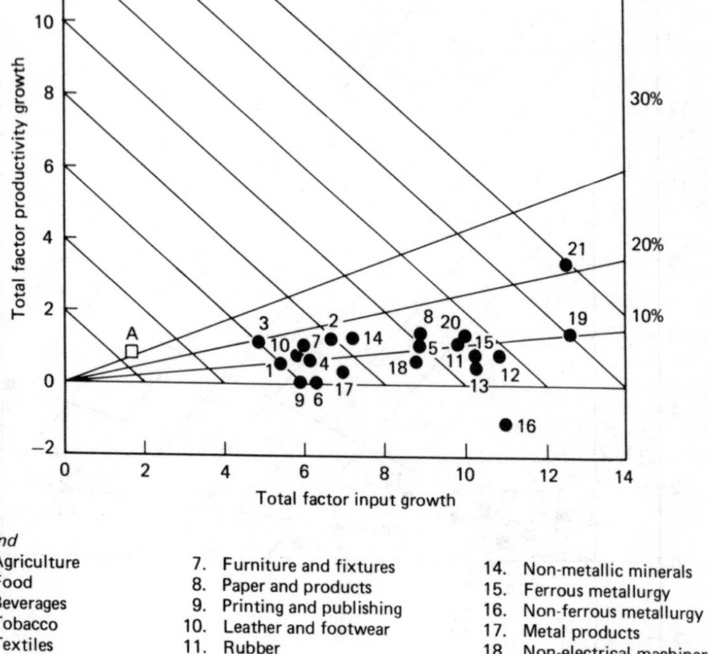

Legend

A. Agriculture
1. Food
2. Beverages
3. Tobacco
4. Textiles
5. Fabricated textiles
6. Wood and products
7. Furniture and fixtures
8. Paper and products
9. Printing and publishing
10. Leather and footwear
11. Rubber
12. Chemicals
13. Petroleum and coal
14. Non-metallic minerals
15. Ferrous metallurgy
16. Non-ferrous metallurgy
17. Metal products
18. Non-electrical machinery
19. Electrical machinery
20. Transport equipment
21. Precision instruments

Figure 18.3 Relationship between total factor productivity growth and total factor input growth by industry

---

and 18.3, three rays out of the origin show the combination of TFP growth and total factor input growth that give the share of the former in output growth. These shares can be observed while at the same time keeping the distinction in the differential growth performance among industries or countries.

Figure 18.2 and 18.3 show a sharp contrast between industry and agriculture in terms of both output growth and productivity change. The country observations for agriculture in Figure 18.2 fall in an output growth range of from 1 to 3 per cent per annum, which is low in comparison with the output growth rates for the industrial sector. The contrast is even more sharply drawn in Figure 18.3; global output growth rates for industrial activities in general lie above 6 per cent per annum, while the global rate of output growth in agriculture is less than 3 per cent. The percentage contribution of TFP to output growth, however, is substantially higher in agriculture than in industry. In general, therefore, output increases have been achieved at a lower cost in terms of factor inputs in agriculture than in the industral sector. However, in Argentina and in the subset of countries with negative productivity growth in agriculture (Turkey, Egypt, and India), the agricultural development process has made less efficient use of resources than industrial growth. Egypt is particularly striking in the sense that it is among the more efficient developing countries in terms of its industrial growth while its performance in agriculture is among the worst.

For the industrial sector the mean rates fall into the interval between a 10 and 30 per cent share of TFP growth in output growth. In Figure 18.2, Sweden is the only country with the share greater than 30 per cent. Yugoslavia's share is below 10 per cent, and in the industrial sector of India, Philippines, and Zambia, negative mean TFP growth has actually reduced output growth below the growth in total factor input. Industries in Figure 18.3 with a share below 10 per cent are: wood products, printing and publishing, chemical products, non-ferrous metals, metal products, and non-electrical machinery.

Three broad groups of economies can be identified from our industrial productivity data. The first group – developed and middle - income countries – is characterised by relatively low factor (capital and labour) growth and TFP growth, but with the latter accounting for a healthy share of output growth. In our data base this group includes the USA, Sweden, Finland, two Latin American countries (Chile and Argentina) and Hungary. The second group – 'typical developing economies' – has somewhat faster growth rates in both factor inputs and TFP compared with the first, but with TFP growth accounting for much smaller shares of output growth. The third group is those countries distinguished by rapid rates of factor input, TFP and GDP growth but with the contibution of TFP to GDP growth falling somewhere between the first and the second group. In our data base, this group includes Japan, Korea, Indonesia and Thailand.

A pattern of agricultural growth distinguished by levels of income also emerges from the data. The developed countries, including

Japan, are all characterised by relative low rates of output growth, low or negative factor input growth, and a high share of TFP growth in the growth of total output (in excess of 50 per cent). Productivity change, arising primarily from input reducing innovations, is clearly driving the process of agricultural output growth in these economies. Among middle-income countries, only Chile demonstrates a similar, though less marked, tendency for agricultural development to proceed largely on the basis of productivity change. The main cluster of middle- and low-income countries is characterised by higher rates of agricultural output growth (in the range of 3.4 per cent per annum) obtained at the cost of more rapid rates of increase in total factor inputs. The percentage contribution of TFP to output growth is correspondingly lower in the less developed countries, but remains in the range of 20-40 per cent. The subset of countries with strongly negative contributions of TFP growth – India, Turkey, Egypt – represent cases in which output growth is lagging behind the growth of inputs. In the cases of Egypt and India this arises largely from continuing increases in the agricultural labour force without compensating increases in agricultural output.

4. AGRICULTURAL GROWTH AND PRODUCTIVITY PERFORMANCE IN INDUSTRY – SOME PRELIMINARY RESULTS

The relationship between agricultural growth and industrial development has been discussed extensively during recent years. The view of proponents of an agriculturally led growth strategy for low income countries is perhaps most succinctly summarised by Mellor (1976):

> Increased agricultural production, based on cost-decreasing technological change, can make large net additions to national income and place that income in the hands of the cultivator classes, who tend to spend a substantial proportion of it on non-agricultural commodities. Agriculture may provide a demand drive for development similar to that often depicted for foreign markets in export led growth.

One obvious linkage between agricultural expansion and productivity performance arises from the Kaldor-Verdoorn principle of a positive relationship between output growth and productivity change, discussed in the preceding section. In developing countries scale economies and size of the market have long been viewed as important determinants of growth and structural change. The existence of economies of scale, or any other justification for the Kaldor-Verdoorn hypothesis, implies that widening of the market should lead to reductions in real production costs.

The results in section 3 tend to support the view that total factor productivity growth in agriculture is sufficiently important as a component of output growth to provide a rationale for such an agriculturally-led strategy. A significant question, however, is to what extent the effects of a strategy based on agriculturally-driven domestic demand expansion might differ, in terms of its effects on productivity change in the industrial sector, from a strategy of export-led industrial growth. Proponents of export-led or 'outward-oriented' strategies frequently emphasise the productivity benefits of competition in external markets in addition to the set of arguments based on expanded market size.[10]

It is important to point out, however, that an agriculturally-led strategy and an industrial export-led growth strategy are not mutually incompatible. Indeed, neutrality of the incentive regime between import substitutes and exports will generally benefit agriculture as well as industrial exports, since import substitution regimes frequently discriminate against both agriculture and exports.

Our data base provides some preliminary indications of the impact of agricultural output growth on industrial productivity change. Table 18.3 presents the results of an initial effort to explore the relationship between components of demand growth which are taken as exogenous (or determined by exogenously determined policy regimes), a measure of the instruments by which the domestic economy is protected from import competition, and a variable representing the absence of market-based allocation of resources, and the rate of TFP growth.

The single equation model estimated is of the form:

$$\begin{aligned} TFPG = a_0 + B_1\, QR + B_2\, NM + B_3\, EX + B_4\, EX^{*}QR + \\ B_5\, EX^{*}NM + B_6\,(M\text{–}Q) + B_7\,(M\text{–}Q)^{*}QR + \\ B_8\,(M\text{–}Q)^{*}NM + B_9\, A + B_{10}\, A^{*}QR + B_{11}\, A^{*}NM + e, \end{aligned}$$

where *TFPG*, the dependent variable is the sector specific rate of TFP change in each industrial activity. *EX*, *M*, *Q*, and *A* are the rates of export, import, output and agricultural growth, respectively. The

Table 18.3 Agricultural growth and productivity change in industry: regression results

TFPG	*Before 1973*	*After 1973*	*Entire period*
a_0	–1.092	–0.202	0.260
	(0.961)	(0.572)	(.428)
QR	1.959*	1.170	0.891
	(1.020)	(1.361)	(.609)
NMKT	0.633	0.537	0.165
	(1.916)	(1.497)	(1.225)
X	0.043***	–0.004	0.006
	(.008)	(.010)	(.006)
*X*QR*	–0.040***	0.000	–0.009
	(.009)	(.013)	(.008)
*X*NMKT*	–0.026	–0.041	–0.040
	(.058)	(.061)	(.051)
(*M–Q*)	0.029	–0.090**	–0.069**
	(.031)	(.031)	(.022)
(*M–Q*)**QR*	–0.000	0.048	0.041
	(.037)	(.035)	(.025)
(*M–Q*)**NMKT*	–0.050	0.079	0.041
	(.046)	(.062)	(.045)
A	0.813**	0.639***	0.560***
	(.327)	(.170)	(.134)
*A*QR*	–0.892***	–0.636**	–0.626***
	(.339)	(.272)	(.158)
*A*NMKT*	–0.751*	–0.535	–0.539**
	(.426)	(.343)	(.255)
R^2	0.291	0.179	0.147
	(2.21)	(3.86)	(3.36)
F-statistic	6.69	4.32	6.10
Number of observations	191	229	400

Notes: Variables are defined as follows:

TFPG = Total Factor Productivity Growth
Q = Output growth
X = Export growth
M = Import growth
A = Agricultural growth
QR = Dummy = 1 for countries with substantial quantitative restrictions
NMKT = Dummy = 1 for non-market economics

Significance levels
*** 99%
** 95%
* 90%

variables *QR* and *NM* are dummy variables, taking a value of one for regimes in which quantity controls of imports and non-market allocation of resources predominate. Because all of the non-market economies in our sample also use quantitative control of imports the variables are defined to be mutually exclusive. A random disturbance, *e*, is added, assumed to be distributed normally with zero mean.

Under these assumptions an OLS regression is applied to the pooled time-series, cross-section of annual observations of TFP growth by industry and country in our data base. The model is estimated for the whole time period and two important sub-periods. The first is the period up to and including 1973, prior to the first oil shock. The second period is that following 1973. Broadly speaking the first period can be thought of as an era of steady state growth in industrial output and trade for the world economy, while the period 1974–83 is one of short-term adjustment to a series of shocks deriving primarily from wide fluctuations in energy prices. The regressions explain between 15 and 30 per cent of the variance in TFP growth rates, and omnibus *F* statistics are significant in all cases.

The results presented in Table 18.3 tend to support the view of a strong complementarity between outward orientation and agriculture-led growth. For economies in which the trade regime is not predominantly subject to quantitative restrictions, the impact of agricultural output expansion of productivity change in industry is positive. Since output growth is determined in part by intermediate and final demand growth arising in agriculture, this is another manifestation of the Verdoorn-Kaldor principle.

The nature of the interaction term between the trade regime and agricultural growth is quite striking, however. Unlike the case for more open economies, agricultural growth appears to have a negative impact on the productivity performance of industry in economies with quantitative trade restrictions. This presumably reflects the ability of domestic producers of industrial products to pass forward cost increases unchecked by import competition during periods of rapid domestic demand expansion. Similar results obtain for non-market-oriented economies.

Results of analysis of the impact of trade related variables on productivity change are mixed. In the period up to 1974 export growth is positively correlated with TFP growth, but in the post-oil-shock adjustment period, and for the whole period, there is apparently no association between export performance and TFP growth.

The variable which represents the difference between the rates of growth of imports and output is a measure of import penetration. The results of the regressions point to quite different relationships between import penetration and TFP performance, depending on the character of the trade regime. No significant relationship exists between import penetration and TFP performance in the period prior to 1974. In the post-oil-shock adjustment, however, open trading economies have a significant negative association between increased rates of import penetration and TFP growth. In economies protected by quotas, however, this result is strongly attenuated, presumably indicating the offsetting effect of increased competition.

In sum our analysis suggests that for the period 1960–80 final demand linkages based on agricultural expansion were a more important source of industrial productivity change *in economies open to external competition* than export growth. During the expansionary phase of world trade prior to 1974, both export growth and agricultural growth were positively correlated with industrial productivity change. In the period after 1974 only agricultural expansion was so correlated. In economies closed to import competition through the use of quantitative import restrictions, however, the relationship between agricultural output expansion and productivity change is reversed. Agricultural growth in such economies has either no relationship or a negative relationship with productivity change. Apparently in these economies the Verdoorn-Kaldor effect of expanding final demand is offset by the relaxation of cost discipline implied by an expanding domestic market and limited foreign competition.

5. CONCLUDING REMARKS.

The results which we have presented tend to raise as many questions as they answer. Broadly, however, they suggest the following generalisations. First, productivity growth is not an exogenously determined variable which proceeds independently of stage of development, phase of the international economic cycle, or choice of policy regime. Secondly, the character of productivity change in agriculture appears to be distinct from that in industry. In particular, TFP growth appears to contribute relatively more to output growth in agriculture than in industry. Thirdly, competition matters in determining productivity performance. The capability to export and the need to maintain a linkage between international and domestic price movements appear crucial in this regard. Finally, there appears to be a significant association between the effect of agriculturally-driven demand expansion of productivity performance in industry and the character of the trade regime. This suggests a strong complementarity between outward-oriented and agriculturally-led growth strategies, and suggests an interdependence of policies in agriculture and external commerce which is somewhat distinct from, but consistent with, the traditional concerns with the allocative effects of interdependence between commercial and agricultural policies.

Appendix Table 18A.1 Gross output growth, aggregate period, by country and sector

Sector / Country	Sweden	USA	Finland	Japan	Yugoslavia	Argentina	Chile	Mexico	Hungary	Korea	Turkey	Philippines	Thailand	Egypt	Zambia	Indonesia	India	Sector average	
Food	1.66	2.81	4.04	5.89	7.20	2.73	2.78	5.11	3.20	16.09	8.47	6.77	9.09	11.41	4.87	5.82	3.70	5.98	(3.69)
Beverages	1.66	2.81	4.04	5.89	7.20	2.73	6.20	7.01	5.40	16.09	8.47	7.29	24.15	11.77	6.80	7.50	10.10	7.95	(5.44)
Tobacco	1.66	0.72	4.04	5.89	5.74	2.54	3.76	2.17	3.80	16.09	8.47	11.19	–	11.77	6.80	10.11	2.10	6.05	(4.38)
Textiles	–2.17	3.45	3.39	3.44	9.77	2.45	1.08	4.84	0.93	18.88	9.47	11.22	18.22	0.48	15.15	11.15	4.40	6.83	(6.38)
Fabricated textiles	–2.17	2.64	3.39	10.15	9.77	2.45	2.71	4.84	2.13	23.34	18.30	6.16	27.75	3.24	8.83	41.24	4.40	9.95	(11.39)
Wood & products	3.20	2.88	3.23	5.78	10.85	1.76	–	6.47	–2.03	16.32	7.35	5.98	4.46	–1.89	3.81	29.16	4.30	6.35	(7.51)
Furniture & fixtures	3.20	3.73	3.23	8.67	10.85	1.76	9.45	–	3.10	13.49	12.37	8.31	12.06	–	3.76	4.31	7.50	7.05	(3.98)
Paper & products	3.36	4.01	4.86	7.61	10.78	–	4.16	7.19	3.70	19.41	13.53	12.63	26.71	9.30	17.33	11.63	8.90	10.32	(6.52)
Printing & publishing	1.17	3.23	4.12	6.56	7.71	–	3.32	–	6.50	–	–	7.73	18.07	–	2.04	6.56	4.70	5.98	(4.39)
Leather & footwear	–2.17	–0.47	3.39	7.30	11.69	–	–1.78	–	–0.10	25.20	6.41	5.12	10.12	–2.30	21.76	7.07	9.70	6.73	(8.27)
Rubber	–	4.77	–	7.02	13.19	4.79	3.20	9.48	1.70	20.90	19.19	13.18	13.43	12.90	10.11	22.97	7.60	10.96	(6.45)
Chemicals & products	6.51	5.91	9.36	10.16	12.14	7.14	4.03	9.03	5.97	21.33	15.23	14.22	16.21	15.76	15.76	18.23	11.00	11.65	(4.98)
Petroleum & coal	–	2.71	–	10.53	5.71	–	7.55	10.52	4.00	22.81	16.60	7.53	–	–	15.91	–	14.20	10.73	(6.12)
Non-metallic mineral	1.28	3.83	7.19	9.30	9.40	4.43	5.66	6.80	4.13	18.93	12.80	9.90	15.75	3.39	2.23	23.88	5.80	8.51	(6.20)
Ferrous metallurgy	2.76	1.28	9.16	9.07	6.08	5.47	6.03	6.98	–1.10	25.68	14.98	10.06	32.63	5.82	–2.81	49.96	6.30	11.08	(13.39)
Non-ferrous metals	2.76	1.28	9.16	7.87	7.54	5.47	–	6.98	–0.45	25.68	14.98	44.21	–	–	–2.81	–	6.30	9.92	(12.57)
Metal products	4.15	3.50	5.83	11.21	12.58	–	7.95	4.94	–1.90	22.19	7.57	6.99	11.63	3.39	–2.55	15.12	4.50	7.32	(6.23)
Non-electrical mach.	4.15	4.17	5.84	10.16	–	10.12	–	9.85	2.80	23.01	17.61	11.69	13.63	3.39	5.38	–	10.40	9.44	(5.83)
Electrical machinery	4.15	5.80	5.84	13.28	15.55	8.05	6.19	9.60	4.55	36.00	19.34	15.74	26.41	16.97	15.43	23.61	12.20	14.04	(8.71)
Transport equipment	4.15	5.05	5.84	10.59	3.09	10.12	–	7.62	2.80	28.68	19.48	6.92	24.83	12.10	14.13	20.50	6.20	11.38	(8.02)
Precision instruments	–	5.69	–	10.94	–	–	–	–	7.50	36.00	19.34	–	–	–	–	–	–	15.89	(12.40)
Other manufacturing	4.48	3.40	5.25	12.77	11.14	–	4.15	–	3.00	–	–	–	13.80	–	17.59	–	6.80	8.24	(5.16)
Average	2.30	3.33	5.33	8.64	9.40	4.80	4.50	7.03	2.71	22.31	13.50	11.14	17.72	7.34	8.55	18.17	7.20	8.29	(7.91)
	(2.39)	(1.64)	(2.07)	(2.52)	(3.06)	(2.89)	(2.70)	(2.22)	(2.63)	(6.12)	(4.65)	(8.34)	(7.69)	(6.15)	(7.43)	(12.84)	(3.14)		

Appendix Table 18A.2 Total factor productivity growth, aggregate period, by country and sector

Sector \ Country	*Sweden*	*USA*	*Finland*	*Japan*	*Yugoslavia*	*Argentina*	*Chile*	*Portugal*	*Mexico*	*Hungary*	*Korea*	*Turkey*	*Philippines*	*Thailand*	*Egypt*	*Zambia*	*Indonesia*	*India*	*Sector average*	
Food	0.09	1.31	0.28	–1.24	–0.57	–0.82	0.12	2.40	0.83	–0.60	5.26	1.91	2.28	0.15	5.55	–4.47	–1.73	–0.24	0.58	(2.37)
Beverages	0.09	1.31	0.28	–1.24	–0.57	–0.82	4.74	–2.70	2.95	–1.30	5.26	1.91	1.37	7.31	5.96	–0.05	–0.73	–0.90	1.27	(2.86)
Tobacco	0.09	–0.68	0.28	–1.24	–1.60	2.83	2.26	–1.50	0.28	1.90	5.26	1.91	3.99	–	5.96	–0.05	0.94	–0.64	1.18	(2.29)
Textiles	1.60	1.92	1.36	0.31	–0.17	–0.27	0.19	0.04	0.69	0.90	4.51	1.44	–1.03	0.36	–2.00	–1.19	3.47	0.30	0.69	(1.58)
Fabricated textiles	1.60	1.09	1.36	1.01	–0.17	–0.27	–1.33	6.02	0.69	0.48	1.62	2.74	2.05	3.76	1.40	–7.40	4.53	0.30	1.08	(2.77)
Wood & products	0.93	0.09	0.70	1.88	–0.60	–1.66	–	–0.92	–0.31	2.67	5.62	–1.20	–0.09	0.15	–4.81	–0.56	–0.58	–0.19	0.07	(2.13)
Furniture & fixtures	0.93	0.26	0.70	0.95	–0.60	–1.66	6.57	7.76	–	1.10	4.88	3.23	–2.65	–2.31	–	–1.80	–1.07	0.96	1.08	(3.09)
Paper & products	0.94	–0.16	0.56	0.84	0.48	–	2.91	–1.70	1.54	0.00	4.52	1.41	–0.94	6.11	2.35	6.47	–1.04	0.12	1.44	(2.37)
Printing & publishing	1.03	0.58	0.47	–0.08	0.67	–	0.08	–0.23	–	1.80	–	–	0.24	1.36	–	–2.90	–1.98	0.01	0.08	(1.28)
Leather & footwear	1.60	0.25	1.36	0.69	0.07	–	–2.07	6.04	–	–0.40	2.80	–0.98	–1.04	1.51	–1.18	8.35	–3.28	0.27	0.87	(2.92)
Rubber	–	2.59	–	0.59	4.70	1.42	–1.50	–3.96	0.79	–0.30	5.88	5.80	–0.32	0.76	–1.57	0.95	3.42	–0.81	1.15	(2.75)
Chemicals & products	1.34	1.21	1.30	2.44	0.10	3.22	–0.18	–7.97	1.48	0.02	4.49	1.62	–0.06	3.94	5.18	–6.59	3.79	–0.36	0.81	(3.41)
Petroleum & coal	–	–1.79	–	–3.16	1.22		3.69	–	–0.19	0.25	0.68	0.45	–0.33	–	–	4.04	–	0.26	0.47	(2.08)
Non-metallic mineral	1.70	0.07	2.01	1.20	1.08	1.52	3.29	3.17	0.84	1.29	4.53	0.26	0.26	2.60	–0.15	–3.75	4.39	–0.50	1.32	(1.95)
Ferrous metallurgy	1.48	–0.59	0.93	0.90	–0.19	0.55	0.35	1.72	–0.04	–0.90	1.87	0.87	–1.54	5.78	0.65	–9.24	13.15	–0.55	0.84	(4.16)
Non-ferrous metals	1.48	–0.59	0.93	0.12	–0.71	0.55	–	–3.36	–0.04	–0.60	1.87	0.87	–5.57	–	–	–9.24	–	–0.55	–1.06	(3.05)
Metal products	1.82	0.50	1.47	1.91	0.54	–	5.04	–0.86	–0.29	–0.60	6.01	1.51	–0.50	–3.26	0.46	–9.17	0.86	–0.55	0.36	(3.25)
Non-electrical mach.	1.82	0.36	1.47	1.29	–	6.49	–	–3.88	2.41	0.30	5.73	1.33	1.44	1.15	0.46	10.65	–	–0.03	0.65	(3.93)
Electrical machinery	1.82	1.58	1.47	3.28	0.25	3.28	–1.91	–0.92	1.61	1.80	7.25	1.83	1.32	2.71	3.81	–9.75	6.20	0.24	1.44	(3.56)
Transport equipment	1.82	0.65	1.47	1.41	0.18	6.49	–	–9.28	–0.33	0.80	5.10	3.33	–1.52	0.61	4.53	3.65	4.67	–0.27	1.37	(3.54)
Precision instruments	–	0.81	–	2.63	–	–	–	5.34	–	2.50	7.25	1.83	–	–	–	–	–	–	3.39	(2.42)
Other manufacturing	1.14	0.46	2.70	2.87	3.88	–	–2.53	–7.40	–	2.20	–	–	–	–3.08	–	–5.68	–	–0.44	–0.53	(3.70)
Average	1.23	0.51	1.11	0.79	0.40	1.39	1.16	–0.58	0.76	0.66	4.52	1.60	–0.13	1.65	1.66	–2.81	–2.06	–0.17	–0.82	(2.90)
	(0.59)	(0.97)	(0.64)	(1.53)	(1.49)	(2.62)	(2.79)	(4.66)	(0.98)	(1.14)	(1.83)	(1.52)	(2.01)	(2.98)	(3.25)	(5.56)	(4.04)	(0.46)		

Sources for Appendix Tables

Sweden & Finland

Wyatt, G. (1983) 'Multifactor Productivity Change in Finnish and Swedish Industries, 1960 to 1980', ETLA Elinkeinoelaman Tutkimuslaitos B 38 (Helsinki).

USA

Gollop, F. M. and Jorgenson, D. W. (1984) *U.S. Economic Growth: 1948–1979*, unpublished manuscript.

Japan

Jorgenson, D. W., Kuroda, M. and Nishimizu, M. (1985) 'Japan-U.S. Industry-level Productivity Comparison, 1960–1979', paper prepared for the US-Japan Productivity Conference, Cambridge, Mass., 26–28 August 1985.

Yugoslavia, Turkey, Korea

Nishimizu, M. and Robinson, S. (1984) 'Trade Policies and Productivity change in Semi-Industrialized Countries', *Journal of Development Economics*, vol. 16, nos 1–2 (September/October).

Argentina

Delfino, J. A. (1984) 'La Productividad En Argentina', mimeo. (Universidad Nacional de Cordobá).

Chile

Mierau, B. (1986) 'The Impact of Trade Policies on Productivity Performance in the Chilean Manufacturing Sector', PhD dissertation, Department of Economics, Georgetown University.

Mexico

World Bank (1986) 'Mexico: Trade Policy Industrial Performance and Adjustment' (restricted distribution).

Portugal

World Bank (1982) 'Portugal: Policies for Restructuring' (restricted distribution).

Hungary

World Bank (1986) 'Hungary: Industrial Policy, Performance and Prospects for Adjustment' (restricted distribution).

Philippines

Hooley, R. (1982) 'Productivity Growth in Philippine Manufacturing: Retrospect and Future Prospects', mimeo.

Thailand

Bateman, D. (1986) 'The Role of Productivity Growth in Thailand's First 20 Years of Industrialization: An Analysis of the Manufacturing Sector', World Bank, Industry Department, mimeo.

Egypt

Handoussa, H., Nishimizu, M. and Page, J. M. Jr (1986) 'Productivity

Change in Egyptian Public Sector Industry After "The Opening", 1973–1979', *Journal of Development Economics*, vol. 21, no. 1, (April).

Zambia

World Bank (1984) 'Zambia: Industrial Policy and Performance' (restricted distribution).

Indonesia

Nehru, V., Kim, H.-S. and Bakeman, D. (1985) 'The Structure and Performance of Indonesian Manufacturing', World Bank mimeo.

India

World Bank (1986) 'India Regulatory Policy Study', (restricted distribution).

Notes

* This paper presents some results of a World Bank Research Project: Productivity Change in Infant Industry. We would like to thank Deborah Bateman for her research assistance and constructive discussions of the issues. The views and interpretations in this paper are those of the authors and should not be attributed to the World Bank, to its affiliated organisations or to any individual acting on their behalf.

** The World Bank does not accept responsibility for the views expressed herein which are those of the authors and should not be attributed to the World Bank or to its affiliated organisations. The findings, interpretations, and conclusions are the results of research supported by the Bank; they do not necessarily represent official policy of the Bank. The designations employed, the presentation of material, and any maps used in this document are solely for the convenience of the reader and do not imply the expression of any opinion whatsoever on the part of the World Bank or its affiliates concerning the legal status of any country, territory, city, area, or of its authorities, or concerning the delimitations of its boundaries, or national affiliation.

1. See Cowing and Stevenson (1981) for representative contributions in these areas, and also for a good survey of related literature and references.
2. A concise summary of the relevant theoretical literature and references are given in Caves, Christensen and Diewert (1981, 1982a, b) (see also Gollop and Jorgenson, 1980).
3. Earlier empirical literature, also included in Chenery (1986), was reviewed by Nadiri (1970, 1972).
4. The data source for each of these countries is given in the Appendix.
5. This distinguishes our estimates from those of Chenery which are based on value added. For a comprehensive theoretical discussion and for the review of relevant literature on this point, see Gollop and Jorgenson (1979).
6. The definitions of the four industrial subsectors are available from the authors.

7. See Verdoorn (1949) and Kaldor (1967).
8. We have adopted the World Bank definition of high income countries to classify countries as developed or developing.
9. Recall that the TFP indices are estimated on the maintained hypothesis of constant returns to scale. Thus, cost reductions arising from movements along a declining long-run average cost curve are indistinguishable empirically from downward shifts in the family of cost curves.
10. For a fuller discussion of trade related issues see Nishimizu and Page (1987).

References

Caves, D. W., Christensen, L. R. and Diewert, W. E. (1981) 'A New Approach to Index Number Theory and the Measurement of Input, Output, and Productivity', *Journal of Political Economy*, vol. 88, no. 5 (October).

Caves, D. W., Christensen, L. R. and Diewert, W. E. (1982a) 'Multilateral Comparisons of Output, Input, and Productivity Using Superlative Index Numbers', *The Economic Journal*, vol. 92, no. 365 (March).

Caves, D. W., Christensen, L. R. and Diewert, W. E. (1982b) 'The Economic Theory of Index Numbers and the Measurement of Input, Output, and Productivity', *Econometrica*, vol. 50, no. 6 (November).

Chenery, H. B. (1986) 'Growth and Transformation', in *Industrialization and Growth: A Comparative Study* (forthcoming).

Cowing, T. G. and Stevenson, R. E. (eds) (1981) *Productivity Measurement in Regulated Industries* (New York: Academic Press)

Gollop, F. M. and Jorgenson, D. W. (1979) *U.S. Economic Growth: 1948–1979*, unpublished manuscript.

Gollop, F. M. and Jorgenson, D. W. (1980) 'U.S. Productivity Growth by Industry, 1947–1973', in Kendrick, J. W. and Vaccara, B. N. (eds) *New Developments in Productivity Measurement and Analysis* (Chicago: University of Chicago Press).

Kaldor, N. (1967) *Strategic Factors in Economic Development* (New York: W. F. Humphrey).

Kawagoe, T. and Hayami, Y. (1985) 'An Intercountry Comparison of Agricultural Production Efficiency', *American Journal of Agricultural Economics*, vol. 67, no. 1 (February).

Mellor, J. W. (1976) *The New Economics of Growth* (Ithaca, New York, and London: Cornell University Press).

Nadiri, M. I. (1970) 'Some Approaches to the Theory and Measurement of Total Factor Productivity: A Survey', *Journal of Economic Literature*, vol. 8, no. 4 (December).

Nadiri, M. I. (1972) 'International Studies of Factor Inputs and Total Factor Productivity: A Brief Survey', *Review of Income and Wealth*, ser. 18, no. 2 (June).

Nishimizu, M. and Robinson, S. (1984) 'Trade Policies and Productivity Change in Semi-Industrialized Countries', *Journal of Development Economics*, vol. 16, nos 1–2 (September/October).

Nishimizu, M. and Page, J. M., jr (1986) 'Productivity Change and Dynamic Comparative Advantage'. *Review of Economics and Statistics*, vol. 68, no. 2 (May).

Nishimizu, M. and Page, J. M, jr (1987) 'Economic Policies and Productivity Change in Industry: An International Comparison', The World Bank, (February) processed.

Verdoorn, P. J. (1949) 'Fattori che regolano lo sviluppo della produttivita del lavoro', *L'Industria*.

World Bank (1985) *World Development Report 1985* (Washington DC.: World Bank and Oxford University Press).

Name Index

Subject Index